Processing and Properties of NANOCOMPOSITES

Processing and Properties of NANOCOMPOSITES

Suresh G. Advani
University of Delaware, USA

World Scientific
NEW JERSEY • LONDON • SINGAPORE • BEIJING • SHANGHAI • HONG KONG • TAIPEI • CHENNAI

Published by

World Scientific Publishing Co. Pte. Ltd.
5 Toh Tuck Link, Singapore 596224
USA office: 27 Warren Street, Suite 401-402, Hackensack, NJ 07601
UK office: 57 Shelton Street, Covent Garden, London WC2H 9HE

British Library Cataloguing-in-Publication Data
A catalogue record for this book is available from the British Library.

PROCESSING AND PROPERTIES OF NANOCOMPOSITES

ISBN 981-270-390-X

Printed by FuIsland Offset Printing (S) Pte Ltd, Singapore

This book is dedicated to the memory of my parents,
Kamla and Gopaldas Advani

Preface

The field on nanotechnology is still in its infancy but continues to progress at a much faster rate that any other field. Many methods to synthesize nano particles, disperse them in a carrying fluid to form a composite and exploit its extraordinary properties is the goal and dream of many researchers engaged in this field. It is not possible to cover every nano particulate matter and its role in materials revolution. The approach adopted here was to focus on carbon nanotubes and nano clays and explore their importance and their role in composites. Hence the chapters presented in this book address processing, rheology, mechanical properties and their interaction with fiber composites. Thus, this book is a collection of nine chapters written by researchers who are at the forefront of their field which address the role of nano particles in composites. The first three chapters focus on the use of Carbon nanotubes in a composite. Chapter 1 is a succinct summary of the state of the art of the carbon nanotubes in composites. Chapter 2 focuses more on the aspects of processing with these nano particles in a suspension. Most research is focused in using these nanotubes sparingly in the composite and exhibit disproportionally better properties. Chapter 3 focuses on how to address higher loadings of these nanotubes and develop nanostructure materials. Chapter 4 explores the interaction between traditional fiber composites and use of nano particles in them in terms of benefits and property enhancement in addition to processing of such materials. Chapter 5 discusses in detail the rheology of suspensions that contain nanofibers and how one can modify existing models to describe their flow behavior. Chapters 6 through 9 address nano clay composites. Chapter 6 is a good overview of the state of the art of the nanoclay usage in various resins and composites. Chapter 7 focuses on the mechanical and physical property characterization of polymer clay nanocomposites.

Chapter 8 discusses further use of nanoclays in thermoplastics and their use in glass fiber composites. Chapter 9 describes methods to prepare nanoclay suspensions with thermosets and the corresponding enhancement in properties. I would like to thank all the authors and reviewers in making this project into reality.

Suresh Advani
GW Laird Professor of Mechanical Engineering
Associate Director, Center for Composite Materials
University of Delaware

Contents

CHAPTER 1

Carbon Nanotube/Nanofibre Polymer Composites

Milo S.P. Shaffer[1] and Jan K.W. Sandler[2]

[1]*Department of Chemistry, Room 103d RCS1*
Imperial College London, SW7 2AZ, UK
Tel: +44(0)20 7594 5825; Fax: +44(0)20 7594 5801
m.shaffer@imperial.ac.uk
[2]*Polymer Engineering, University of Bayreuth*
Universitätsstrasse 30, D-95447 Bayreuth, Germany
jan.sandler@uni-bayreuth.de

1. Introduction

Although the terms *nanomaterial* and *nanocomposite* represent new and exciting fields in materials science, such materials have actually been used for centuries and have always existed in nature. However, it is only recently that the means to characterise and control structure at the nanoscale have stimulated rational investigation and exploitation. A nanocomposite is defined as a composite material where at least one of the dimensions of one of its constituents is on the nanometre size scale [1]. The term usually also implies the combination of two (or more) distinct materials, such as a ceramic and a polymer, rather than spontaneously phase-segregated structures. The challenge and interest in developing nanocomposites is to find ways to create macroscopic components that benefit from the unique physical and mechanical properties of very small objects within them.

Natural materials such as bone, tooth, and nacre are very good examples of the successful implementation of this concept, offering excellent mechanical properties compared to those of their constituent materials. Such composites actually exhibit beautifully organised levels of hierarchical structure from macroscopic to microscopic length scales, and provide a powerful motivation for improving our processing control. Currently, we are striving to understand the behaviour of just the smallest building blocks in such materials which are the natural versions of nanocomposites. Significantly, two contrasting phases are often combined: a hard nanoscale reinforcement (such as hydroxyapatite or calcium carbonate) is embedded in a soft, usually protein-based, matrix. Although the composite character of these materials itself plays a crucial role, the question remains, why the nanometre scale is so important.

From a simple mechanical point of view, the situation in such biocomposites is quite familiar: the matrix transfers the load via shear to the nanoscale reinforcement [2]. A large length-to-diameter (aspect) ratio of the mineral reinforcement compensates for the low modulus of the soft protein matrix, leading to an optimised stiffness of the composite. The fracture toughness of such biocomposites, on the other hand, hinges on the ultimate tensile strength, σ_f, of the reinforcement. Crucially, the use of a nanomaterial allows access to the maximum theoretical strength of the material, since mechanical properties become increasingly insensitive to flaws at the nanoscale [2]. This observation is an extension of the classic approach to strong materials, namely to reduce the dimensions until critical flaws are excluded. At the nanoscale, highly crystalline reinforcements are used in which all but the smallest atomistic defects can be eliminated. It is clear that a high aspect ratio must be maintained in order to ensure suitable stress transfer. This general concept of exploiting the inherent properties of nanoscaled materials is not limited to the mechanical properties of a material, since a wide range of physical properties also depend on defect concentrations. In addition, the small size scale can generate inherently novel effects through, for example, quantum confinement, or through the dramatic increase in interfacial area.

The concept of creating both structural and functional multi-phase nanocomposites with improved performance is currently under development in a wide variety of metallic, ceramic, and polymeric matrices, although the emphasis to date has been on polymeric systems. Similarly, the filler particles can be organic or inorganic with a wide range of material compositions and structures. The resulting composites generally exhibit a number of enhanced properties, so that the material cannot easily be classified as a structural or functional composite. The term *reinforcement*, as opposed to plain *filler*, is equally frequently used for the nanoscale component, without a clear distinction.

Carbon nanotubes (CNTs) have attracted particular interest because they are predicted, and indeed observed, to have remarkable mechanical and other physical properties. The combination of these properties with very low densities suggests that CNTs are ideal candidates for high-performance polymer composites; in a sense they may be the next generation of carbon fibres. Although tens or hundreds of kilograms of carbon nanotubes are currently produced per day, the development of high-strength and high-stiffness polymer composites based on these carbon nanostructures has been hampered so far by the lack of availability of high quality (high crystallinity) nanotubes in large quantities. In addition, a number of fundamental challenges arise from the small size of these fillers. Although significant advances have been made in recent years to overcome difficulties with the manufacture of polymer nanocomposites, processing remains a key challenge in fully utilising the properties of the nanoscale reinforcement. A primary difficulty is achieving a good dispersion of the nanoscale filler in a composite, independent of filler shape and aspect ratio. Without proper dispersion, filler aggregates tend to act as defect sites which limit the mechanical performance; such agglomerates also adversely influence physical composite properties such as optical transmissivity.

When dispersing small particles in a low viscosity medium, diffusion processes and particle-particle and particle-matrix interactions play an increasingly important role as the diameter drops below 1 μm. It is not only the absolute size but rather the specific surface area of the filler, and the resulting interfacial volumes, which significantly influence the dispersion process. These regions can have distinctly different properties from the bulk polymer and can represent a substantial volume fraction of the matrix for nanoparticles with surface areas of the order of hundreds of m^2/g. The actual interphase volume depends on the dispersion and distribution of the filler particles, as well as their surface area.

In traditional fibre composites, the interfacial region is defined as the volume in which the properties deviate from those of the bulk matrix or filler [3]. However, it is simpler, in the case of small particles, to consider a straightforward calculation [4] of the interparticle distance s:

$$s = d^* \left[\left(\frac{\pi}{6\phi_V} \right)^{1/3} - 1 \right], \quad (1)$$

where d is the diameter and ϕ_V is the volume fraction of uniformly-sized, spherical particles on a lattice. As an example, a 15 vol% loading of 10 nm diameter particles leads to an interparticle distance of only 5 nm. This figure, which is similar to the radius of gyration of a typical polymer molecule, shows that essentially the entire polymer matrix in a nanocomposite can behave as if it were part of an 'interphase'. In other words, the properties of the whole matrix may differ from the pure bulk polymer in terms of the degree of cure, chain mobility, chain conformation, and degree of chain ordering or crystallinity. These effects may influence both processing and final properties of the polymeric phase. One simple but important consequence is that it becomes increasingly difficult to 'wet' adequately all of the nanofiller surface with polymer; thus, the volume fraction of nanomaterial that can be uniformly dispersed using conventional processing techniques is decreased. In addition, the strong influence of interfacial interactions during processing can alter the matrix microstructure which, particularly in the case of semicrystalline polymers, can significantly affect the mechanical behaviour of the nanocomposite independently of direct load-bearing by the filler [5]. Therefore, the matrix microstructure must be critically assessed when evaluating the performance of carbon nanotube-polymer composites. Although the high aspect ratio of carbon nanotubes generally appears to be a clear benefit for the exploitation of their mechanical as well as physical properties such as electrical conductivity, it is as yet not established which of the many different types of nanotubes will yield the ultimate performance in a polymer composite. Nanotubes have shown a remarkable range of structural features, and the resulting structure-property-relationships are only slowly emerging.

2. Carbon Nanotubes and Nanofibres

Carbon nanotubes are often seen as the intersection of traditional carbon fibres with the fullerene family [6]. It was only realised relatively recently that solids of pure elemental carbon with sp^2-hybridisation can form a wide variety of well-defined crystalline structures. The first recognition of the fullerenes as closed structures [7] in 1985, and their subsequent synthesis in a carbon arc [8], stimulated enormous new interest in carbon materials. In 1991, Iijima observed a graphitic tubular structure in an arc discharge apparatus that was used to produce C_{60} and other fullerenes. His realisation [9] of the structural richness of these particles, which came to be known as nanotubes, generated enormous interest; indeed, there are now in excess of 10,000 papers discussing the science of CNTs, including a large fraction on polymer composites. Although Iijima is often credited with the discovery of CNTs, there are earlier reports in the literature, notably by Endo in 1976, of the synthesis of tubular carbon structures using hydrocarbon decomposition [10]. There are even reports in the catalysis literature of the 1950s of attempts to remove troublesome fibrous carbon deposits. The electric arc method was actually used as early as 1960 by Roger Bacon [11] to produce graphite whiskers, and one can also speculate that nanotubes were likely present in his experiments as a by-product, although unobserved. The explosion of interest in 1991 was driven by the renewed interest in carbon following the discovery of fullerenes, and the structural perfection of Iijima's nanotubes which implied a whole range of exciting properties.

CNTs have typical diameters in the range of ~1–50 nm and lengths of many microns (even centimetres in special cases). They can consist of one or more concentric graphitic cylinders. In contrast, commercial (PAN and pitch) carbon fibres are typically in the 7–20 μm diameter range, while vapour-grown carbon fibres (VGCFs) have intermediate diameters ranging from a few hundred nanometres up to around a millimetre. The variation in diameter of fibrous graphitic materials is summarised in Figure 1. Crucially, conventional carbon fibres do not have the same potential for structural perfection that can be observed in CNTs. Indeed, there is a general question as to whether the smallest

CNTs should be regarded as very small fibres or heavy molecules, since the diameters of the smallest nanotubes are similar to those of common polymer molecules. This ambiguity is characteristic of nanomaterials, and it is not yet clear to what extent conventional fibre composite understanding can be extended to CNT composites.

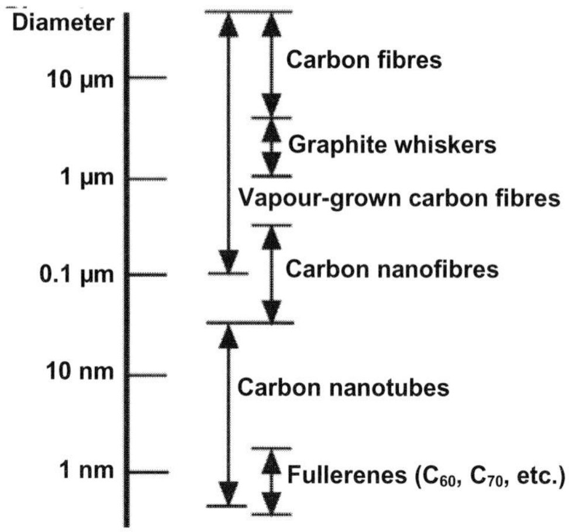

Fig. 1. Comparison of diameters of various fibrous carbon-based materials.

The nanotube structure can be defined more exactly by considering a single-wall carbon nanotube (SWCNT) as a conformal mapping of the two-dimensional hexagonal lattice of a single graphene sheet onto the surface of a cylinder. The graphite sheet may be 'rolled' in different orientations along any two-dimensional lattice vector (m,n) which then maps onto the circumference of the resulting cylinder; the orientation of the graphite lattice relative to the axis defines the *chirality* or *helicity* of the nanotube [12]. As-grown, each nanotube is closed at both ends by a hemispherical cap formed by the replacement of hexagons with pentagons in the graphite sheet which induces curvature. SWCNTs are usually obtained in the form of so-called ropes or bundles, containing between 20 and 100 individual tubes packed in a hexagonal array [13–16]. Rope formation is energetically favourable due to the Van der Waals attractions between isolated nanotubes [17]. Multi-wall carbon nanotubes (MWCNTs) provide an alternative route to stabilisation. They consist of two or more coaxial cylinders, each rolled out of single sheets, separated by approximately the interlayer spacing in graphite. The outer diameter of such MWCNTs can vary between 2 and a somewhat arbitrary upper limit of about 50 nm; the inner hollow core is often (though not necessarily) quite large with a diameter commonly about half of that of the whole tube. Carbon nanofibres (CNFs) are mainly differentiated from nanotubes by the orientation of the graphene planes: whereas the graphitic layers are parallel to the axis in nanotubes, nanofibres can show a wide range of orientations of the graphitic layers with respect to the fibre axis. They can be visualised as stacked graphitic discs or (truncated) cones, and are intrinsically less perfect as they have graphitic edge terminations on their surface. Nevertheless, these nanostructures can be in the form of hollow tubes with an outer diameter as small as ~5 nm, although 50–100 nm is more typical. The stacked cone geometry is often called a 'herringbone fibre' due to the appearance of the longitudinal cross-section. Slightly larger (100–200 nm) fibres are also often called CNFs, even if the graphitic orientation is approximately parallel to the axis. Representative transmission electron micrographs of commercially available nanofibres with a mean outer diameter of around 150 nm are shown in Figure 2. The schematic diagram highlights the orientation of the graphitic planes in the inner and outer regions of the nanofibre wall, and is intended to illustrate the structural complexity that can arise in these materials.

A variety of synthesis methods now exist to produce carbon nanotubes and nanofibres. However, these carbon nanostructures differ greatly with regard to their diameter, aspect ratio, crystallinity, crystalline orientation, purity, entanglement, surface chemistry, and straightness. These structural variations dramatically affect intrinsic properties, processing, and behaviour in composite systems. However, it is not yet clear which type of nanotube material is most suitable for composite applications, nor is there much theoretical basis for rational design. Ultimately, the selection will depend on the matrix material, processing technology, and the property enhancement required. Thus, in order to interpret the data obtained for nanotube composites, and to develop the required understanding, it is essential to appreciate the range of nanotube materials available.

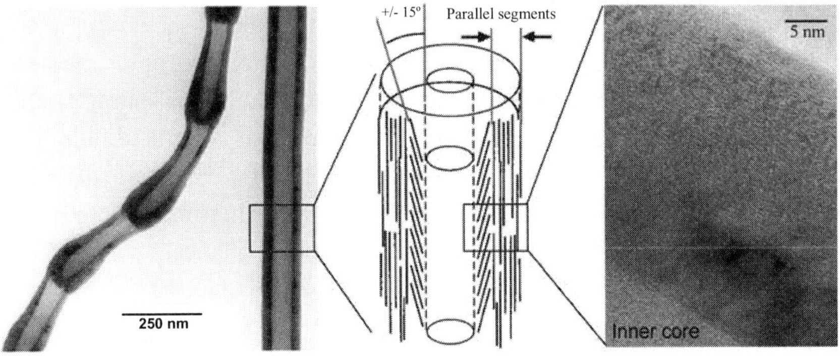

Fig. 2. Representative transmission electron micrographs of commercial carbon nanofibres, highlighting structural variations both in overall morphology and in the orientation of the graphitic planes. The leftmost image shows a 'bamboo' and a 'cylindrical' CNF, whilst the rightmost image shows a high magnification image of one wall of the cylindrical fibre which reveals the graphitic arrangement sketched in the central panel.

2.1. *Production methods*

Both MWCNTs and SWCNTs can be produced by a variety of different processes which can broadly be divided into two categories: high-temperature evaporation using arc-discharge [13,14,18–20] or laser ablation [15,21], and various chemical vapour deposition (CVD) or catalytic growth processes [16,22–24]. In the high-temperature methods MWCNTs can be produced from the evaporation of pure carbon, but the synthesis of SWCNTs requires the presence of a metallic catalyst. The CVD approach requires a catalyst for both types of CNTs but also allows the production of CNFs. Many variants exist, but the basic methods can be described as follows.

The electric arc method is based on the generation of a DC arc plasma between two carbon electrodes in an inert (usually helium) atmosphere. While the anode is consumed, a soft, dark black, fibrous deposit forms on the cathode which consists of about 50 vol% straight MWCNTs, often arranged in a fractal structure. Addition of a suitable catalyst such as Ni-Co, Co-Y or Ni-Y leads to the formation of interconnected web-like SWCNT bundles on the walls of the reaction chamber [13,14,25,26]. Macroscopically long ropes of well-aligned SWCNTs can be synthesised by the arc-discharge method using hydrogen [27]. Laser ablation is a similar process, using a different technique to generate the carbon vapour; although the method can generate MWCNTs, it is usually used for the production of SWCNTs at yields higher than 70%. A graphite target, containing a 1–2% metal catalyst, is held in a furnace at 1200°C in an inert atmosphere and is evaporated using a high power laser [15,28]. The resulting products are swept from the high-temperature zone by the flowing inert gas and are deposited on a conical water-cooled copper collector.

The products of both high-temperature routes tend to be highly crystalline, with low defect concentrations, but are relatively impure, containing other, unwanted carbonaceous impurities. These methods usually work on the gram scale and are, therefore, relatively expensive. For the use of nanotubes in composites, large quantities of nanotubes are required at low cost, ideally without the requirement for complicated purification. At present, only CVD-grown nanotubes satisfy these requirements and, as such, tend to be the materials of choice for composite work, both in academia and in industry. The main contaminants in CVD materials are residual catalyst particles which are mostly incorporated into the nanotubes. On the other hand, these gas-phase processes operate at lower temperatures and lead to structurally more imperfect nanotubes, as shown by the deviation from the ideal cylindrical structure in Figure 3.

Carbon filaments can form when small metal particles, almost always containing iron, nickel, or cobalt, are exposed to CO or hydrocarbon gases at temperatures between 500 and 1200°C. The carbonaceous feedstock decomposes on the catalyst, generating carbon which diffuses through or around the catalyst to produce a fibre with a diameter similar to the metal particle [29]. Originally, the method was developed for growing VGCFs with diameters up to several hundred micrometres through simultaneous pyrolytic overgrowth on the primary fibre. Later, various orientations of the graphitic planes with respect to the fibre axis were observed depending on the crystallographic orientation of the catalyst particle [30]. The arrangement of the graphitic planes can vary from perpendicular to the fibre axis to parallel, thus generating a range of

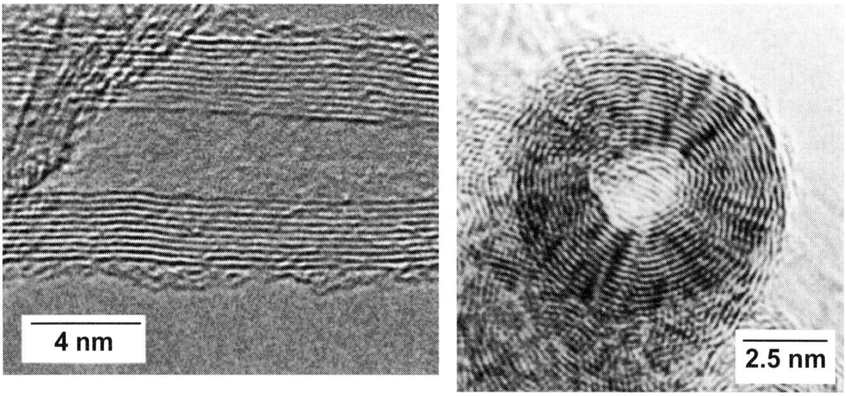

Fig. 3. Transmission electron micrographs of commercial MWCNTs grown by CVD methods, with the beam perpendicular (left) and parallel (right) to the axis.

CNF structures [31]. Twisted, helical, and branched nanofibres have also been reported [32].

Under the right conditions, entangled mats of catalytically-grown CNTs can be produced [22]. Generally, the experiment is carried out in a flow furnace at atmospheric pressure [33,34]. In perhaps the simplest embodiment, the catalyst is placed in a ceramic boat which is then put into a quartz tube. A reaction mixture consisting of, for example, acetylene and argon is passed over the catalyst bed for several hours at temperatures ranging from 500 to 1100°C [34]. In fact, there are many options for introducing the catalyst ranging from injection of organometallic vapours and metallic colloids, to pre-deposition of metal films or particles on ceramic supports. Injecting the catalyst into the gas stream allows for the continuous production of nanotubes. A number of commercial routes to the production of vapour-grown CNFs and CNTs are based on a 'floating catalyst' carried in the gas stream inside a continuous flow reactor [35].

The CVD products discussed above tend to be highly entangled; however, aligned nanotube arrays, as shown in Figure 4a, can be obtained under conditions that lead to rapid and dense nucleation on flat substrates [36–39]. With sufficient growth time, lengths in the millimetre range have been observed [40]. Such aligned MWCNT films can be easily removed from the substrate and might be considered as the ideal nanotube material for composite applications, as the low degree of nanotube entanglement should allow straightforward dispersion in a polymer matrix. The synthesis of well-aligned and comparatively straight MWCNTs on substrates can be further enhanced by the application of a plasma during growth [41,42]. This approach has recently been shown to allow the production of aligned nanotube films at temperatures as low as 120°C [43], opening the door for direct nanotube growth on polymer substrates. The use of plasma accelerates growth, and increases alignment, but tends to reduce crystallinity.

Although a number of commercial CVD routes to SWCNTs exist, the so-called 'HiPco' process has received particular attention. The gas-phase growth of SWCNTs using high-pressure carbon monoxide as the carbon source was developed by Nikolaev *et al.* [44] but is now commercially exploited by Carbon Nanotechnologies Inc., USA. The product is shown in Figure 4b and is widely used for research purposes, although it is, at present, still too expensive for large-scale commercial composite applications.

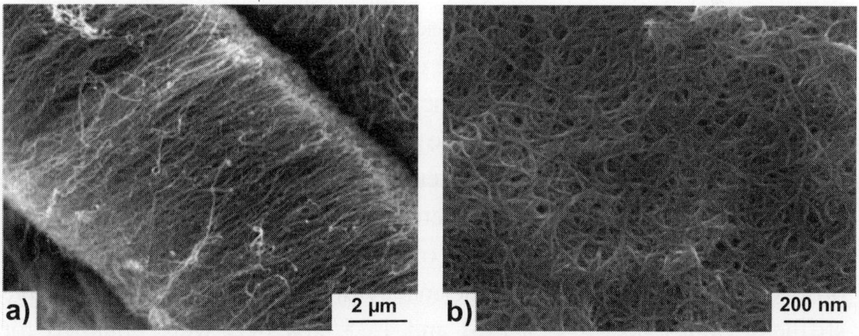

Fig. 4. Scanning electron micrographs of (a) aligned multi-wall carbon nanotubes and (b) HiPco single-wall carbon nanotubes produced by CVD methods.

In summary, the quality and yield of carbon nanotubes depend on the synthesis technique and the specific growth conditions used. Catalytic processes generally involve lower growth temperatures which lead to both variations in the orientation of the graphitic planes with respect to the tube axis and to an increased concentration of structural defects. It is not surprising that most studies aimed at investigating the fundamental mechanical and physical properties of individual nanotubes have been performed on high-temperature materials with few structural defects. Relatively little effort has been applied to CNFs, although given their structural appearance, their properties should, at best, resemble those of very defective catalytic nanotubes.

2.2. *Mechanical properties of carbon nanotubes*

Although challenging, a number of experimental studies have focussed on the direct determination of the mechanical properties of individual carbon nanotubes. Experimental measurements of nanotube deformations have mostly been analysed by assuming nanotubes to be elastic beams. The resulting elastic constants belong to the framework of continuum elasticity, and any estimate of these material parameters for nanotubes therefore implies the continuum assumption. There is also a general question as to what should be taken as the cross-section of a nanotube. For MWCNTs, values have usually been calculated based on a thick-walled cylinder, ignoring the cross-section of the hollow core, although it might be argued that the area of the core should be included since it is an intrinsic feature of the structure. The question is even more difficult with SWCNTs which are only a single layer of atoms thick. Usually, the thickness, t, is taken to be the interlayer spacing of graphite, although difficulties with bending stiffness remain [45].

The original determinations of CNT stiffness were based on observing the amplitude of thermal vibrations in a TEM; average stiffness values of 1.8 TPa [46] and 1.25 TPa [47] were obtained for MWCNTs and SWCNTs, respectively. For MWCNTs, the estimated nanotube stiffness appeared to depend on the diameter [48], an effect that was explained by the occurrence of wave-like distortions for multi-wall carbon nanotubes with diameters of greater than 12 nm, as predicted by a combination of finite element analysis and non-linear vibration analysis

[49]. Falvo *et al.* [50] showed that MWCNTs could be repeatedly bent to large angles (> 120°) with an AFM tip without undergoing catastrophic failure; this observation was supported by high-resolution TEM studies that indicated reversible buckling as a mechanism for stress relief [51–53]. Lourie *et al.* captured the buckling of SWCNTs in compression and bending by embedding the nanotubes in a polymer film [54].

Static models of beam bending have also been used to quantify mechanical properties of nanotubes. AFM measurements led to an average bending stiffness of arc-grown MWCNTs of about 1 TPa [55,56]; however, catalytic nanofibres with a higher defect concentration were found to have a substantially lower stiffness of only 10 to 50 GPa [56]. Whereas point defects do not affect the nanotube stiffness greatly, deviations from a perfectly parallel alignment of the graphitic layers to the axis have a significant detrimental effect on properties, due to the high anisotropy of graphite.

More recently, a mechanical loading stage operating inside an SEM was used to perform the first *in-situ* tensile tests on individual MWCNTs and ropes of SWCNTs. Individual arc-grown MWCNTs were found to fracture by a so-called *sword-in-sheath* mechanism in the outermost shell. Strength values for the outer shell ranged from 11 to 63 GPa at fracture strains of up to 12% and modulus values ranged from 270 to 950 GPa [57]. By assuming that the load is carried only by the SWCNTs on the perimeter of the rope, fracture strengths ranging from 13 to 52 GPa and moduli between 320 and 1470 GPa were obtained [58]. It is interesting to note that the maximum fracture strain was found to be 5.3% which is close to the theoretical value of ~5% for defect nucleation in individual SWCNTs [59].

The experimental results for highly crystalline nanotubes (produced by high-temperature methods) show that such nanotubes can indeed have a Young's modulus approaching the theoretical value of 1.06 TPa [52], the in-plane modulus of graphite, in agreement with theoretical studies [60,61]. (Young's modulus values of around 5.5 TPa [46] relate to an assumed effective SWCNT wall thickness of 0.066 nm). It should be borne in mind that a single value of Young's modulus should not be uniquely used to describe both the tension/compression *and* bending behaviour of carbon nanotubes. Tension and compression are mostly governed by the in-plane σ-bonds, while pure bending is affected by the out-of-plane π-bonds. However, within a given mode, continuum elasticity does seem to be applicable to the elastic properties of such nanostructures, up to the point at which local instabilities occur, as long as the geometry of the nanotubes is properly taken into account.

A number of theoretical studies have addressed the structural stability of nanotubes in tension, compression, bending, and torsion. Under axial loads, abrupt changes in nanotube morphology were observed which depended on the nanotube length [62,63]. Nanotube buckling due to bending has also been demonstrated [62,64] and is characterised by a collapse of the cross-section in the middle of the tube, in agreement with experimental observations [51–53].

The strength of nanotubes depends on the distribution of defects, as well as interlayer interactions in MWCNTs and bundles of SWCNTs. The defect density is potentially low in these nanostructures and defect sites may be distributed over large distances due to the small diameter and high aspect ratio. However, defect density will depend on the growth process, and the strength should only approach the theoretical limit for nanotubes grown at high-temperatures. There are relatively few experimental results but those mentioned above are in good agreement with theoretical predictions [65] and indeed indicate that the strength of nanotubes can be one order of magnitude higher than that of current high-strength carbon fibres. Further evidence for the high strength of high temperature nanotubes has been found in other tensile tests [66], although the strength decreased significantly for 2 mm long bundles of MWCNTs grown in a CVD process [67]. In this case, the average strength of about 1.7 GPa might be related to the higher defect concentration as a result of the lower growth temperature, but could also be attributed to gauge length effects, with individual nanotubes being shorter than the overall length of the bundle. Initial fragmentation tests of nanotubes embedded in thin film polymer composite films also led to an estimated high tensile strength of nanotubes [68,69], although an accurate determination of the fragment length of an embedded nanotube in a TEM sample is challenging.

Although the high nanotube strength is derived from the strong in-plane graphitic bonds, the weak interlayer interactions may cause problems. MWCNTs appear to fall victim to their own in-plane structural perfection which minimises load transfer to the inner shells when the outermost shell is strained in tension. Whether end effects, high aspect ratios, or modest defect concentrations can alleviate this problem remains to be seen. Similarly, the inner SWCNTs in a bundle may not contribute to the overall mechanical performance. For these reasons, some have suggested that individually dispersed SWCNTs should be the ideal reinforcement. Experimentally, these ideas are supported by the failure mechanisms observed in individual CNT tensile tests and by some composite data. For example, macroscopic epoxy samples containing 5 wt% of dispersed MWCNTs subjected to tensile and compressive loads showed a more pronounced modulus enhancement in compression; the result is consistent with the idea that only the outer nanotube layers are stressed in tension, whereas all layers contribute under compression [70].

2.3. *Transport properties of carbon nanotubes*

The electronic properties of nanotubes are another area of great scientific interest, yet they are also challenging to measure directly on individual CNTs. Structural defects as well as bends or twists are again thought to have a strong effect on the transport properties [71]. Initial theoretical studies of the electronic properties of SWCNTs, based on band-folding, indicated that nanotube shells can be either metallic or semiconducting depending critically on helicity [72–74], with a small or moderate band gap (for semiconducting tubes) inversely proportional to the tube radius [75,76]. On average, approximately 1/3 of SWCNTs are metallic and 2/3 semiconductors [72]; this ratio tends to be observed in real samples because current synthesis methods offer little, if any, control over helicity. Since MWCNTs have larger diameters, confinement effects disappear, and the transport properties approach those of turbostratic graphite [77]. Interlayer interactions which might be important in small diameter MWCNTs appear to be weak; theoretical studies of double-wall nanotubes indicate that the overall behaviour is determined by the electronic properties of the external shell [78,79].

Results of experimental transport measurements of nanotubes, therefore, vary strongly between individual SWCNTs, SWCNT bundles, individual MWCNTs, or MWCNT bundles or mats. Connections to individual CNTs are usually made either by random deposition of CNTs on pre-patterned electrodes [80,81], or subsequent deposition of contacts using focussed ion beam techniques [82]. The first experimental transport measurement of individual SWCNTs was carried out by Tans *et al.* [83] and showed that there are indeed metallic and semiconducting SWCNTs, verifying the theoretical predictions. The room temperature conductivity was about 10^5 to 10^6 S/m for the metallic nanotubes and about 10 S/m for semiconducting tubes. Scanning tunnelling spectroscopy verified experimentally that the electronic properties sensitively depend on nanotube diameter and helicity [84,85], and that there is no preferred helicity in laser-grown SWCNT material [84]. The presence of a large fraction of semiconducting tubes should therefore be considered when interpreting transport measurements on bundles of SWCNTs, as verified experimentally by a temperature-dependent resistivity of bundles [86]. The conductivity of SWCNT bundles was found to vary between 1×10^4 [87] and 3×10^6 S/m [88,89] at room temperature, depending on sample type and entanglement state. These values approach that for the in-plane conductivity of graphite (2.5×10^6 S/m [90]). Conductivities of individual MWCNTs have been reported to range between 20 and 2×10^7 S/m [82], depending on the helicities of the outermost shells [91] or the presence of defects [92]. The electronic properties of larger diameter MWCNTs approach those of graphite.

Remarkable similarities between the conductivity behaviour of nanotube networks and conducting polymers have been pointed out by Kaiser *et al.* [93]. In analogy to conducting polymers, a good description of the experimentally observed conductivity behaviour is given by a simple model of metallic conduction with hopping or tunnelling through small electrical barriers, e.g. tangled regions, inter-rope or intertube contacts, or tube defects.

Lastly, the axial thermal conductivity of individual, perfect CNTs is expected to be very high [94], greater than that of diamond, with experimental values for MWCNTs reaching 3300 W/m/K [95].

3. Carbon Nanotube/Nanofibre-reinforced Polymer Composites

There are at least three general experimental methods to produce polymer nanocomposites: mixing in the liquid state, solution-mediated processes and *in-situ* polymerisation techniques. The direct melt-blending approach is much more commercially attractive than the latter two methods, as both solvent processing and *in-situ* polymerisation are less versatile and more environmentally contentious.

The literature on processing and evaluating macroscopic nanotube/nanofibre-polymer composites is still in its infancy but developing rapidly. This situation is not surprising, given that initial attempts to produce such nanocomposites were hindered by the small quantities of nanotubes available; although, more recently, the focus on CVD synthesis techniques has enabled the manufacture of large-scale polymer nanocomposites. As yet, no standard approach has been established to assess the resulting nanocomposite properties or to correlate them with the intrinsic nanotube characteristics. For example, a large number of studies have focussed on the effect of nanotubes and nanofibres on the composite stiffness, failing to report other, more relevant, properties such as strength and strain to failure. Such mechanical properties are more dependent on filler dispersion, alignment, and interface than the stiffness, and are more difficult both to improve and to analyse. Nevertheless, a number of interesting studies have been reported that illuminate the potential of nanotube and nanofibre composites.

3.1. *Thermosetting carbon nanotube/nanofibre composites*

Thermosetting epoxies have found a widespread use in applications ranging from household glues to high-performance composites. Increases in toughness, glass transition temperature (T_g), and mechanical properties above T_g would be of particular technological benefit. Initial experimental studies of nanotube-based thermosetting nanocomposites focussed on the production of thin epoxy films which required only small amounts of filler and provided information about dispersion and interfacial properties. When using thermosetting matrices, as-received nanotubes are often directly mixed with the liquid matrix (especially epoxy) precursors. Mechanical mixing can be aided by ultrasonication and vacuum-assisted processing is often applied to ensure defect-free composite samples for mechanical testing. Chemically-treated nanotubes are often first dispersed in surfactants or solvents to which the epoxy is added.

A first example of this approach was published by Ajayan *et al.* who dispersed arc-grown MWCNTs by mechanical mixing in an epoxy resin [96]. The composites were microtomed for investigation by TEM, resulting in individual MWCNTs aligned parallel to the cutting direction. This alignment, and the lack of nanotube fracture, was interpreted as an indication that the nanotubes are strong and that the nanotube-matrix interface is weak. However, samples prepared with harder epoxy resins did allow cutting of nanotubes for direct cross-sectional examination [97]. Similar epoxy samples have been used to determine intrinsic mechanical nanotube characteristics [54,68,98–100]; interfacial bonding was generally assumed to be good because the epoxy was observed to wet both MWCNTs and bundles of SWCNTs.

Only relatively low concentrations (< ~5 wt%) of nanotubes can easily be incorporated in thermosetting composites, due to rapidly increasing viscosity and subsequent processing difficulties, at higher loadings. Even well-dispersed, shortened nanotubes can form a stiff gel, simply in solvent, due to their high aspect ratio and resulting network-forming ability [101]; the large interaction volume may also increase the background viscosity of the solvent/matrix. A further problem is aggregation of the nanofiller, which is a major issue in all nanocomposites, even at modest loading fractions. The introduction of pure nanotubes into thermosetting resins tends to yield only moderate increases in stiffness, whilst the strength and strain to failure of the matrix are usually degraded. This poor performance is often attributed to poor dispersion and various processing remedies have been explored. One tactic is to use continual mechanical stirring or ultrasonication to prevent reagglomeration of nanotubes [102]. This approach has been shown to reduce the average MWCNT cluster size in epoxy composites and hence to improve the composite tensile stiffness, although the strain to failure still tends to be reduced [103]. An alternative to mechanical agitation is to introduce a surfactant [104]. Composites containing as little as 1 wt% of surfactant-dispersed MWCNTs show an improved thermomechanical behaviour as compared to the pure nanotube material, although the surfactant itself decreases the storage modulus of the epoxy significantly. The combination of MWCNTs and surfactant can increase the composite T_g, providing evidence for an improved interaction between filler and matrix.

A third strategy for improved dispersion is to use chemical routes to directly functionalise the CNTs [105]. The end caps of both single and multi-wall nanotubes can be opened under oxidising conditions, leading to carboxyl, carbonyl, and hydroxyl groups both at the opened ends and at defects on the side walls. Such oxidised nanotubes show a better solubility and can form electrostatically stabilised colloidal dispersions in water as well as alcohols [101,106], leading to improved dispersions in epoxy systems [107]. An example of well-dispersed, oxidised multi-wall nanotubes in an epoxy is shown in Figure 5. One drawback of such acid treatments is that they degrade the length of individual nanotubes [107], although an associated reduction in SWCNT bundle diameter may be an advantage [108,109]. Further improvements in solubility can be achieved by fluorination [110], again leading to improvements in both the stiffness and strength, on the addition of 1 wt% of oxidised and fluorinated SWCNTs [105]. In addition to improving dispersion, chemical functionalisation can encourage direct covalent coupling between the CNTs and the matrix; one example is the use of amino-functionalised MWCNTs in epoxy systems to yield improved properties [111]. The improved mechanical performance in these functionalised systems may reflect both the enhanced dispersion and an increased interaction of the nanotube surface groups with the polymer.

Using larger CNFs allows the processing of thermosetting composites with higher loadings, up to about 20 wt%, with relatively little void content [112,113]; the increase most likely reflects the smaller surface area of nanofibres compared to nanotubes, as well as their greater tendency to break during shear processing. At these filler fractions, randomly oriented CNFs in epoxy have been found to be an as effective

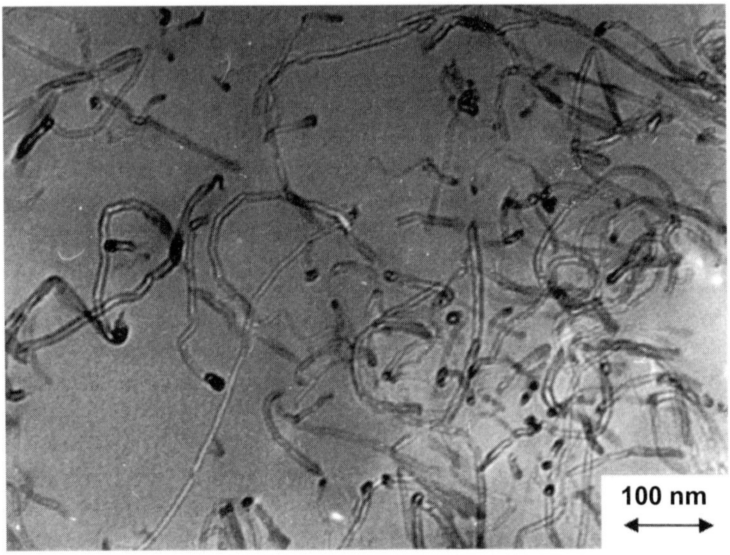

Fig. 5. Transmission electron micrograph of well-dispersed catalytically-grown multi-wall carbon nanotubes in an epoxy matrix as a result of a chemical oxidation treatment.

reinforcement as aligned short macroscopic VGCFs; a simple rule-of-mixture approach classified the CNFs as comparable to low modulus, medium-strength carbon fibres [112]. Interestingly, the CNFs were not as effective in a phenolic resin, possibly as a result of a better interfacial bonding in the epoxy composites [113].

The effect of nanotubes on either the curing reaction or the thermal degradation of thermosets has not yet been fully established. A change in curing is particularly relevant as the cross-link density has a pronounced influence on the mechanical performance, and nanotubes, particularly functionalised ones, both affect stoichiometry and act as massively parallel cross-linking sites. Acceleration of the epoxy cure reaction has been observed for untreated SWCNT bundles [114], the effect being most pronounced for low curing temperatures. This increase in reaction rate arises partly from the high thermal conductivity of the nanotubes but also depends on the specific surface area and surface chemistry, as shown by a comparative study of CNFs and carbon black [115]. The higher the degree of graphitisation of the filler surface, the less pronounced the effect on the curing rate. In addition, the presence of

SWCNTs degraded the thermal stability of the composite slightly [114]. Such effects still need to be addressed for chemically modified CNTs.

Both nanofibres and nanotubes have been used as electrically conductive fillers in epoxy composites [103,116–120]. Such composites display a characteristic percolation behaviour (see section 3.3.3 for a more detailed discussion in the context of thermoplastic nanocomposites). Both the percolation threshold and the maximum composite conductivity appear to depend on the type of nanoscale carbon filler and the degree of dispersion. In general, CNT-based composites have higher conductivities and lower percolation thresholds than either carbon black or CNF-based systems. Indeed, a CNT-filled epoxy currently shows the lowest percolation threshold observed in any system, at around 0.0025 wt% [121]. Figure 6 compares the best results achieved using MWCNTs from an aligned CVD-process with those obtained using entangled nanotubes and carbon black, in an aerospace grade epoxy

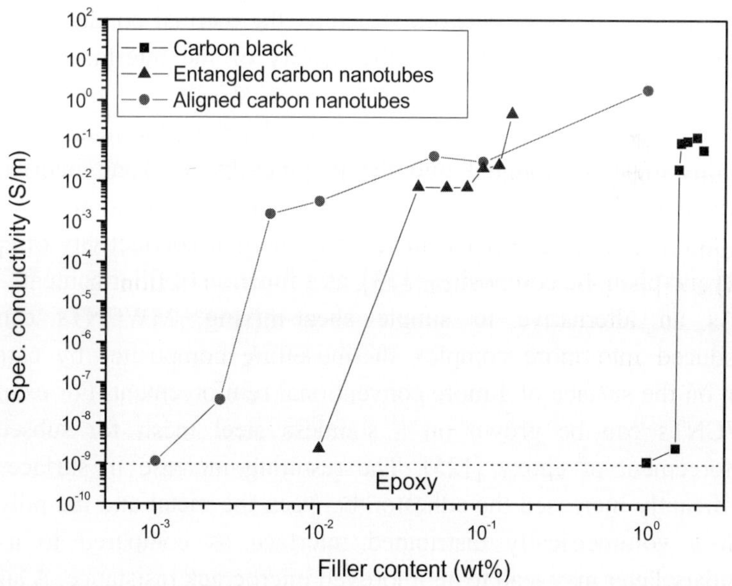

Fig. 6. Epoxy composite conductivity as a function of filler weight fraction for aligned CVD-grown multi-wall carbon nanotubes, compared to results obtained with commercially-available entangled nanotubes and carbon black particles.

system. It is tempting to attribute this low threshold simply to the high aspect ratio of the conductive filler. However, such a low value can only be explained in the light of a complicated dispersion and reaggregation behaviour during processing; in essence, well-dispersed nanotubes are destabilised and trapped just as a network forms [122]. The network formation behaviour can be manipulated not only by temperature and shear rate, but also by the application of external electrical fields [123], an approach that offers the possibility of achieving bulk conductive nanotube-polymer composites with anisotropic electrical properties, whilst maintaining a high degree of optical transparency.

In contrast, the thermal conductivity of cured nanotube-epoxy composites shows a minimal, or at best linear, increase with nanotube content. The enhancement appears to be greater for SWCNTs than CNFs [124], probably reflecting the intrinsic properties of the fillers, although it is impossible to rule out the effects of specific surface area and chemistry. In contrast to the electrical percolation behaviour, which is dominated by the filler network structure, the thermal conductivity of a composite is more sensitive to the quality of the interfacial bonding between filler and matrix. Whereas electrons travel along filler particles and can tunnel through remaining polymer barriers, thermally-activated phonons must be coupled into the polymer by a strong interface of intermediate thermal impedance. Even CNF loadings up to 40 wt% only lead to a moderate and linear increase in thermal conductivity of epoxy [112] and phenolic composites [113], as a function of filler content.

As an alternative to simple shear-mixing, MWCNTs can be introduced into more complex thermosetting composites by carrying them on the surface of a more conventional reinforcement. For example, MWCNTs can be grown on a stainless steel mesh for subsequent reinforcement of epoxy [125]. The resulting increase in surface area significantly increased the adhesion between the metal and the polymer. Such a volumetrically distributed interface as compared to a thin boundary layer may lead to an improved microcrack resistance. A similar effect has also been shown for catalytically-grown MWCNTs [126] and CNFs [127] on carbon fibre surfaces. Although the macroscopic fibre surface was somewhat damaged at the high temperatures used for the nanotube/nanofibre growth, single fibre fragmentation tests showed reduced fragment lengths for the coated carbon fibres compared to the as-received fibres in epoxies, implying an improvement of interfacial shear strength of up to 500%.

3.2. *Elastomeric carbon nanotube/nanofibre composites*

Surprisingly little effort has been directed towards using nanotubes and/or nanofibres as a reinforcement in elastomeric polymers. A comparative study on silicone-based nano-composites containing bundles of SWCNTs and CNFs has been reported by Frogley *et al.* [128]. Filler loading fractions of 1 and 4 wt%, respectively, were achieved. An evaluation of the mechanical tensile properties showed an approximately linear increase in composite stiffness, but also a reduction in strength and strain to failure for both types of filler, with increasing loading fraction. The increase in initial modulus (a linear stress-strain behaviour up to an elongation of 10% was assumed) was more pronounced for the bundles of SWCNTs compared to the nanofibres, although the (modest) CNF composite stiffnesses compare well with data for similar nanofibres in a rubbery epoxy [129]. However, at a higher strain, around 80%, the stiffness of the rubber composites was found to be similar, within experimental accuracy, independent of filler type and loading fraction. On the other hand, the addition of up to 10 wt% CNFs in rubbery epoxy did lead to a significant increase in composite strength and strain to failure [129]. Such increases in properties are not observed for comparable loading fractions of spherical carbon black particles in the same matrix, highlighting the influence of an increased aspect ratio. Catalytically-grown MWCNTs dramatically increased the modulus and strength of a similar flexible epoxy, but reduced the strain to failure at a loading of 4 wt% [117]. These differences in composite performance most likely reflect variations in the quality of the filler dispersion, since the entangled MWCNTs formed clusters which reduced the deformability of the composite. In general, the nanofillers have proportionately greater influence on the stiffness of rubbery matrices than hard ones, due to the lower intrinsic modulus.

The stiffness of elastomeric composites generally depends on the distribution of particles, whereas the dispersion is critical for the strength and strain to failure [130]. Fundamental issues regarding the matrix chemistry, especially the degree of cross-linking as a function of nanotube type and loading fraction, as well as the direct interfacial strength have not yet been established. However, it is well known that the surface area surface chemistry of carbon black has a pronounced influence on the resulting properties of elastomeric nanocomposites [131,132] and some similar effects may be anticipated.

3.3. *Thermoplastic carbon nanotube/nanofibre composites*

Approaches to the manufacture of nanotube/nanofibre thermoplastic composites cover a very broad range of processing technologies, including, in some cases, combinations of different methods. Standard techniques such as extrusion and injection-moulding are preferred for economical reasons, but are often inapplicable due to limited quantities of nanofiller or the desire for high volume fractions. A concise review of the literature poses some difficulties due to the vast number of experimental parameters that have been explored. The studies, although mostly consistent in themselves, generally evaluate rather different composite systems and seldom present both an assessment of mechanical performance and matrix morphology. Thus, a complete, consistent picture has not yet emerged.

3.3.1. *Thermoplastic nanotube/nanofibre composites processing: dispersion and alignment*

Composite processing machines ranging from bench-top, custom-made injection-moulding machines, suitable for small-volume composites, to large-scale extruders have been used to produce nanotube/nanofibre-filled compounds. In general, straightforward addition of nanotube/polymer mixtures to processing machines is complicated by the low apparent density (typically around 0.1 g cm^{-3}) of as-produced filler materials. Both solution- and dry-blending of the nanotube-polymer components have been used prior to the extrusion step [133–135]. In addition, this type of initial solution-compounding can lead to polymer-coated nanotubes, which may assist subsequent dispersion in other thermoplastic matrices [136].

Substantial shear forces appear to be necessary during the first composite processing step, in order to disperse nanotubes and nanofibres in the polymer, especially in the as-produced state and at high filler loadings. For example, filler contents up to 60 wt% of nanofibres [137] and about 30 wt% of MWCNTs [138] in thermoplastics have been realised using melt-compounding. In the case of nanotubes especially, the degree of nanotube dispersion depends on both the entanglement state of the as-received material and the particular processing technology. In general, the degree of dispersion is reasonable in thermoplastic systems, and better than in thermosets. As an example, Figure 7 shows the dispersion of commercial carbon nanofibres in a thermoplastic poly(ether ether ketone) (PEEK) matrix as a result of twin-screw extrusion at a nanofibre loading fraction of 15 wt%. The high intrinsic viscosity of thermoplastic matrices in general, has the dual advantage of

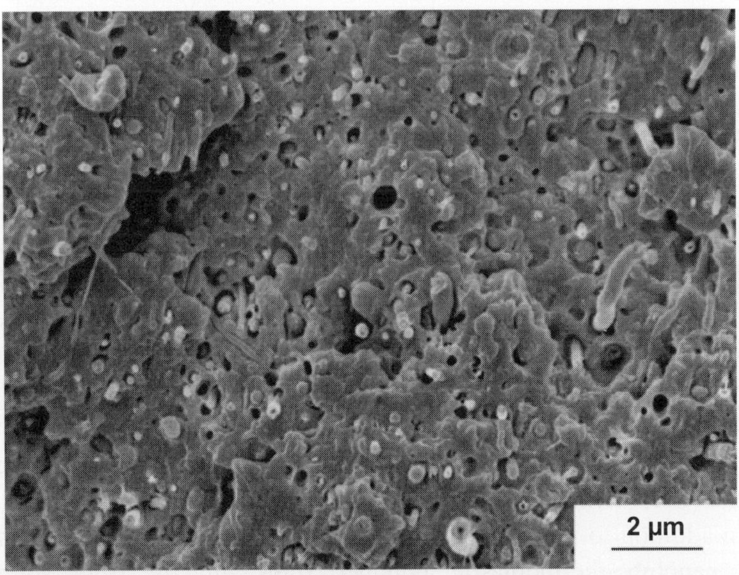

Fig. 7. Representative scanning electron micrograph showing dispersed carbon nanofibres in a poly(ether ether ketone) matrix at a filler content of 15 wt% after twin-screw extrusion.

increasing the shear applied to the aggregates (even breaking the individual CNTs/CNFs) and minimising the opportunity for reaggregation. However, even extensive twin-screw extrusion does not lead to a complete break-up of the entanglements in commercially available catalytically-grown MWCNTs [139–141]; nanotube clusters are observed even at low filler concentrations. Prolonged mixing times simply lead to an improved distribution of nanotube aggregates [140]. In contrast, nanotube weight fractions of up to 30 wt% of arc-grown as well as of CVD-grown MWCNTs were shown to disperse in poly(methyl methacrylate) (PMMA) [138] and polystyrene (PS) [135] matrices, and the use of aligned, rather than entangled CVD nanotubes, also appears to be advantageous [142]. Although filler dispersion can be aided by ball milling of the raw filler material prior to processing [143], this approach degrades the aspect ratio of individual particles significantly more than shear-intensive melt processing [144].

Interestingly, nanofibre loading fractions up to 10 wt% were found to have no significant influence on the shear viscosity of polypropylene (PP) composites in the shear rate regime typically encountered during thermoplastic processing [145]. In case of a polycarbonate (PC) matrix, the shear viscosity was even reduced with increasing nanofibre content up to 10 wt% [146], most likely as a result of pronounced shear alignment of the filler, a well known behaviour for short fibre-filled polymers [147]. At higher nanofibre loadings, however, the rheological behaviour of such polymer nanocomposites changes. A pronounced increase in shear viscosity, especially at low shear rates, reflects the presence of a nanofibre network structure and/or aggregates [145]. Similar rheological thresholds were observed for oxidised CNFs, at even lower loadings, indicating a greater interfacial interaction during processing [146,148] and for compounds based on entangled MWCNTs [139]. This rheological threshold depends on the nanotube/nanofibre type and treatment as well as on the polymer matrix and generally indicates the onset of interactions between individual filler particles or clusters.

In addition to melt-processing, significant efforts have been made to cast nanotube-thermoplastic polymer films directly from solution. However, most systems require large volumes of solvents in order to fully solubilise both the polymer and then the nanotubes. Common solvents are organic liquids of high toxicity such as toluene, chloroform, tetrahydrofurane (THF), or dimethyl formamide (DMF), although a number of aqueous systems have also been explored. Solution-casting has been used to manufacture MWCNT-containing PS [149–151], poly(hydroxyaminoether) (PHAE) [152,153], poly(vinyl alcohol) (PVA) [106,154], ultra-high molecular weight polyethylene (UHMWPE) [155] and PP [156] composite films with homogeneous nanotube dispersions. Similarly, SWCNT-containing PP [157], PVA [158,159] and PVA/PVP (poly(vinyl pyrrolidone)) [160] composite films have been prepared. In many cases, ultrasonication is used to aid nanotube dispersion in the liquid state, although prolonged high-energy sonication has the potential to introduce defects into nanotubes [161]; the ultrasonic treatment may also stabilise the dispersion by grafting polymer onto the CNT surface through trapping of radicals generated as a result of chain scission [162]. In addition, surfactants [160,163], polymer-functionalised nanotubes [158,164], and other chemical treatments of the constituents [165] are often employed. Last but not least, a number of studies have investigated the creation of nanotube-polymer composites [166–170] by *in-situ* polymerisation.

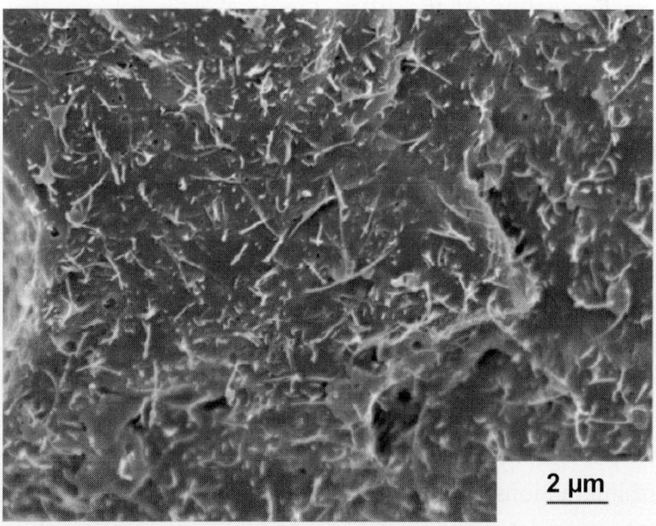

Fig. 8. Representative scanning electron micrograph showing dispersion of 5 wt% of CVD-grown MWCNTs in solution-cast and subsequently hot-pressed polypropylene.

The quality of the nanotube/nanofibre dispersion appears to strongly depend on the initial degree of entanglement of the as-prepared materials, as well as on the strength of the shear forces experienced during processing. Ideally, the individual nanotubes should form an inherently electrostatically or sterically stabilised dispersion, with a long lifetime relative to the casting process. The nature of the interaction between nanofiller and polymer is particularly important in relatively low viscosity solutions in which reaggregation can be rapid. A range of amphiphilic, water-soluble polymers have been found to solubilise SWCNTs efficiently in aqueous solution [171], effectively acting as surfactants. Alternatively, CNTs can be functionalised in order to produce stable dispersions in the desired solvent; for example, oxidisation produces electrostatically-stabilised MWCNTs in water [101]. Once is the solvent is removed, and the nanofillers are homogeneously dispersed, subsequent sample processing does not necessarily require shear forces. For example, the dispersion can be maintained during subsequent compression-moulding of solution-blended material [135,138,144]; the comparatively high melt viscosity of thermoplastics, as compared to epoxies for example, appears sufficient to prevent reaggregation of the filler. Figure 8 demonstrates that a good dispersion can be retained in a system of 5 wt% CVD-grown MWCNTs in polypropylene processed by solution-casting and subsequent hot-pressing.

Processing of compounds under conditions involving both shear and elongational flows, such as injection-moulding, can be used to induce alignment of the nanofiller [172]. Similarly, Kuriger *et al.* showed that flow-induced nanofibre alignment occurred during extrusion [144]; the degree of nanofibre alignment was improved by optimisation of the extruder die geometry. Keeping the extruded strand under tension minimised die swell effects and the orientation of the filler was maintained. However, a decreasing degree of alignment with increasing nanofibre content was observed, most likely as a result of nanofibre-nanofibre interactions altering the flow field [144]. Similarly, drawing of composite extrudates was shown to induce significant nanotube alignment [135] as did mechanical stretching of solid nanocomposites above the glass transition temperature [152,153], spin-casting of nanotube-polymer solutions [151], and the application of magnetic fields during *in-situ* polymerisation [173].

In addition, a number of studies have investigated the direct melt-spinning of composite fibres [142,174–181]. Similar to hot-drawing of nanocomposites [152], such melt-spinning leads to an improved alignment of the nanoscale filler and the polymer matrix, thereby maximising composite performance. As shown in Figure 9 for a melt-spun polyamide-12 nanocomposite fibre containing 5 wt% of CVD-grown MWCNTs, such melt-spinning leads to aligned nanotubes in the polymer fibre while maintaining a good quality of the overall fibre surface finish. In the case of SWCNT bundles in an isotropic petroleum pitch matrix, a pronounced influence of the filler on the melt elongation behaviour of the composite was observed [174]. Such an increase in melt elongation properties, with nanotube addition, enhances the spinnability of polymers. For example, fine CNF-polyester (PET) fibres with diameters as low as 25 μm can be produced under stable conditions [178]. Measurements of the extensional rheology of PEEK/CNF blends have quantitatively demonstrated improvements in extensional viscosity and melt strength, and have shown that the resulting stabilisation of the melt allows the production of novel PEEK foams [182]. In fact, as shown in Figure 10, at high elongation shear rates, there is a crossover leading to reduced viscosity, but at rates relating to foaming processes [183],

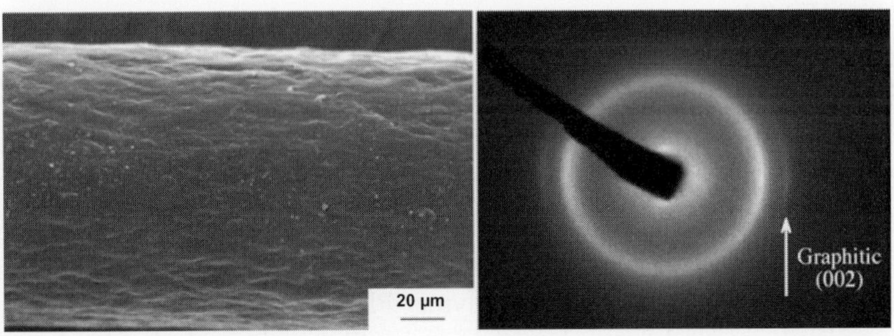

Fig. 9. Scanning electron micrograph of surface finish of melt-spun polyamide-12 nanocomposite fibre containing 5 wt% of CVD-grown MWCNTs and the corresponding 2D WAXS fibre diffraction pattern highlighting the nanotube alignment (evidenced by the anisotropic arc in the graphitic (002) peak arising from the intershell spacing).

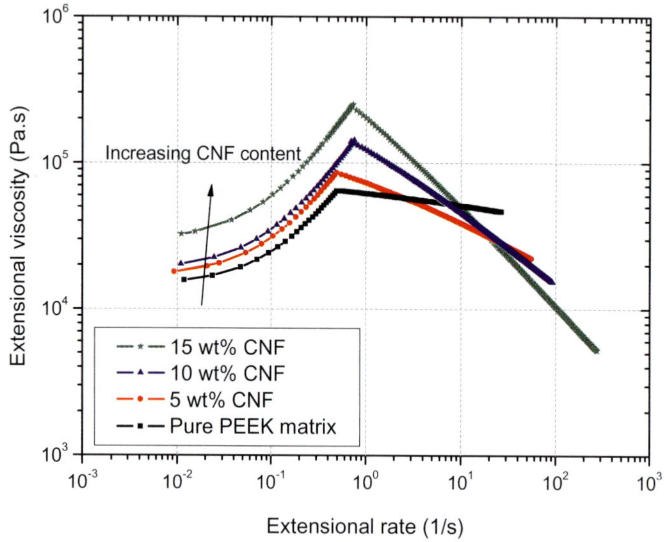

Fig. 10. Apparent elongational viscosity of a PEEK matrix at 360°C as a function of the strain rate with increasing nanofibre content as determined by unaxial Rheotens melt elongation experiments.

bubble expansion is stabilised. Carbon nanostructures can thus beneficially alter the processing behaviour of polymers, as well as enhancing the properties of the resulting composite solids [184].

In addition to melt processing, solution-spinning has also been successfully used to produce high loading fraction SWCNT-PVA fibres. Surfactant-stabilised dispersions of CNTs are injected into a PVA bath, forming a fibre that can be handled and drawn [185,186]. There is a desire to exploit lyotropic liquid crystalline phases to produce fibres with improved alignment. Some promising steps have been taken in this direction using pure CNT dispersions [187,188], as well as combinations of SWCNTs with the rigid rod polymer PBO [167]. Solution-processing has also been used to produce composite films using a Layer-By-Layer technique to deposit alternating thin layers of negatively charged SWCNTs and positively charged polyelectrolyte [189].

3.3.2. *Mechanical properties of thermoplastic nanotube/nanofibre composites*

The published mechanical data show that the tensile modulus of nanotube/nanofibre-thermoplastic composites is generally improved, although a detailed comparison of the data is difficult due to the different types of fillers, surface treatments, matrices, processing techniques, and test methods that have been used. In general, the stiffening effect of nanotubes and nanofibres appears to be more prominent in semicrystalline rather than amorphous thermoplastics, possibly due to the nucleation effects discussed below. Although a linearly increasingly composite stiffness has been observed up to 15 wt% of nanofibres [172], in many cases, a non-linear relationship is observed with increasing filler loading fraction. Even when a homogeneous dispersion of the nanoscale filler is claimed for all concentrations, the stiffness enhancement is usually most prominent for low filler weight fractions, with the critical concentration depending on the specific materials and processing conditions used. This observation might relate to the decreasing nanofiller alignment with increasing weight fraction [144,151], although alignment variations of the polymer matrix have not properly been taken into account. In semicrystalline matrices the (often unanalysed) increases in crystallinity may be the source of the non-linearity. Furthermore, in most cases, there are probably changes in dispersion (which are notoriously hard to quantify), as the larger surface areas associated with high loading fractions become increasingly difficult to accommodate within the polymer; similar effects have been seen in nanoclay-filled polymers [190]. However, even in the presence of nanotube clusters, enhancements in composite stiffness can be observed [134]. Overall, some interesting trends in composite stiffness can be distinguished. For example, Figure 11 shows a comparative plot of the tensile modulus of melt-compounded nanofibre-reinforced (a) amorphous PC [146] and PMMA [133], and (b) semicrystalline PP [143,144,191] and PEEK [192] composites as a function of reported nanofibre weight and volume fraction, respectively.

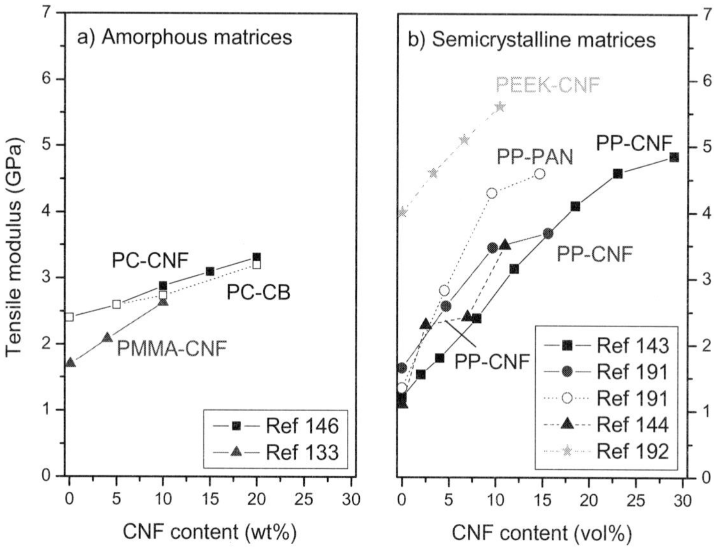

Fig. 11. Comparative evaluation of composite stiffness increase for nanofibre-reinforced (a) amorphous and (b) semicrystalline matrices as a function of nanofibre loading.

Considering the case of the amorphous matrices, in Figure 11a, both studies revealed a linear increase in composite stiffness with increasing nanofibre weight fraction. A good dispersion of nanofibres was claimed [133,146], independent of filler weight content. Furthermore, partial nanofibre alignment in the direction of the tensile axis can be assumed as a result of the shear flow conditions during production, and was verified by extensive image analysis of TEM images in the case of the PMMA composites [133]. More disappointingly, comparable PC composites containing dispersed equiaxed carbon black particles showed an identical increase in composite tensile modulus to the CNF composites [146]. The relative increase in composite tensile modulus with increasing nanofibre content is only slightly higher for the PMMA composites but agrees well with other data for melt-spun composite fibres [180]. Turning to the melt-compounded, semicrystalline, PP nanocomposites shown in Figure 11b, more pronounced increases in composite tensile stiffness, with the addition of nanofibres, are apparent, as compared to the amorphous composites. This increase is linear only up to a critical nanofibre volume fraction which varies between the studies. A direct comparison to values for short PAN-based carbon fibres in PP (processed under similar conditions) indicates a smaller reinforcement for the nanofibres. In the case of the PEEK data, the increase in stiffness is linear (a constant degree of crystallinity was confirmed); the slope can be used to extract a modulus of the CNFs if a simple short fibre model is assumed and if the aspect ratio and orientation are known [172]. Experimentally, these factors are hard to establish accurately, but estimates put the CNF tensile modulus at around 100 GPa, lower than that of CNTs grown at high temperature but consistent with expectations for CVD-grown CNFs.

Comparable loading fractions of MWCNTs, rather than CNFs, in both solution-cast [149,151] and extruded PS films, produced a more pronounced stiffness enhancement, indicating that the crystalline quality of the nanoscale filler contributes to the overall performance of the nanocomposite. Similarly, comparisons of different types of filler in PA-12 composite fibres showed that well-dispersed CNTs produce a greater improvement in stiffness than CNFs, but that the type of nanotube also has an effect [142]. In addition, a direct comparison of aligned versus unoriented nanotubes showed a significant enhancement of the composite stiffness for the aligned nanotubes [135].

Dynamic mechanical testing as a function temperature has shown that the stiffening effect of the nanotubes is more pronounced above the softening point of the matrix, in both MWCNT-PMMA [138] and MWCNT-PVOH [106] composites. In addition, the glass transition temperature is often increased. This observation indicates that the matrix mobility is influenced by the presence of the nanotubes, a common effect for polymer systems containing finely dispersed particles [193].

Turning to other mechanical properties, enhancements in composite yield stress, strength, and toughness generally appear more difficult to achieve, especially for filler loading fractions exceeding about 10 wt%. Improvements in impact strength of nanofibre and nanotube-reinforced PMMA composites [133] and in tensile strength of amorphous MWCNT-PS nanocomposites [135,149,151] have been reported. These properties depend on the homogeneity of specimens achieved during processing as well as on interfacial issues relating to the specific filler types and matrices. For example, the impact properties of nanofibre-PC composites were significantly decreased, even at low nanofibre contents, most likely as a result of aromatic hydrocarbons on the nanofibre surface enhancing chemical stress cracking of the polycarbonate [146]. The most prominent strength enhancements of bulk nanocomposites have been achieved for well-dispersed and aligned nanofibres in PEEK [172], up to filler loading fractions of 10 vol%, and in PP [144], at nanofibre loading fractions as low as 5 vol%. In the PP case, a further sub-linear strength increase up to 11 vol% CNFs was reported. This deviation from a linear relationship most likely reflects the decreasing nanofibre alignment at higher filler contents, but might also be a consequence of microstructural variations and an increasing void content at higher concentrations [112]. Improvements in yield strength have also been reported in a number of nanocomposite fibre systems, as discussed later.

CNTs and CNFs are also interesting additives for tribological applications [194]; they can significantly reduce the wear rate of polymers, apparently independently of the degree of nanotube dispersion achieved during processing. Although the exact wear reducing mechanism of the nanoscale constituent is as yet not clear, there appear to be number of beneficial effects contributing to the overall performance increase. Firstly, the basic strength and stiffness of the nanocomposites is enhanced; secondly, the small size of the nanomaterials results in smaller wear debris particles and therefore less pronounced roughening of the bearing counterparts; thirdly, the carbon-based fillers may act, or break up to act, as solid lubricants under dry sliding conditions. These nanoscale fillers can be easily compounded together with standard, more established tribological aids (such as PTFE or carbon fibres), to allow fine-tuning of the resulting overall mechanical performance. As an example, the observed improvements in specific wear rate of a range of commercial high-performance PEEK compounds with the addition of 10 wt% of carbon nanofibres are shown in Figure 12. The wear-reducing potential of carbon nanostructures is currently being studied in a range of ultra-high molecular weight polyethylene composites for medical joint replacement applications [195]. In addition, of course, the nanofillers provide the means to improve the wear

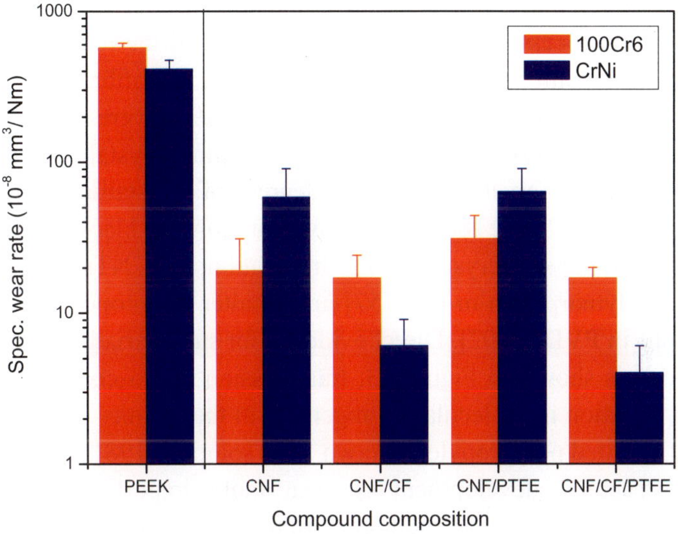

Fig. 12. Influence of a 10 wt% carbon nanofibre loading on the wear performance of a PEEK matrix and a range of commercial PEEK compounds for tribological applications against two different steels.

properties of microstructured parts in which more conventional fillers cannot be accommodated, and in which problems of erosion are particularly problematic. As an example, Endo and colleagues helped Seiko develop a CNF-filled nylon watch gear, less than 200 μm in diameter.

A number of studies have shown a significant influence of the nanoscale fillers on the resulting morphology of a semicrystalline matrix. The microstructure, in terms of crystal structure, crystallinity, and crystal orientation, of a semicrystalline polymer is dependent on its thermal history, the manufacturing process employed, and the presence of possible nucleation sites. The combination of high shear and elongational flows, occurring during processing under non-isothermal conditions, leads to complex variations in molecular orientation and crystal morphology. Furthermore, there is a direct influence of reinforcing fibres on the surrounding matrix which depends on the interfacial interactions, the fibre volume fraction, as well as the aspect ratio and orientation of the fibres [196,197]. Finally, relative movements between filler and matrix during processing result in a modification of the local flow field and molecular conformation. In the case of PP, nanofibres [137,198], MWCNTs [156,198], and SWCNTs [157,179,199] were all found to alter the crystallisation kinetics and the resulting crystal structure in terms of average crystal size and degree of crystallinity. Similar nucleation effects have been observed for carbon black [200] and nanoclays [201] in semicrystalline PP. In addition, SWCNTs were shown to induce polymorphism in PP [157] and to alter the proportions of the polymorphs in PVDF [202]. Direct evidence for such variations in matrix morphology is shown in Figure 13; transmission microscopy highlights the clear variation in molecular arrangement in the vicinity of nanofibres in highly oriented polypropylene drawn from solution [198].

Microstructural development is even more important for highly oriented polymers, such as melt-spun fibres. There are two important deformation processes which govern the resulting mechanical and physical properties of melt-spun thermoplastic fibres: melt- drawing and subsequent drawing in the solid state. The final orientation of polymer molecules depends on the relative draw ratios applied during the two process stages [203] and such orientations influence the crystallization kinetics [204]. Filler particles acting as nucleation sites can further alter

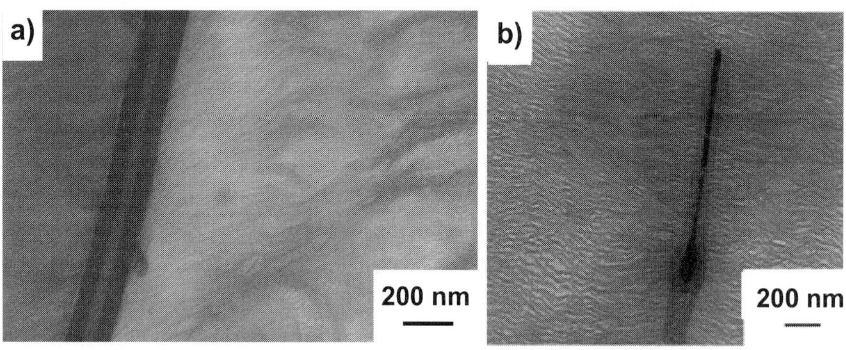

Fig. 13. Transmission electron micrographs of variations in local polymer crystallinity in the vicinity of nanofibres in highly oriented PP, drawn from solution (with permission from [198]). In (a) note the debonded interface on the left, and the associated lack of matrix modification.

the structural development [204]. One would therefore expect nanofibres and nanotubes to have a pronounced effect on the microstructure of spun composite fibres. Such effects are likely to prove even more prominent than those observed for colouring pigments and common nucleating agents such as talc [205], since the high aspect ratio of dispersed nanotubes and nanofibres could lead to a significant load transfer in the molten state, as observed experimentally [174,182].

Melt-spinning of nanofibre-reinforced PP composite fibres significantly improved the relative modulus increase, most likely as a result of enhanced nanofibre alignment [175,176]. This improvement strongly depended on the draw ratio of the composite fibres and on the filler content [175]. Increasing the CNF content above 5 vol% did not further increase the composite modulus at higher draw ratios, as in the case of amorphous polymer composite fibres. On the other hand, the initial stiffening effect of small additions of SWCNTs to PP fibres dwarfs the impact of CNFs but appears to saturate at loading fractions as low as 1–2 wt% [177]. The greater impact of the SWCNTs may be attributed to their intrinsically greater perfection and their flexibility which may allow them to align better during processing. The saturation effect is not yet fully understood, although, again, it may relate to the difficulty of accommodating the very high surface area of SWNCTs. On the other hand, the possible microstructural variations were not considered and initial increases in crystallinity, due to nucleation effects, might contribute to the rapid improvements in composite stiffness. In contrast, a modest stiffening effect but a large increase in nanocomposite strength and toughness has been reported for entangled MWCNTs in UHMWPE films [155]. The toughness of these nanocomposites was further improved by a subsequent drawing process at elevated temperatures which led to an increased ductility compared to the pure polymer. Here, electron microscopy provided direct evidence for nanotube-nucleated shish-kebab PE crystals, an effect that might account for the observed ductility improvement in these semicrystalline nanocomposites. Attempts to improve the performance of existing high performance fibres has met with mixed success, but the introduction of SWCNTs into PBO has been shown to improve both stiffness and especially strength (by 50%), demonstrating that nanotubes do have the potential to extend the ultimate mechanical performance of materials [167].

Some of the other most promising improvements in mechanical properties have appeared in solution-processed PVA nanocomposite fibres [185,186] which contain high loading fractions of SWCNTs, up to 60 wt%. The strength and stiffness of these fibres, of around 1 GPa and 100 GPa, respectively, are relatively high, although still modest compared to carbon fibres (and theoretical predictions); however, the strain to failure can be large, leading to a very high-energy absorption; Figure 14 shows that the combined strength and strain to failure of these fibres lies outside the envelope of conventional materials. Similar, basic fibre properties have been obtained from pure SWCNT fibres, either as-grown [206] or spun from a lyotropic superacid solution [188]. There is still plenty of scope for improving the alignment and density of nanotubes within these fibres. The Layer-By-Layer assembly of SWCNT films produced a 2D tensile strength of 220 MPa, comparing favourably with engineering ceramics [189].

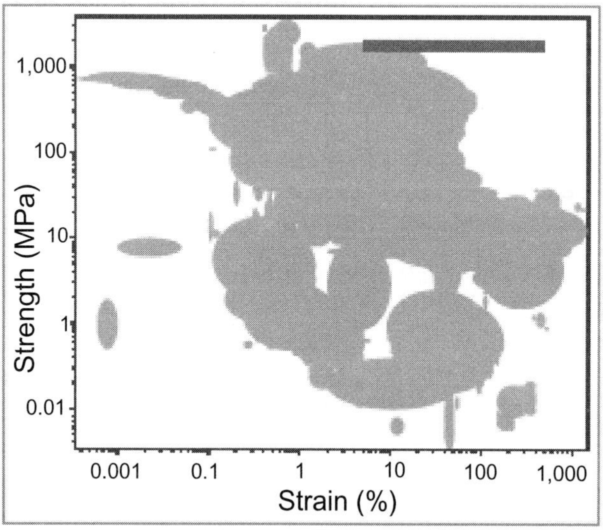

Fig. 14. Comparison of the strength and failure strain for carbon nanotube composite fibres for different degrees of initial pre-draw (red line) and the 3,000 materials of all types (grey field) in the Cambridge Materials Selector database. Taken from [186].

3.3.3. *Transport properties of thermoplastic nanotube/nanofibre composites*

As well as mechanical reinforcement, there is considerable interest in functional nanocomposites in order to exploit the unique physical properties, such as high thermal or electrical conductivity, of the constituents. Electrically conductive polymer composites, for example, are used in anti-static packaging applications, as well as in specialised components in the electronics, automotive, and aerospace sector. The incorporation of conductive filler particles into an insulating polymer host leads to bulk conductivities at least exceeding the anti-static limit of 10^{-6} S/m. Common conductive fillers are metallic or graphitic particles in any shape (spherical, platelet-like or fibrous) and size. However, the incorporation of CNTs allows for a low percolation threshold, a high quality surface finish, a robust network, and good mechanical properties – a combination not obtained with any other filler. The use of CNTs/CNFs as a conductive filler in thermoplastics is their biggest current application, and is widespread across the automotive and electronic sectors.

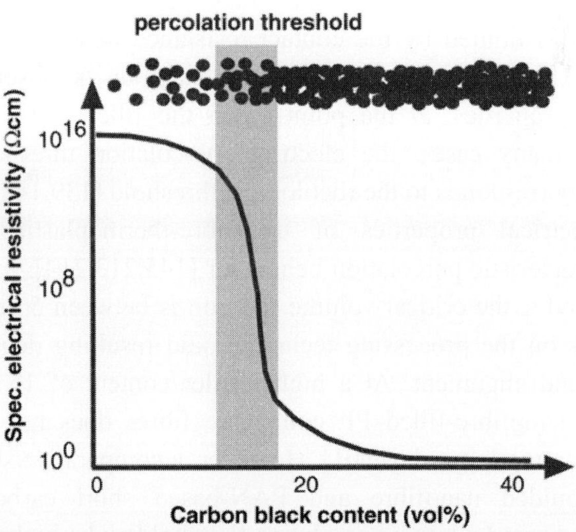

Fig. 15. Schematic representation of the electrical resistivity in a carbon black-filled functional polymer composite with increasing filler loading fraction (according to [207]).

In general, the electrical conductivity of a particulate composite reveals a non-linear increase with the filler concentration, passing through a percolation threshold, as shown in Figure 15. At low filler concentrations, the conductive particles are separated from each other and the electrical properties of the composite are dominated by the matrix. With increasing filler concentration local clusters of particles are formed. At the percolation threshold, ϕ_c, these clusters form a connected three-dimensional network through the component, resulting in a jump in the electrical conductivity.

Close to the percolation threshold, the electrical conductivity follows a power-law of the form

$$\sigma_0 \propto \left(\phi_V - \phi_c\right)^t \qquad \phi_V > \phi_c, \tag{2}$$

where ϕ_V is the volume fraction of the filler [208]. The exponent t in this equation was found to be surprisingly uniform for systems of the same dimensionality. For three-dimensional percolating systems t varies between 1.6 and 2, in simulations [209,210]. The percolation threshold is reduced on increasing the aspect ratio [211], but the maximum conductivity is limited by the contact resistance between neighbouring particles [212]. A related percolation behaviour is observed in the rheological properties, at the point when the filler particles begin to interact. In many cases, the electrical percolation threshold of bulk composites corresponds to the rheological threshold [139,141].

The electrical properties of nanofibre-thermoplastic composites exhibit characteristic percolation behaviour [145,213,214]. In the case of untreated CNFs, the critical volume fraction is between 5 and 10 vol%, but depends on the processing technique and resulting degree of CNF dispersion and alignment. At a higher filler content of 15 vol%, even drawing of nanofibre-filled PP composite fibres does not destroy the conductive network [gordeyev01]. However, a comparative study of bulk injection-moulded nanofibre and PAN-based short carbon fibre-PP composites showed a lower percolation threshold and a higher maximum bulk conductivity for the macroscopic filler [175].

The electrical properties of CNF-composites are influenced by surface and thermal treatments [214]. Oxidation increased the percolation threshold and decreased the maximum bulk conductivity, whereas high-temperature graphitisation produced the opposite trends. These results verify that the electrical performance of the composite depends on the intrinsic conductivity as well as the dispersion and alignment of the filler. Oxidation encourages interaction with the polymer, increasing the contact resistance, whereas graphitisation both reduces polymer interactions and improves the intrinsic conductivity. As an alternative to graphitisation, electro-deposition of copper on nanofibre surfaces has been shown to improve the maximum bulk composite conductivity [148].

Similarly, the electrical percolation threshold of thin MWCNT-thermoplastic films also depends on the type of nanotube and surface treatment. Threshold values from around 5 wt% for oxidised catalytic MWCNTs in PVA [106] to around 0.06 wt% and 0.5 wt% for arc-discharge MWCNTs in PVA [215] and PMMA [216], respectively, have been reported. Interestingly, the optical transparency of such conductive thin film composites for anti-static applications can be significantly improved by using SWCNTs [169], even in bundled form. As discussed previously for epoxy systems, these very low percolation thresholds are far below the expected values for randomly distributed fibres, and are the result of active aggregation processes.

A significant reduction in the critical nanotube volume fraction for electrical percolation can be achieved by exploiting the concept of double percolation through the formation of a co-continuous morphology in nanotube-filled polymer blends [141,217,218]. This concept of double percolation was introduced by Sumita *et al.* [219] who achieved percolation of carbon black in the continuous phase of a polymer blend. Similar success has been demonstrated for nanofibres in a PE/PMMA blend [220].

As in the case of the thermosetting systems, the thermal conductivity of nanofibre-thermoplastic composites does not show a percolation transition, even at higher filler volume fractions [112,144,213]. A linear increase in thermal conductivity is observed, although the magnitude depends to some extent on the alignment of the filler [144], in agreement with data for short carbon fibre composites [221]. The overall performance increase for nanofibres was similar to that observed for short carbon fibres in a similar system [213].

4. Conclusions

Although commercial nanotube-polymer composites exist today, they almost exclusively employ relatively low loadings (3–5 wt%) within thermoplastic matrices for the purposes of anti-static dissipation, particularly in the automotive and electronics industries [222]. Such applications exploit bulk quantities of relatively defective catalytically-grown materials. On the other hand, individual perfect nanotubes appear to have axial stiffnesses towards that of diamond, and strengths ten times that of any other available material. There are, therefore, considerable efforts underway to exploit these properties in macroscopic structural composites. In addition to these remarkable headline mechanical properties, there is interest in thermal conductivity, thermal stability, flame retardance, wear resistance, and so on. However, the successful exploitation of the promising mechanical and other properties of CNTs and CNFs in polymer composites is as yet hindered by a number of fundamental issues.

It has become clear that issues of dispersion, alignment, and stress transfer are crucial, and often problematic at this size scale. Dispersion is often obtained by using unentangled nanotubes, high viscosities, and high shear rates. However, a more subtle approach uses surface modifications or coatings on the nanofiller to stabilise individual particles. Surface modification has the added advantage of improving stress transfer to the matrix (although it tends to increase contact resistance). The drawback with the direct modification of the filler surface is that it will damage the properties of SWCNTs and the outer shell(s) of MWCNTs. The area of surface chemistry of nanotubes is therefore an important area for future development. A degree of alignment has been successfully obtained using shear and elongation, as well as, to a lesser extent, magnetic and electrical fields. However, as with the development of high-performance polymers, producing materials that approach perfect orientation will be challenging.

One major uncertainty is the type and quality of nanotubes that should be used. A wide variety of synthesis methods have been employed, yielding nanotubes of different size, aspect ratio, crystallinity, crystalline orientation, purity, entanglement, and straightness. All these factors affect the processing and properties of the resulting composites but it has not yet emerged what the 'ideal' carbon nanotube would be; the answer may vary with the matrix and application. As an example, consider even the question of the ideal nanotube diameter. Very small diameters, particularly single-wall nanotubes, are relatively flexible, potentially leading to lower viscosities and greater robustness during processing, but also to potentially persistent entanglements. The higher the surface area, the greater the impact of the nanotubes on matrix conformation and crystallinity; the nature and significance of such effects are not yet clear but depend on the polymer used. For the smallest nanotube sizes, where the diameters of the nanotubes and polymer molecules are similar, particulate concepts from composite or nucleation theory may no longer be helpful. The composite is essentially a polymer-polymer blend, although one with rather unusual characteristics which may give rise to new behaviours.

One major difficulty with small diameter nanotubes is that they become increasingly difficult to wet. By trivial estimation, even a 1 vol% loading of single-wall nanotubes ensures that all of the polymer molecules are within one radius of gyration (say 5 nm) of a nanotube. This result implies that a complete wetting of high loading fractions of single-wall nanotubes will be difficult, at least by conventional means, and that even more modest concentrations may be brittle and hard to process due to the constraint of the matrix. Alternative approaches, based on layer-by-layer assembly and possibly lyotropic spinning of single-wall nanotube solutions have already been explored and proven to be interesting routes to thin films and fibres. However, for the purpose of simple, bulk composites, intrinsically straight, highly crystalline, multi-wall nanotubes might be expected to yield the best mechanical properties, as long as internal shear failures can be minimised. It remains to be seen whether there is an optimal defect concentration that prevents internal sliding without harming the intrinsic properties excessively. Strain arguments would suggest that smaller multi-wall nanotubes will be less prone to shear failure than their larger relatives. When taken together with the supposed greater perfection of higher curvature nanotubes due to self-selection during growth, it might be hypothesised that small diameter multi-wall nanotubes, with say 2–4 shells, should be preferred.

5. Outlook

Whichever type of nanotube is selected, it will be necessary to develop new synthesis routes. At present, catalytic MWCNTs and CNFs appear as the optimum choice, given that such materials can be most readily obtained in large quantities with a high purity. However, these materials are intrinsically defective and wavy, both of which are expected to be highly detrimental to the mechanical performance [223]. Somehow nanotubes with a crystalline quality closer to arc-grown nanotubes need to be obtained at a cost similar, or indeed below, current CVD-grown products.

As the absolute size of the reinforcement decreases towards the size of the polymer molecules, interactions between filler and matrix become more important, both during processing as well as in the solid-state. Significant progress has been made in under-standing the interactions between polymers and flat surfaces, but the interactions of polymers with highly curved surfaces at the molecular scale, are still largely unknown. Variations in crystallinity are important but the effects of constraint and other changes in polymer morphology in the vicinity of highly curved surfaces may also have significant effects on the deformation behaviour of the composite. As in biological nanocomposites, a high strength might not be exclusively linked to the intrinsic strength of the filler but might reflect increased yield stresses in the vicinity of the filler, as recently observed during pull-out experiments [224]. Given these issues and the change in scale towards molecular dimensions, it is not surprising that concepts of traditional fibre-reinforced composites are often not applicable to nanocomposites. Further work is required to provide a sound theoretical basis which will allow the successful prediction of the resulting mechanical and physical properties of nanotube/nanofibre-based polymer composites.

Carbon nanotubes and nanofibres may not produce practical replacements for existing high-performance materials in the near future. However, there is continuing market for electrically conducting polymer compounds, and immediate potential to develop the reinforcement of delicate composite structures such as thin films, fibres, and the matrices of conventional fibre composites. Indeed, market projections for polymer nanocomposite technology show a 160 million lb market for carbon nanotube-filled products by 2009 [225]. Although the full potential of nanotube composites remains to be realised, much progress has been made, and these nanocomposite systems have a bright future once the fundamental questions are resolved.

References

1. P.M. Ajayan, P. Redlich, M. Rühle; "Structure of carbon nanotube-based nanocomposites", J. Microsc.-Oxford 185(2) (1997), 275–282.
2. H.J. Gao, B.H. Ji, I.L. Jäger, E. Arzt, P. Fratzl; "Materials become insensitive to flaws at nanoscale: Lessons from Nature", Proc. Nat. Acad. Sci. USA 100(10) (2003), 5997–5600.
3. L.T. Drzal, M.J. Rich, M.F. König, P.F. Lloyd; "Adhesion of graphite fibers to epoxy matrices. 2. The effect of fiber finish", J. Adhesion 16(2) (1983), 133–152.
4. S.H. Wu; "A generalized criterion for rubber toughening – the critical matrix ligament thickness", J. Appl. Polym. Sci. 35(2) (1988), 549–561.
5. A. Galeski; "Strength and toughness of crystalline polymer systems", Prog. Polym. Sci. 28(12) (2003), 1643–1699.
6. T. Ebbesen; "Carbon Nanotubes, Preparation and Properties", CRC Press USA (1997).
7. H.W. Kroto, J.R. Heath, S.C. O'Brian, R.F. Curl, R.E. Smalley; "C_{60}: Buckminsterfullerene", Nature 318(6042) (1985), 162–163.
8. W. Krätschmer, L.D. Lamb, K. Fostiropoulos, D.R. Huffmann; "Solid C_{60}: A new form of carbon", Nature 347(6291) (1990), 354–358.
9. S. Iijima; "Helical microtubules of graphitic carbon", Nature 354(6348) (1991), 56–58.
10. A. Oberlin, M. Endo, T. Koyama; "Filamentous growth of carbon through bezene decomposition", J. Cryst. Growth 32(3) (1976), 335–349.
11. R. Bacon; "Growth, structure, and properties of graphite whiskers", J. Appl. Phys. 31(2) (1960), 283–290.
12. P.J.F. Harris; "Carbon nanotubes and related structures", Cambridge University Press (2001).
13. D.S. Bethune, C.H. Kiang, M.S. de Vries, G. Gorman, R. Savoy, J. Vazquez, R. Beyers; "Cobalt-catalysed growth of carbon nanotubes with single-atomic-layer walls", Nature 363(6430) (1993), 605–607.

14. C. Journet, W.K. Master, P. Bernier, A. Loiseau, M. Lamy de la Chapelle, S. Lefrant, P. Deniard, R. Lee, J.E. Fischer; "Large-scale production of single-walled carbon nanotubes by the electric-arc technique", Nature 388(6644) (1997), 756–758.
15. A. Thess, R. Lee, P. Nikolaev, H. Dai, P. Petit, J. Robert, C. Xu, Y.H. Lee, S.G. Kim, A.G. Rinzler, D.T. Colbert, G.E. Scuseria, D. Tomanek, J.E. Fischer, R.E. Smalley; "Crystalline ropes of metallic carbon nanotubes", Science 273(5274) (1996), 483–487.
16. H.M. Cheng, F. Li, G. Su, H.Y. Pan, L.L. He, X. Sun, M.S. Dresselhaus; "Large-scale and low-cost synthesis of single-walled carbon nanotubes by the catalytic pyrolysis of hydrocarbons", Appl. Phys. Lett. 72(25) (1998), 3282–3284.
17. J. Tersoff, R.S. Ruoff; "Structural properties of a carbon-nanotube crystal", Phys. Rev. Lett. 73(5) (1994), 676–679.
18. T.W. Ebbesen, P.M. Ajayan; "Large-scale synthesis of carbon nanotubes", Nature 358(6383) (1992), 220–222.
19. D.T. Colbert, J. Zhang, S.M. McClure, P. Nikolaev, Z. Chen, J.H. Hafner, D.W. Owens, P.G. Kotula, C.B. Carter, J.H. Weaver, A.G. Rinzler, R.E. Smalley; "Growth and sintering of fullerene nanotubes", Science 266(5188) (1994), 1218–1222.
20. M. Cadek, R. Murphy, B. McCarthy, A. Drury, B. Lahr, R.C. Barklie, M. in het Panhuis, J.N. Coleman, W.J. Blau; "Optimisation of the arc-discharge production of multi-walled carbon nanotubes", Carbon 40(6) (2002), 923–928.
21. A.G. Rinzler, J. Liu, H. Dai, P. Nikolaev, C.B. Huffman, F.J. Rodriguez-Macias, P.J. Boul, A.H. Lu, D. Heymann, D.T. Colbert, R.S. Lee, J.E. Fischer, A.M. Rao, P.C. Eklund, R.E. Smalley; "Large-scale purification of single-wall carbon nanotubes: Process, product, and characterization", Appl. Phys. A 67(1) (1998), 29–37.
22. M. Endo, K. Takeuchi, S. Igarashi, K. Kobori, M. Shiraishi, H.W. Kroto; "The production and structure of pyrolytic carbon nanotubes (PCNTs)", J. Phys. Chem. Solids 54(12) (1993), 1841–1848.
23. C.N.R. Rao, A. Govindaraj, R. Sen, B.C. Satishkumar; "Synthesis of multi-walled and single-walled nanotubes, aligned-nanotube bundles and nanorods by employing organometallic precursors", Mater. Res. Innovat. 2(3) (1998), 128–141.
24. R. Andrews, D. Jacques, A.M. Rao, F. Derbyshire, D. Qian, X. Fan, E.C. Dickey, J. Chen; "Continuous production of aligned carbon nanotubes: A step closer to commercial realization", Chem. Phys. Lett. 303(5–6) (1999), 467–474.
25. Z. Shi, Y. Lian, F.H. Liao, X. Zhou, Z. Gu, Y. Zhang, S. Iijima; "Large scale synthesis of single-wall carbon nanotubes by arc-discharge method", J. Phys. Chem. Solids 61(7) (2000), 1031–1036.
26. Y. Saito, K. Nishikubo, K. Kawabata, T. Matsumoto; "Carbon nanocapsules and single-layered nanotubes produced with platinum-group metals (Ru, Rh, Pd, Os, Ir, Pt) by arc discharge", J. Appl. Phys. 80(5) (1996), 3062–3067.
27. C. Liu, H.-M. Cheng, H.T. Cong, F. Li, G. Su, B.L. Zhou, M.S. Dresselhaus; "Synthesis of macroscopically long ropes of well-aligned single-wall carbon nanotubes", Adv. Mater. 12(16) (2000), 1190–1192.

28. T. Guo, P. Nikolaev, A. Thess, D.T. Colbert, R.E. Smalley; "Catalytic growth of single-walled nanotubes by laser vaporization", Chem. Phys. Lett. 243(1–2) (1995), 49–54.
29. A. Oberlin, M. Endo, T. Koyama; "High resolution electron microscope observations of graphitized carbon fibers", J. Cryst. Growth 14 (1976), 133–135.
30. N. Krishnankutty, N.M. Rodriguez, R.T.K. Baker; "Effect of copper on the decomposition of ethylene over an iron catalyst", J. Catal. 158(1) (1996), 217–227.
31. F.W.J. Van Hattum, J.M. Benito-Romero, A. Madronero, C.A. Bernado; "Morphological, mechanical and interfacial analysis of vapour-grown carbon fibres", Carbon 35(8) (1997), 1175–1183.
32. N.M. Rodriguez; "A review of catalytically grown carbon nanofibres", J. Mater. Res. 8(12) (1993), 3233–3250.
33. A.M. Benito, Y. Maniette, E. Munoz, M.T. Martinez; "Carbon nanotubes production by catalytic pyrolysis of benzene", Carbon 36(5) (1998), 681–683.
34. V. Ivanov, J.B. Nagy, P. Lambin, A. Lucas, X.B. Zhang, X.F. Zhang, D. Bernaerts, G. van Tendeloo, S. Amelinckx, J. van Landuyt; "The study of carbon nanotubules produced by catalytic method", Chem. Phys. Lett. 223(4) (1994), 329–335.
35. G.G. Tibbetts, D.W. Gorkiewicz, R.L. Alig; "A new reactor for growing carbon-fibers from liquid-phase and vapor-phase hydrocarbons", Carbon 31(5) (1993), 809–814.
36. W.Z. Li, S.S. Xie, L.X. Qian, B.H. Chang, B.S. Zou, W.Y. Zhou, R.A. Zhao, G. Wang; "Large-scale synthesis of aligned carbon nanotubes", Science 274(5293) (1996), 1701–1703.
37. M. Terrones, N. Grobert, J. Olivares, J.P. Zhang, H. Terrones, K. Kordatos, W.K. Hsu, J.P. Hare, P.D. Townsend, K. Prassides, A.K. Cheetham, H.W. Kroto, D.R.M. Walton; "Controlled production of aligned-nanotube bundles", Nature 388(6637) (1997), 52–55.
38. Z.F. Ren, Z.P. Huang, J.W. Xu, J.H. Wang, P. Bush, M.P. Siegel, P.N. Provenico; "Synthesis of large arrays of well-aligned carbon nanotubes on glass", Science 282(5391) (1998), 1105–1107.
39. C. Singh, M.S.P. Shaffer, I.A. Kinloch, A.H. Windle; "Production of aligned carbon nanotubes by the CVD injection method", Physica B 323(1–4) (2002), 339–340.
40. Z.W. Pan, S.S. Xie, B.H. Chang, C.Y. Wang, L. Lu, W. Liu, W.Y. Zhou, W.Z. Li, L.X. Qian; "Very long carbon nanotubes", Nature 394(6694) (1998), 631–632.
41. Z.P. Huang, J.W. Xu, Z.F. Ren, J.H. Wang, M.P. Siegel, P.N. Provenico; "Growth of highly oriented carbon nanotubes by plasma-enhanced hot filament chemical vapour deposition", Appl. Phys. Lett. 73(26) (1998), 3845–3847.
42. C. Bower, W. Zhu, D.J. Werder, O. Zhou; "Plasma-induced alignment of carbon nanotubes", Appl. Phys. Lett. 77(6) (2000), 830–832.
43. S. Hofmann, C. Ducati, J. Robertson, B. Kleinsorge; "Low-temperature growth of carbon nanotubes by plasma-enhanced chemical vapour deposition", Appl. Phys. Lett. 83(1) (2003), 135–137.

44. P. Nikolaev, M.J. Bronikowski, R.K. Bradley, F. Rohmund, D.T. Colbert, K.A. Smith, R.E. Smalley; "Gas-phase catalytic growth of single-walled carbon nanotubes from carbon monoxide", Chem. Phys. Lett. 313(1–2) (1999), 91–97.
45. S. Govindjee, J.L. Sackman; "On the use of continuum mechanics to estimate the properties of nanotubes", Solid State Commun. 110(4) (1999), 227–230.
46. M.M.J. Treacy, T.W. Ebbesen, J.M. Gibson; "Exceptionally high Young's modulus observed for individual carbon nanotubes", Nature 381(6584) (1996), 678–680.
47. A. Krishnan, E. Dujardin, T.W. Ebbesen, P.N. Yianilos, M.M.J. Treacy; "Young's modulus of single-walled nanotubes", Phys. Rev. B 58(20) (1998), 14013–14019.
48. P. Poncharal, Z.L. Wang, D. Ugarte, W.A. de Heer; "Electrostatic deflections and electromechanical resonances of carbon nanotubes", Science 283(5407) (1999), 1513–1516.
49. J.Z. Liu, Q. Zheng, Q. Jiang; "Effect of a rippling mode on resonances of carbon nanotubes", Phys. Rev. Lett. 86(21) (2001), 4843–4846.
50. M.R. Falvo, G.J. Clary, R.M. Taylor, V. Chi, F.P. Brooks Jr., S. Washburn, R. Superfine; "Bending and buckling of carbon nanotubes under large strain", Nature 389(6651) (1997), 582–584.
51. J.F. Despres, E. Daguerre, K. Lafdi; "Flexibility of graphene layers in carbon nanotubes", Carbon 33(1) (1995), 87–92.
52. R.S. Ruoff, D.C. Lorents; "Mechanical and thermal properties of carbon nanotubes", Carbon 33(7) (1995), 925–930.
53. S. Iijima, C. Brabec, A. Maiti, J. Bernholc; "Structural flexibility of carbon nanotubes", J. Chem. Phys. 104(5) (1996), 2089–2092.
54. O. Lourie, D.M. Cox, H.D. Wagner; "Buckling and collapse of embedded carbon nanotubes", Phys. Rev. Lett. 81(8) (1998), 1638–1641.
55. E.W. Wong, P.E. Shehan, C.M. Lieber; "Nanobeam mechanics: Elasticity, strength and toughness of nanorods and nanotubes", Science 277(5334) (1997), 1971–1975.
56. J.-P. Salvetat, A.J. Kulik, J.-M. Bonard, G.A.D. Briggs, T. Stökli, K. Metenier, S. Bonnamy, F. Beguin, N.A. Burnham, L. Forro; "Elastic modulus of ordered and disordered multiwalled carbon nanotubes", Adv. Mater. 11(2) (1999), 161–165.
57. M.-F. Yu, O. Lourie, M.J. Dyer, K. Moloni, T.F. Kelly, R.S. Ruoff; "Strength and breaking mechanism of multiwalled carbon nanotubes under tensile load", Science 287(5453) (2000), 637–640.
58. M.-F. Yu, B.S. Files, A. Arepalli, R.S. Ruoff; "Tensile loading of ropes of single wall carbon nanotubes and their mechanical properties", Phys. Rev. Lett. 84(24) (2000), 5552–5555.
59. M.B. Nardelli, B.I. Yakobson, J. Bernholc; "Brittle and ductile behavior in carbon nanotubes", Phys. Rev. Lett. 81(21) (1998), 4656–4659.
60. J.P. Lu; "Elastic properties of carbon nanotubes and nanoropes", Phys. Rev. Lett. 79(7) (1997), 1297–1300.
61. N. Yao, V. Lordi; "Young's modulus of single-walled carbon nanotubes", J. Appl. Phys. 84(4) (1998), 1939–1943.
62. B.I. Yakobson, C.J. Brabec, J. Bernholc; "Nanomechanics of carbon tubes: Instabilities beyond linear response", Phys. Rev. Lett. 76(14) (1996), 2511–2514.

63. B.I. Yakobson, R.E. Smalley; "Fullerene nanotubes: $C_{1,000,000}$ and beyond", Am. Sci. 85(6) (1997), 324–337.
64. J. Bernholc, C.J. Brabec, M.B. Nardelli, A. Maiti, C. Roland, B.I. Yakobson; "Theory of growth and mechanical properties of nanotubes", Appl. Phys. A 67(1) (1998), 39–46.
65. T. Belytschko, S.P. Xiao, G.C. Schatz, R.S. Ruoff; "Atomistic simulation of nanotube fracture", Phys. Rev. B 65(235430) (2002).
66. D.A. Walters, L.M. Ericson, M.J. Casavant, J. Liu, D.T. Colbert, K.A. Smith, R.E. Smalley; "Elastic strain of freely suspended single-wall carbon nanotube ropes", Appl. Phys. Lett. 74(25) (1999), 3803–3805.
67. Z.W. Pan, S.S. Xie, L. Lu, B.H. Chang, L.F. Sun, W.Y. Zhou, G. Wang, D.L. Zhang; "Tensile tests of ropes of very long aligned multiwall carbon nanotubes", Appl. Phys. Lett. 74(21) (1999), 3152–3154.
68. H.D. Wagner, O. Lourie, Y. Feldmann, R. Tenne; "Stress-induced fragmentation of multiwall carbon nanotubes in a polymer matrix", Appl. Phys. Lett. 72(2) (1998), 188–190.
69. F. Li, H.M. Chai, S. Bai, G. Su, M.S. Dresselhaus; "Tensile strength of single-walled carbon nanotubes directly measured from their macroscopic ropes", Appl. Phys. Lett. 77(20) (2000), 3161–3163.
70. L.S. Schadler, S.C. Giannaris, P.M. Ajayan; "Load transfer in carbon nanotubes epoxy composites", Appl. Phys. Lett. 73(26) (1998), 3842–3844.
71. C.L. Kane, E.J. Mele; "Size, shape, and low energy electronic structure of carbon nanotubes", Phys. Rev. Lett. 78(10) (1997), 1932–1935.
72. R. Saito, M. Fujita, G. Dresselhaus, M.S. Dresselhaus; "Electronic structure of chiral graphene tubules", Appl. Phys. Lett. 60(18) (1992), 2204–2206.
73. N. Hamada, S.-I. Sawada, A. Oshiyama; "New one-dimensional conductors: Graphitic microtubules", Phys. Rev. Lett. 68(10) (1992), 1579–1581.
74. J.W. Mintmire, B.I. Dunlop, C.T. White; "Are fullerene tubules metallic?", Phys. Rev. Lett. 68(5) (1992), 631–634.
75. C.T. White, D.H. Robertson, J.W. Mintmire; "Helical and rotational symmetries of nanoscale graphitic tubules", Phys. Rev. B 47(9) (1993), 5485–5488.
76. J.W. Mintmire, C.T. White; "Electronic and structural properties of carbon nanotubes", Carbon 33(7) (1995), 893–902.
77. L.X. Benedict, V.H. Crespi, S.G. Louie, M.L. Cohen; "Static conductivity and superconductivity of carbon nanotubes: Relations between tubes and sheets", Phys. Rev. B 52(20) (1995), 14935–14940.
78. K. Tanaka, H. Aoki, H. Ago, T. Yamabe, K. Okahara; "Interlayer interaction of two graphene sheets as a model of double-layer carbon nanotubes", Carbon 35(1) (1997), 121–125.
79. P. Lambin, V. Meunier, A. Rubio; "Electronic structure of polychiral carbon nanotubes", Phys. Rev. B 62(8) (2000), 5129–5135.
80. L. Langer, L. Stockman, J.P. Heremans, V. Bayot, C.H. Olk, C. van Haesendonck, Y. Bruynseraede, J.P. Issi; "Electrical resistance of a carbon nanotube bundle", J. Mater. Res. 9(4) (1994), 927–932.

81. J.P. Heremans, C.H. Olk, D.T. Morelli; "Magnetic susceptibility of carbon structures", Phys. Rev. B 49(21) (1994), 15122–15125.
82. T.W. Ebbesen, H.J. Lezec, H. Hiura, J.W. Bennett, H.F. Ghaemie, T. Thio; "Electrical conductivity of individual carbon nanotubes", Nature 382(6586) (1996), 54–56.
83. S.J. Tans, M.H. Devoret, H. Dai, A. Thess, R.E. Smalley, L.J. Geerligs, C. Dekker; "Individual single-wall carbon nanotubes as quantum wires", Nature 386(6624) (1997), 474–477.
84. T.W. Odom, J.-L. Huang, P. Kim, C.M. Lieber; "Atomic structure and electronic properties of single-walled carbon nanotubes", Nature 391(6662) (1998), 62–64.
85. J.W.G. Wildöer, L.C. Venema, A.G. Rinzler, R.E. Smalley, C. Dekker; "Electronic structure of atomically resolved carbon nanotubes", Nature 391(6662) (1998), 59–62.
86. M. Bockrath, D.H. Cobden, P.L. McEuen, N.G. Chopra, A. Zettl, A. Thess, R.E. Smalley; "Single-electron transport in ropes of carbon nanotubes", Science 275(5308) (1997), 1922–1925.
87. G.T. Kim, E.S. Choi, D.C. Kim, D.S. Suh, Y.W. Park, K. Liu, G. Duesberg, S. Roth; "Magnetoresistance of an entangled single-wall carbon-nanotube network", Phys. Rev. B 58(24) (1998), 16064–16069.
88. J.E. Fischer, H. Dai, A. Thess, R. Lee, N.M. Hanjani, D.L. Dehaas, R.E. Smalley; "Metallic resistivity in crystalline ropes of single-wall carbon nanotubes", Phys. Rev. B 55(8) (1997), 4921–4924.
89. A.D. Bozhko, D.E. Sklovsky, V.A. Nalimova, A.G. Rinzler, R.E. Smalley, J.E. Fischer; "Resistance vs. pressure of single-wall carbon nanotubes", Appl. Phys. A 67(1) (1998), 75–77.
90. J.R. Charlier, J.-P. Issi; "Electrical conductivity of novel forms of carbon", J. Phys. Chem. Solids 57(6–8) (1996), 957–965.
91. J.-O. Lee, C. Park, J.-J. Kim, J. Kim, J.W. Park, K.-H. Yoo; "Formation of low-resistance ohmic contact between carbon nanotubes and metal electrodes by a rapid thermal annealing method", J. Phys. D: Appl. Phys. 33(16) (2000), 1953–1956.
92. H. Dai, E.W. Wong, C.M. Lieber; "Probing electrical transport in nanomaterials: Conductivity of individual carbon nanotubes", Science 272(5261) (1996), 523–526.
93. A.B. Kaiser, G.S. Duesberg, S. Roth; "Heterogeneous model for conduction in carbon nanotubes", Phys. Rev. B 57(3) (1998), 1418–1421.
94. J. Hone, M. Whitny, A. Zettl; "Thermal conductivity of single-walled carbon nanotubes", Synth. Metals 103(1–3) (1999), 2498–2499.
95. P. Kim, L. Shi, A. Majumdar, P.L. McEuen; "Thermal transport measurements of individual multiwalled nanotubes", Phys. Rev. Lett. 87(215502) (2001).
96. P.M. Ajayan, O. Stephan, C. Colliex, D. Trauth; "Aligned carbon nanotube arrays formed by cutting a polymer resin-nanotube composite", Science 265(5176) (1994), 1212–1214.
97. L.A. Bursill, J.L. Peng, X.D. Fan; "Cross-sectional high-resolution transmission electron-microscopy study of the structures of carbon nanotubes", Phil. Mag. A 71(5) (1995), 1161–1176.

98. O. Lourie, H.D. Wagner; "Transmission electron microscopy observations of fracture of single-wall carbon nanotubes under axial tension", Appl. Phys. Lett. 73(24) (1998), 3527–3529.
99. O. Lourie, H.D. Wagner; "Evidence of stress transfer and formation of fracture clusters in carbon nanotube-based composites", Comp. Sci. Tech. 59(6) (1999), 975–977.
100. P.M. Ajayan, L.S. Schadler, C. Giannaris, A. Rubio; "Single-walled carbon nanotube-polymer composites: strength and weaknesses", Adv. Mater. 12(10) (2000) 750–752
101. M.S.P. Shaffer, X. Fan, A.H. Windle; "Dispersion and packing of carbon nanotubes", Carbon 36(11) (1998), 1603–1612.
102. K.T. Lau, S.-Q. Shi, H.-M. Cheng; "Micro-mechanical properties and morphological observation on fracture surfaces of carbon nanotube composites pre-treated at different temperatures", Comp. Sci. Tech. 63(8) (2003), 1161–1164.
103. J.B. Bai, A. Allaoui; "Effect of the length and the aggregate size of MWCNTs on the improvement efficiency of the mechanical and electrical properties of nanocomposites – experimental investigation", Comp. Part A 34(8) (2003), 689–694.
104. X. Gong, J. Liu, S. Baskaran, R.D. Voise, J.S. Young; "Surfactant-assisted processing of carbon nanotube/polymer composites", Chem. Mater. 12(4) (2000), 1049–1052.
105. J. Zhu, J.D. Kim, H. Peng, J.L. Margrave, V.N. Khabashesku, E.V. Barrera; "Improving the dispersion and integration of single-wall carbon nanotubes in epoxy composites through functionalisation", Nano Lett. 3(8) (2003), 1107–1113.
106. M.S.P. Shaffer, A.H. Windle; "Fabrication and characterization of carbon nanotube/poly(vinyl alcohol) composites", Adv. Mater. 11(11) (1999), 937–941.
107. M.S.P. Shaffer; "Carbon nanotubes", PhD thesis, University of Cambridge, UK (1998).
108. J. Liu, A.G. Rinzler, H.J. Dai, J.H. Hafner, R.K. Bradley, P.J. Boul, A. Lu, T. Iverson, K. Shelimov, C.B. Huffman, F. Rodriguez-Macias, Y.S. Shon, T.R. Lee, D.T. Colbert, R.E. Smalley; "Fullerene pipes", Science 280(5367) (1998), 1253–1256.
109. N. Yao, V. Lordi, S.X.C. Ma, E. Dujardin, A. Krishnan, M.M.J. Treacy, T.W. Ebbesen; "Structure and oxidation patterns of carbon nanotubes", J. Mater. Res. 13(9) (1998), 2432–2437.
110. E.T. Mickelson, I.W. Chiang, J.L. Zimmermann, P.J. Boul, J. Lozano, J. Liu, R.E. Smalley, R.H. Hauge, J.L. Margrave; "Solvation of fluorinated single-wall carbon nanotubes in alcohol solvents", J. Phys. Chem. B 103(21) (1999), 4318–4322.
111. F.H. Gojny, J. Nastalczyk, Z. Roslaniec, K. Schulte; "Surface modified multi-wall carbon nanotubes in CNT/epoxy-composites", Chem. Phys. Lett. 370(5–6) (2003), 820–824.
112. R.D. Patton, C.U. Pittman Jr., L. Wang J.R. Hill; "Vapor grown carbon fiber composites with epoxy and poly(phenylene sulfide) matrices", Comp. Part A 30(9) (1999), 1081–1091.

113. R.D. Patton, C.U. Pittman Jr., L. Wang. J.R. Hill, A. Day; "Ablation, mechanical and thermal conductivity of vapor grown carbon fiber/phenolic matrix composites", Comp. Part A 33() (2002), 243–251.
114. D. Puglia, L. Valentini, J.M. Kenny; "Analysis of the cure reaction of carbon nanotubes/epoxy resin composites through thermal analysis and Raman spectroscopy", J. Appl. Polym. Sci. 88(2) (2003), 452–458.
115. M. Yin, J.A. Koutsky, T.L. Barr, N.M. Rodriguez, R.T.K. Baker, L. Klebanov; "Characterization of carbon microfibers as a reinforcement for epoxy resins", Chem. Mater. 5(7) (1993), 1024–1031.
116. J. Sandler, M.S.P. Shaffer, T. Prasse, W. Bauhofer, K. Schulte, A.H. Windle; "Development of a dispersion process for carbon nanotubes in an epoxy matrix and the resulting electrical properties", Polymer 40(21) (1999), 5967–5971.
117. A. Allaoui, S. Bai, H.M. Cheng, J.B. Bai; "Mechanical and electrical properties of a MWNT/epoxy composite", Comp. Sci. Tech. 62(15) (2002), 1993–1998.
118. S. Barrau, P. Demont, A. Peigny, C. Laurent, C. Lacabanne; "DC and AC conductivity of carbon nanotubes-polyepoxy composites", Macromolecules 36(14) (2003), 5187–5194.
119. T. Prasse, J.-Y. Cavaille, W. Bauhofer; "Electric anisotropy of carbon nanofibre/epoxy resin composites due to electric field induced alignment", Comp. Sci. Tech. 63(13) (2003), 1835–1841.
120. L. Valentini, D. Puglia, E. Frulloni, I. Armentano, J.M. Kenny, S. Santucci; "Dielectric behavior of epoxy matrix/single-walled carbon nanotube composites", Comp. Sci. Tech. 64(1) (2004), 23–33.
121. J.K.W. Sandler, J.E. Kirk, I.A. Kinloch, M.S.P. Shaffer, A.H. Windle; "Ultra-low electrical percolation threshold in carbon-nanotube-epoxy composites", Polymer 44(19) (2003), 5893–5899.
122. C.A. Martin, J.K.W. Sandler, M.S.P. Shaffer, M.-K. Schwarz, W. Bauhofer, K. Schulte, A.H. Windle; "Formation of percolating networks in multi-wall carbon-nanotube-epoxy composites", Comp. Sci. Tech. 64(15) (2004), 2309–2316.
123. C.A. Martin, J.K.W. Sandler, A.H. Windle, M.-K. Schwarz, W. Bauhofer, K. Schulte, M.S.P. Shaffer; "Electric field-induced aligned multi-wall carbon nanotube networks in epoxy composites", Polymer 46(3) (2005), 877–886.
124. M.J. Biercuk, M.C. Llaguno, M. Radosavljevic, J.K. Hyun, A.T. Johnson, J.E. Fischer; "Carbon nanotube composites for thermal management", Appl. Phys. Lett. 80(15) (2002), 2767–2769.
125. R.L. Vander Wal, L.J. Hall; "Nanotube coated metals: New reinforcement materials for polymer matrix composites", Adv. Mater. 14(18) (2002), 1304–1308.
126. E.T. Thostenson, W.Z. Li, D.Z. Wang, Z.F. Ren, T.-W. Chou; "Carbon nanotube/carbon fiber hybrid multiscale composites", J. Appl. Phys. 91(9) (2002), 6034–6037.
127. W.B. Downs, R.T.K. Baker; "Modification of the surface properties of carbon fibers via the catalytic growth of carbon nanofibres", J. Mater. Res. 10(3) (1995), 625–633.
128. M.D. Frogley, D. Ravich, H.D. Wagner; "Mechanical properties of carbon nanoparticle-reinforced elastomers", Comp. Sci. Tech. 63(11) (2003), 1647–1654.

129. P. Richard, T. Prasse, J-Y. Cavaille, L. Chazeau, C. Gauthier, J. Duchet; "Reinforcement of rubbery epoxy by carbon nanofibres", Mater. Sci. Eng. A 352(1–2) (2003), 344–348.
130. M.A. Osman, A. Atallah, M. Müller, U.W. Sluter; "Reinforcement of poly(dimethylsiloxane) networks by mica flakes", Polymer 42(15) (2001), 6545–6556.
131. J.B. Donnet; "Nano and microcomposites of polymers, elastomers and their reinforcement", Comp. Sci. Tech. 63(8) (2003), 1085–1088.
132. G. Heinrich, M. Klüppel, T.A. Vilgis; "Reinforcement of elastomers", Curr. Opin. Solid State Mater. Sci. 6(3) (2002), 195–203.
133. C.A. Cooper, D. Ravich, D. Lips, J. Mayer, H.D. Wagner; "Distribution and alignment of carbon nanotubes and nanofibrils in a polymer matrix", Comp. Sci. Tech. 62(7–8) (2002), 1105–1112.
134. W. Tang, M.H. Santare, S.G. Advani; "Melt processing and mechanical property characterization of multi-walled carbon nanotube/high density polyethylene (MWNT/HDPE) composite films", Carbon 41(14) (2003), 2779–2785.
135. E.T. Thostenson, T.-W. Chou; "Aligned multi-walled carbon nanotube-reinforced composites: processing and mechanical characterization", J. Phys. D: Appl. Phys. 35(16) (2002), 77–80.
136. Z. Jin, K.P. Pramoda, S.H. Goh, G. Xu; "Poly(vinylidene fluoride)-assisted melt-blending of multi-walled carbon nanotube/poly(methyl methacrylate) composites", Mater. Res. Bull. 37(2) (2002), 271–278.
137. K. Lozano, E.V. Barrera; "Nanofiber-reinforced thermoplastic composites I: Thermoanalytical and mechanical analyses", J. Appl. Polym. Sci. 79(1) (2001), 125–133.
138. Z. Jin, K.P. Pramoda, G. Xu, S.H. Goh; "Dynamic mechanical behavior of melt-processed multi-walled carbon nanotube/poly(methyl methacrylate) composites", Chem. Phys. Lett. 337(1–3) (2001), 43–47.
139. P. Pötschke, T.D. Fornes, D.R. Paul; "Rheological behavior of multiwalled carbon nanotube/polycarbonate composites", Polymer 43(11) (2002) 3247–3255.
140. P. Pötschke, S.M. Dudkin, I. Alig; "Dielectric spectroscopy on melt processed polycarbonate-multiwalled carbon nanotube composites", Polymer 44(17) (2003), 5023–5030.
141. O. Meincke, D. Kaempfer, H. Weickmann, C. Friedrich, M. Vathauer, H. Warth; "Mechanical properties and electrical conductivity of carbon-nanotube filled polyamide-6 and its blends with acrylonitrile/butadiene/styrene" Polymer 45(3) (2004), 739–748.
142. J.K.W. Sandler, S. Pegel, M. Cadek, F. Gojny, M. van Es, J. Lohmar, W.J. Blau, K. Schulte, A.H. Windle, M.S.P. Shaffer; "A comparative study of melt-spun polyamide-12 fibres reinforced with carbon nanotubes and nanofibres", Polymer 45(6) (2004), 2001–2015.
143 G.G. Tibbetts, J.J. McHugh; "Mechanical properties of vapor-grown carbon fiber composites with thermoplastic matrices", J. Mater. Res. 14(7) (1999), 2871–2880.

144. R.J. Kuriger, M.K. Alam, D.P. Anderson, R.L. Jacobsen; "Processing and characterization of aligned vapor grown carbon fiber reinforced polypropylene", Comp. Part A 33(1) (2002), 53–62.
145. K. Lozano, J. Bonilla-Rios, E.V. Barrera; "A study on nanofiber-reinforced thermoplastic composites II: Investigation of the mixing rheology and conduction properties", J. Appl. Polym. Sci. 80(8) (2001), 1162–1172.
146. O.S. Carneiro, J.A. Covas, C.A. Bernado, G. Caldeira, F.W.J. Van Hattum, J.-M. Ting, R.L. Alig, M.L. Lake; "Production and assessment of polycarbonate composites reinforced with vapour-grown carbon fibres", Comp. Sci. Tech. 58(3–4) (1998), 401–407.
147. A. Ramazani, A. Ait-Kadi, M. Grmela; "Rheological modelling of short fiber thermoplastic composites", J. Non-Newton. Fluid Mech. 73(3) (1997), 241–260.
148. P. Cortes, K. Lozano, E.V. Barrera, J. Bonilla-Rios; "Effects of nanofiber treatments on the properties of vapor-grown carbon fiber reinforced polymer composites", J. Appl. Polym. Sci. 89(9) (2003), 2527–2534.
149. D. Qian, E.C. Dickey, R. Andrews, T. Rantell; "Load transfer and deformation mechanisms in carbon nanotube-polystyrene composites", Appl. Phys. Lett. 76(20) (2000), 2868–2870.
150. P.C. Watts, W.K. Hsu, G.Z. Chen, D.J. Fray, H.W. Kroto, D.R.M. Walton; "A low resistance boron-doped carbon nanotube-polystyrene composite", J. Mater. Chem. 11(10) (2001), 2482–2488.
151. B. Safadi, R. Andrews, E.A. Grulke; "Multiwalled carbon nanotube polymer composites: synthesis and characterization of thin films", J. Appl. Polym. Sci. 84(14) (2002), 2660–2669.
152. L. Jin, C. Bower, O. Zhou; "Alignment of carbon nanotubes in a polymer matrix by mechanical stretching", Appl. Phys. Lett. 73(9) (1998), 1197–1199.
153. C. Bower, R. Rosen, J. Han, O. Zhou; "Deformation of carbon nanotubes in nanotube-polymer composites", Appl. Phys. Lett. 74(22) (1999), 3317–3319.
154. M. Cadek, J.N. Coleman, V. Barron, K. Hedicke, W.J. Blau; "Morphological and mechanical properties of carbon-nanotube-reinforced semicrystalline and amorphous polymer composites", Appl. Phys. Lett. 81(27) (2002), 5123–5125.
155. S.L. Ruan, P. Gao, X.G. Yang, T.X. Yu; "Toughening high performance ultrahigh molecular weight polyethylene using multiwalled carbon nanotubes", Polymer 44(19) (2003), 5643–5654.
156. E. Assouline, A. Lustiger, A.H. Barber, C.A. Cooper, E. Klein, E. Wachtel, H.D. Wagner; "Nucleation ability of multiwall carbon nanotubes in polypropylene composites", J. Polym. Sci. B: Polym. Phys. 41(5) (2003), 520–527.
157. B.P. Grady, F. Pompeo, R.L. Shambaugh, D.E. Resasco; "Nucleation of polypropylene crystallization by single-walled carbon nanotubes", J. Phys. Chem. B 106(23) (2002), 5852–5858.
158. M.C. Paiva, B. Zhou, K.A.S. Fernando, Y. Lin, J.M. Kennedy, Y.-P. Sun; "Mechanical and morphological characterization of polymer-carbon nanocomposites from functionalized carbon nanotubes", Carbon 42(14) (2004), 2849–2854.

159. O. Probst, E.M. Moore, D.E. Resasco, B.P. Grady; "Nucleation of polyvinyl alcohol crystallization by single-walled carbon nanotubes", Polymer 45(13) (2004), 4437–4443.
160. X. Zhang, T. Liu, T.V. Sreekumar, S. Kumar, V.C. Moore, R.H. Hauge, R.E. Smalley; "Poly(vinyl alcohol)/SWCNT composite film", Nano Lett. 3(9) (2003), 1285–1288.
161. K.L. Lu, R.M. Lago, Y.K. Chen, M.L.H. Green, P.F. Harris, S.C. Tsang; "Mechanical damage of carbon nanotubes by ultrasound", Carbon 34(6) (1996), 814–816.
162. A. Koshio, M. Yudasaka, M. Zhang, S. Iijima; "A simple way to chemically react single-wall carbon nanotubes with organic materials using ultrasonication", Nano Lett. 1(7) (2001), 361–363.
163. A. Dufresne, M. Paillet, J.L. Putaux, R. Canet, F. Carmona, P. Delhaes, S. Cui; "Processing and characterization of carbon nanotube/poly(styrene-co-butyl acrylate) nanocomposites", J. Mater. Sci. 37(18) (2002), 3915–3923.
164. Y. Lin, B. Zhou, K.A.S. Fernando, P. Liu, L.A. Allard, Y.-P. Sun; "Polymeric carbon nanocomposites from carbon nanotubes functionalised with matrix polymer", Macromolecules 36(19) (2003), 7199–7204.
165. C.A. Mitchell, J.L. Bahr, S. Arepalli, J.M. Tour, R. Krishnamoorti; "Dispersion of functionalized carbon nanotubes in polystyrene", Macromolecules 35(23) (2002), 8825–8830.
166. Z.J. Jia, Z.Y. Wang, C.L. Xu, J. Liang, B.Q. Wei, D.H. Wu, S.W. Zhu; "Study on poly(methyl methacrylate)/carbon nanotube composites", Mater. Sci. Eng. A 271(1–2) (1999), 395–400.
167. S. Kumar, T.D. Dang, F.E. Arnold, A.R. Bhattacharyya, B.G. Min, X. Zhang, R.A. Vaia, C. Park, W.W. Adams, R.H. Hauge, R.E. Smalley, S. Ramesh, P.A. Willis; "Synthesis, structure, and properties of PBO/SWNT composites", Macromolecules 35(24) (2002), 9039–9043.
168. H.J. Barraza, F. Pompeo, E.A. O'Rear, D.E. Resasco; "SWNT-filled thermoplastic and elastomeric composites prepared by miniemulsion polymerization", Nano Lett. 2(8) (2002), 797–802.
169. Z. Ounaies, C. Park, K.E. Wise, E.J. Siochi, J.S. Harrison; "Electrical properties of single wall carbon nanotube reinforced polyimide composites", Comp. Sci. Tech. 63(11) (2003), 1637–1646.
170. Z. Roslaniec, G. Broza, K. Schulte; "Nanocomposites based on multiblock polyester elastomers (PEE) and carbon nanotubes (CNT)", Comp. Interfaces 10(1) (2003), 95–102.
171. M.J. O'Connell, P. Boul, L.M. Ericson, C. Huffman, Y. Wang, E. Haroz, C. Kuper, J. Tour, K.D. Ausman, R.E. Smalley; "Reversible water-solubilization of single-walled carbon nanotubes by polymer wrapping", Chem. Phys. Lett. 342(3–4) (2001), 265–271.
172. J. Sandler, P. Werner, M.S.P. Shaffer, V. Demchuk, V. Altstädt, A.H. Windle; "Carbon-nanofibre-reinforced poly(ether ether ketone) composites", Comp. Part A 33(8) (2002), 1033–1039.

173. T. Kimura, H. Ago, M. Tobita, S. Ohshima, M. Kyotani, M. Yumura; "Polymer composites of carbon nanotubes aligned by a magnetic field", Adv. Mater. 14(19) (2003), 1380–1383.
174. R. Andrews, D. Jacques, A.M. Rao, T. Rantell, F. Derbyshire, Y. Chen, J. Chen, R.C. Haddon; "Nanotube composite carbon fibers", Appl. Phys. Lett. 75(9) (1999), 1329–1331.
175. S.A. Gordeyev, J.A. Ferreira, C.A. Bernado, I.M. Ward; "A promising conductive material: highly oriented polypropylene filled with short vapour-grown carbon fibres", Mater. Lett. 51(1) (2001), 32–36.
176. S. Kumar, H. Doshi, M. Srinivasarao, J.O. Park, D.A. Schiraldi; "Fibers from polypropylene/nano carbon fiber composites", Polymer 43(5) (2002), 1701–1703.
177. J.C. Kearns, R.L. Shambaugh; "Polypropylene fibers reinforced with carbon nanotubes", J. Appl. Polym. Sci. 86(8) (2002), 2079–2084.
178. H. Ma, J. Zeng, M.L. Realff, S. Kumar, D.A. Schiraldi; "Processing, structure, and properties of fibers from polyester/carbon nanofiber composites", Comp. Sci. Tech. 63(11) (2003), 1617–1628.
179. A.R. Bhattacharyya, T.V. Sreekumar, T. Liu, S. Kumar, L.M. Ericson, R.H. Hauge, R.E. Smalley; "Crystallization and orientation studies in polypropylene/single wall carbon nanotube composites", Polymer 44(8) (2003), 2373–2377.
180. J. Zeng, B. Saltysiak, W.S. Johnson, D.A. Schiraldi , S. Kumar; "Processing and properties of poly(methyl methacrylate)/carbon nano fiber composites", Comp. Part B 35(2) (2004), 173–178.
181. J. Sandler, A.H. Windle, P. Werner, V. Altstädt, M. van Es, M.S.P. Shaffer; "Carbon-nanofibre-reinforced poly(ether ether ketone) fibres", J. Mater. Sci. 38(10) (2003), 2135–2141.
182. P. Werner, R. Verdejo, F. Wöllecke, V. Altstädt, J.K.W. Sandler, M.S.P. Shaffer; "Carbon nanofibers allow foaming of semicrystalline poly(ether ether ketone)", Adv. Mater. 17(23) (2005), 2864–2869.
183. N.S. Ramesh, S.T. Lee; "Blowing agent effect on extensional viscosity calculated from fiber spinning method for foam processing", J. Cell. Plast. 36(5) (2000), 374–385.
184. D.S. Bangarusampath, H. Ruckdäschel, J.K.W. Sandler, V. Altstädt, M.S.P. Shaffer; "Rheological properties of carbon nanofibre reinforced poly(ether ether ketone) composites under shear and elongational flow", Polymer (2006), submitted.
185. B. Vigolo, A. Penicaud, C. Coulon, C. Sauder, R. Pailler, C. Journet, P. Bernier, P. Poulin; "Macroscopic fibers and ribbons or oriented carbon nanotubes", Science 290(5495) (2000), 1331–1334.
186. A.B. Dalton, S. Collins, E. Munoz, J.M. Razal, V.H. Ebron, J.P. Ferraris, J.N. Coleman, B.G. Kim, R.H. Baughman; "Super-tough carbon-nanotube fibres – These extraordinary composite fibres can be woven into electronic textiles", Nature 423(6941) (2003), 703.
187. L.M. Eriscon, H. Fan, H.Q. Peng, V.A. Davis, W. Zhou, J. Sulpizio, Y.H. Wang, R. Booker, J. Vavro, C. Guthy, A.N.G. Parra-Vasquez, M.J. Kim, S. Ramesh, R.K. Saini, C. Kittrell, G. Lavin, H. Schmidt, W.W. Adams, W.E. Billups, M. Pasquali,

W.F. Hwang, R.H. Hauge, J.E. Fischer, R.E. Smalley; "Macroscopic, neat, single-walled carbon nanotube fibers"", Science 305(5689) (2004), 1447–1450.
188. W. Zhou, J. Vavro, C. Guthy, K.I. Winey, J.E. Fischer, L.M. Ericson, S. Ramesh, R. Saini, V.A. Davis, C. Kittrell, M. Pasquali, R.H. Hauge, R.E. Smalley; "Single wall carbon nanotube fibers extruded from super-acid suspensions: Preferred orientation, electrical, and thermal transport", J. Appl. Phys. 95(2) (2004), 649–655.
189. A.A. Mamedov, N.A. Kotov, M. Prato, D.M. Guldi, J.P. Wicksted, A. Hirsch; "Molecular design of strong single-wall carbon nanotube/polyelectrolyte multilayer composites", Nature Mat. 1(3) (2002), 190–194.
190. S.S. Ray, M. Okamoto; "Polymer/layered silicate nanocomposites: a review from preparation to processing", Prog. Polym. Sci. 28(11) (2003), 1539–1641.
191. F.W.J. Van Hattum, C.A. Bernado, J.C. Finegan, G.G. Tibbetts, R.L. Alig, M.L. Lake; "A study of the thermomechanical properties of carbon fiber-reinforced polypropylene composites", Polym. Comp. 20(5) (1999), 683–688.
192. J.K.W. Sandler; "Structure-property-relationships of carbon nanotubes / nanofibres and their polymer composites", PhD thesis, University of Cambridge, UK (2005).
193. G. Tsagaropoulos, A. Eisenberg; "Dynamic mechanical study of the factors affecting the two glass transition behavior of filled polymers. Similarities and differences with random ionomers", Macromolecules 28(18) (1995), 6067–6077.
194. P. Werner, V. Altstädt, R. Jaskulka, O. Jacobs, J.K.W. Sandler, M.S.P. Shaffer, A.H. Windle; "Tribological behaviour of carbon-nanofibre-reinforced poly(ether ether ketone)", Wear 257(9–10) (2004), 1006–1014.
195. Y.S. Zoo, J.W. An, D.P. Lim, D.S. Lim; "Effect of carbon nanotube addition on tribological behavior of UHMWPE", Tribol. Lett. 16(4) (2004), 305–309.
196. M.J. Folkes, D.A.M. Russell; "Orientation effects during the flow of short-fibre reinforced thermoplastics", Polymer 21() (1980), 1252–1258.
197. S.E. Barbosa, D.R. Ercoli, M.A. Bibbo, J.M. Kenny; "Rheology of short-fiber composites: A systematic approach", Comp. Struct. 27(1–2) (1994), 83–91.
198. J. Sandler, G. Broza, M. Nolte, K. Schulte, Y.-M. Lam, M.S.P. Shaffer; "Crystallization of carbon nanotube and nanofiber polypropylene composites", J. Macromol. Sci. B: Phys. B42(3–4) (2003), 479–488.
199. L. Valentini, J. Biagiotti, J.M. Kenny, S. Santucci; "Morphological characterization of single-walled carbon nanotubes-PP composites", Comp. Sci. Tech. 63(8) (2003), 1149–1153.
200. M. Mucha, J. Marszalek, A. Fidrych; "Crystallization of isotactic polypropylene containing carbon black as a filler", Polymer 41(11) (2000), 4137–4142.
201. W. Xu, M. Ge, P. He; "Nonisothermal crystallization kinetics of polypropylene/montmorrillonite nanocomposites", J. Polym. Sci. B: Polym. Phys. 40(5) (2002), 408–414.
202. N. Levi, R. Czerw, S.Y. Xing, P. Iyer, D.L. Carroll; "Properties of polyvinylidene difluoride-carbon nanotube blends", Nano Lett. 4(7) (2004), 1267–1271.
203. T. Kitao, S. Ohya, J. Furukawa, S. Yamashita; "Orientation of polymer molecules during melt spinning. II. Orientation of crystals in as-spun polyolefin fibers", J. Polym. Sci. - Polym. Phys. Ed. 11 (1973), 1091–1109.

204. F.M. Lu, J.E. Spruiell; "The role of crystallization kinetics in the development of the structure and properties of polypropylene filaments", J. Appl. Polym. Sci. 49(4) (1993), 623–631.
205. A. Marcincin; "Modification of fiber-forming polymers by additives", Prog. Polym. Sci. 27(5) (2002), 853–913.
206. Y.L. Li, I.A. Kinloch, A.H. Windle; "Direct spinning of carbon nanotube fibers from chemical vapor deposition synthesis", Science 304(5668) (2004), 276–27.
207. H.J. Mair, S. Roth, Eds.; "Elektrisch leitende Kunststoffe", Carl Hanser Hamburg, Germany, (1989).
208. R. Zallen; "The physics of amorphous solids", John Wiley & Sons, New York (1983).
209. D. Stauffer; "Introduction to percolation theory", Taylor & Francis, London (1984).
210. F. Lux; "Models proposed to explain the electrical conductivity of mixtures made of conductive and insulating materials", J. Mater. Sci. 28(2) (1993), 285–301.
211. F. Carmona, P. Prudhorn, F. Barreau; "Percolation in short fibers epoxy-resin composites – conductivity behaviour and finite size effects near threshold", Solid State Commun. 51(4) (1984), 255–257.
212. M.T. Connor, S. Roy, T.A. Ezquerra, F.J. Balta-Calleja; "Broadband AC conductivity of conductor-polymer composites", Phys. Rev. B 57(4) (1998), 2286–2294.
213. S.A. Gordeyev, F.J. Macedo, J.A. Ferreira, F.W.J. Van Hattum, C.A. Bernado; "Transport properties of polymer-vapour-grown carbon fibre composites", Physica B 279(1–3) (2000), 33–36.
214. I.C. Finegan, G.G. Tibbetts; "Electrical conductivity of vapor-grown carbon fiber/thermoplastic composites", J. Mater. Res. 16(6) (2001), 1668–1674.
215. B.E. Kilbride, J.N. Coleman, J. Fraysse, P. Fournet, M. Cadek, A. Drury, S. Hutzler, S. Roth, W.J. Blau; "Experimental observation of scaling laws for alternating and direct current conductivity in polymer-carbon nanotube composite thin films", J. Appl. Phys. 92(7) (2002), 4024–4030.
216. C. Stephan, T.P. Nguyen, B. Lahr, W.J. Blau, S. Lefrant, O. Chauvet; "Raman spectroscopy and conductivity measurements on polymer-multiwalled carbon nanotubes composites", J. Mater. Res. 17(2) (2002), 396–400.
217. P. Pötschke, A.R. Bhattacharyya, A. Janke; "Morphology and electrical resistivity of melt mixed blends of polyethylene and carbon nanotube filled polycarbonate", Polymer 44(26) (2003), 8061–8069.
218. M. Wu, L.L. Shaw; "On the improved properties of injection-molded, carbon nanotube-filled pet/pvdf blends", J. Power Sources 136(1) (2004), 37–44.
219. M. Sumita, K. Sakata, S. Asai, K. Miyasaka, H. Nakagawa; "Dispersion of fillers and the electrical-conductivity of polymer blends filled with carbon black", Polym. Bull. 25(2) (1991), 265–271.
220. C. Zhang, X.S. Yi, H. Yui, S. Asai, M. Sumita; "Morphology and electrical properties of short carbon fiber-filled polymer blends: High-density polyethylene poly(methyl methacrylate)", J. Appl. Polym. Sci. 69(9) (1998), 1813–1819.
221. S.-Y. Fu, Y.-W. Mai; "Thermal conductivity of misaligned short-fiber-reinforced polymer composites", J. Appl. Polym. Sci. 88(6) (2003), 1497–1505.

222. V. Jamieson; "Whatever happened to nanotubes, the hollow threads of carbon that were going to change the world?", New Scientist 2386, 15 March (2003).
223. F.T. Fisher, R.D. Bradshaw, L.C. Brinson; "Effects of nanotube waviness on the modulus of nanotube-reinforced polymers", Appl. Phys. Lett. 80(24) (2002), 4647–4649.
224. A.H. Barber, S.R. Cohen, H.D. Wagner; "Measurement of carbon nanotube-polymer interfacial strength", Appl. Phys. Lett. 82(23) (2003), 4140–4142.
225. Principa Partners; "Nanocomposites 1999: Polymer technology for the next century", Techn. Report, Principa Partners Consulting (1999).

CHAPTER 2

Dispersion, Bonding and Orientation of Carbon Nanotubes in Polymer Matrices

Suresh G. Advani* and Zhihang Fan

Department of Mechanical Engineering
and Center for Composite Materials
University of Delaware, Newark, DE 19711, USA
**advani@udel.edu*

This chapter will report on the progress of various researchers in addressing the key challenges in manufacturing polymer nanocomposites with carbon nanotubes. Unique properties of the carbon nanotubes (CNT) offer the potential for them to serve as excellent reinforcements for advanced nanocomposites with better mechanical and multi functional properties. One critical step that will affect the outcome is the processing method and approach that is used to manufacture such CNT/Polymer composites. The processing method will influence the final micro structure of CNT in the composite which determines the CNT effectiveness in the nanocomposite. This chapter will focus on reviewing the current state of research on CNT dispersion, interfacial bonding between the CNT and the polymer matrix, and CNT orientation during flow. All these parameters influence the evolution of the microstructure and hence play a key role in tailoring the properties of the polymer nanocomposites.

1. Introduction

Since the discovery of the carbon nanotubes (CNT) [1], many researchers have devoted their efforts to understand and explore the use of CNT in various fields. Carbon nanotubes (CNT) can be created by rolling a graphite sheet, as shown in the Figure 1. As we know, the

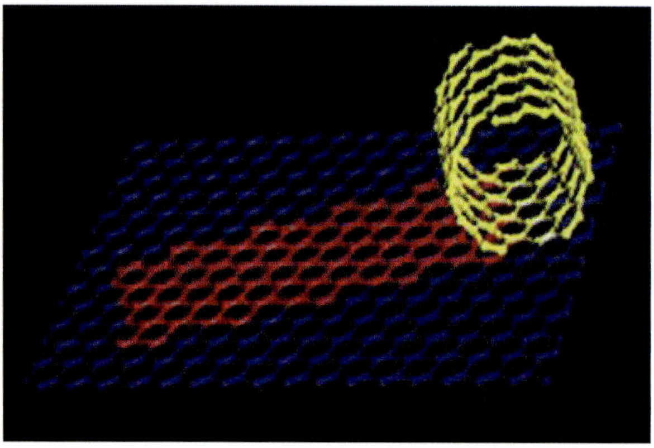

Fig. 1. Carbon Nanotubes may be considered as a rolling cylinder from graphite sheet. (Berkeley research review.)

graphite has C-C sp^2 bonds, which are stronger than the C-C sp^3 bonds in diamond because the C-C sp^2 bond length is shorter than the C-C sp^3. However, the molecular force between the graphite layers is van der Waals force, which allows them to slide with relative ease. The CNT is formed with rolled graphite sheets forming a closed cylindrical structure. There are two types of carbon nanotubes, Single-Walled Carbon Nanotubes (SWNT) and Multi-Walled Carbon Nanotubes (MWNT). SWNT is much stronger than MWNT, because SWNT is one cylindrical layer of graphite while MWNT is formed by many co-axial cylindrical layers of graphite which deteriorates the strength due to the possibility of slip between the layers [2]. One can envision many uses for such structure due to their superior properties. Table 1 lists the properties of CNT along with some comparisons. However, the properties of specific nanotubes depend on the diameter, length, chirality, morphology, and the structure.

The exceptional mechanical and physical properties of the CNT make them likely candidates not only for composite structures but also for multi-functional CNT composites with tailored electrical and thermal properties.

However, to realize the potential application of CNT in composites and polymers to form useful and multifunctional objects, two difficulties

Table 1. Properties of CNT.

Property	CNT	Comments
Young's Modulus	~1 TPa [3]	10–100 times higher than the strongest steel
Maximum Tensile Strength	30 GPa [4]	
Thermal Conductivity	~ 2000 W/m/K [5]	Twice as high as diamond
Maximum Current Density	10^{13} A/m^2 [6]	1000 times higher than copper wires
Aspect Ratio	~1000	

in processing need to be addressed: homogeneous dispersion and creating strong interfacial bonding between CNT and polymer matrix. These are the two requirements that provide maximum effectiveness in load transfer [2,7–9]. To achieve certain anisotropic functionality, such as strong electrical conductivity in specific direction, we must also be able to control orientation of CNTs in that specified direction to tailor the CNT benefits. Hence CNT dispersion, CNT/Matrix interfacial bonding, and CNT orientation state are three important parameters in CNT composites processing. This chapter will review the current state of the art in processing of CNT/Polymer composites. The discussion will be presented in three sections: CNT dispersion, CNT/Polymer Matrix interfacial bonding and CNT orientation. All these parameters influence the processing of such materials and hence the formation of the microstructure which will influence the properties of the manufactured nanocomposite.

2. CNT Dispersion in Polymer

2.1. *Classification*

The aim is to create homogeneously separated CNT in the polymer suspension, as shown in Figure 2a. However, CNTs due to their nano scale, have strong van der Waals forces between them. In addition, CNTs can be envisioned as long flexible nano fibers which can entangle very easily with each other. Moreover, CNT formation process creates them in agglomeration form requiring a separation process. These hurdles in CNT composite processing raise the following question: how to separate or exfoliate CNTs and disperse them in polymer homogeneously? Directly mixing such CNTs into the polymer will result in CNT aggregates of different sizes and non-uniform distribution, as shown in Figure 2b.

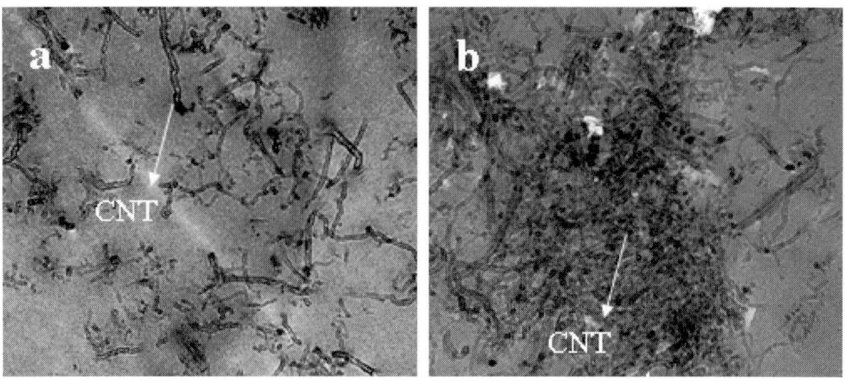

Fig. 2. TEM images of (a) Separate and homogeneous CNT dispersion; (b) CNT aggregates.

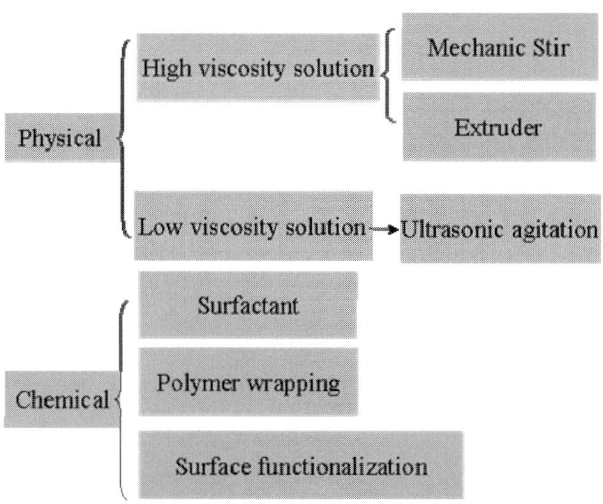

Fig. 3. Classification of methods to disperse Carbon Nanotubes in polymers.

Hence many researchers have focused on improving the CNT dispersion quality. Dispersion methods can be divided into two categories: Physical approaches and chemical approaches. In each category, several methods have been attempted, as shown in Figure 3.

2.2. *Physical approaches*

Usually physical approaches are simple and straightforward, which relies on directly mixing the nanotubes with the aid of mechanical force. One such physical approach was attempted by Haggenmuellera *et al.* [10]. They cut a MWNT/PMMA composite sample in which the MWNT were visibly not well dispersed into small pieces. They mixed these pieces and hot pressed the mixture to create a new sample. They repeated this cutting and mixing procedure 20 times and the authors found that the MWNT dispersion quality improved. Their comment about the quality was based on observing sections of the sample under an optical microscope [10].

To mix CNT with highly viscous fluids, especially thermoplastic melts such as polyethlene (PE) and Polystyrene (PS), one could rely on high shear forces to exfoliate the CNT aggregates and improve the CNT dispersion. Extruder is the most popular equipment choice to disperse and process CNT based composites using high shear mixing rates. Usually, pellets containing clustered CNTs are fed into the extruder hopper. A twin screw rotates at high speed and creates high shear flow. The hypothesis is that the high shear rate will exfoliate the CNT clusters and mix them uniformly with the resin [11–14]. Figure 4 shows the dispersion state of MWNT in Polycarbonate mixed in the DACA Micro Compunder by Pötschk *et al.* [14] .

Mechanical Stir or Magnetic Stir could also be used to generate shear flow to disperse CNT for lower viscosity solutions usually thermosets or thermoplastics containing solvents. Moniruzzaman *et al.* mixed SWNT in epoxy at 100 rpm for 1 h and claimed that a uniform dispersion was achieved [15]. However, Sandler *et al.* used a stir to mix MWNTs in epoxy at 2000 rpm for 5 minutes and still found MWNT aggregates dispersed in the resin [16]. The discrepancy points to the fact that the high speed mechanical stir or magnetic stir can not generate sufficiently

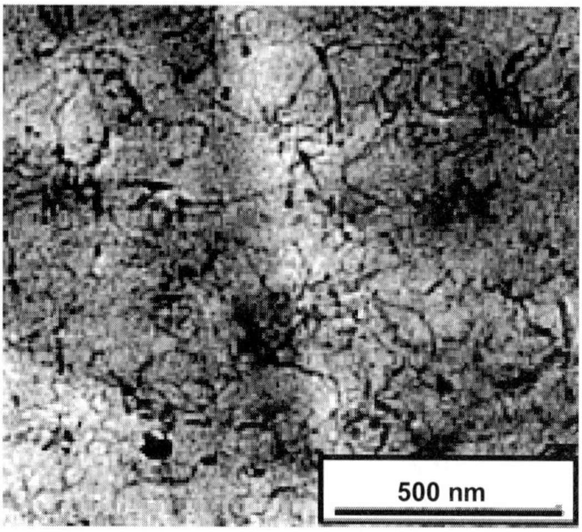

Fig. 4. TEM image of MWNT dispersion state in Polycarbonate mixed in an extruder [14].

high shear stress to exfoliate the CNT aggregates for better dispersion as in an extruder.

In case of low viscosity fluids, ultra sonic agitation is effective. The ultrasonic agitation water bath has a high agitation frequency of 40–50 kHz. The high frequency agitation induces bubbles to nucleate on the CNT surfaces and when the bubbles burst or collapse, the discharged energy generates sufficient forces to push the CNTs apart. Safadi *et al.* found that the ultrasonic agitation improved the CNT dispersion by decreasing the CNT aggregates and in some cases even separated the CNTs [17]. But the usefulness of ultrasonic agitation is limited to low viscosity suspensions, such as water, acetone and Dimethylformamide (DMF). Thus modus operandi has been to mix CNT and polymer melt with a low viscosity solvent using ultrasonic water bath and then evaporate the solvent resulting in hopefully a well dispersed CNT/Polymer suspension. This method has been coined as solution sonication. Following this way, Vaccarini et al. were able to obtain uniform SWNT dispersion in epoxy by sonicating SWNT and polymer with Dichlodo Methylene Chloride (CH_2C_{12}) and then evaporating

CH_2C_{12} [18]. Very few researchers were able to obtain dispersed CNT suspension only using ultrasonic agitation. Usually solution sonication is an important step combined with other methods to disperse CNT in polymer melts and solutions. It should be noticed that the two popular methods for dispersion: extruder and ultrasonic agitation both do cause some damage to CNT surface and may also lead to attrition in CNT length which may be partially responsible for degradation of CNT's mechanical properties.

2.3. *Chemical approaches*

Since CNT surface is hydrophobic and inert to many polymers, physical approaches are usually not very effective or efficient. The recent interest in chemical modification and functionalization of the CNT surface has shown to disperse CNT in selected polymers and aqueous media. The latter is very important for application of CNT in biology, which had been discussed in detail by Lin *et al.* [9]. Many chemical modification methods have been attempted in the past decade. These methods could be grouped into three categories: Surfactant Assistance, Polymer Wrapping and Surface Functionalization. It should be noticed, however, there is no clear boundary between the three categories.

Surfactant Assistance

There are many types of surfactants, but to date, there is no universal one that can achieve good CNT dispersion in all polymeric resins. Surfactant facilitates the link between CNT and the polymer when the CNTs were not compatible with the suspending polymer [19–27]. As shown in Figure 5, Gong *et al.* [21] found that Polyoxyethylene 8 lauryl ($C_{12}EO_8$) could improve the compatibility between CNT with epoxy, so that the CNT aggregates size decreased and uniformly dispersed. Sodium dodecylbenzene sulfonate (SDBS) was found efficient to assist CNT dispersion in water even at a relatively high concentration [19]. The advantage of surfactant is its ease in facilitating dispersion but the dispersion stability and interfacial bonding between CNT and the matrix are not that high.

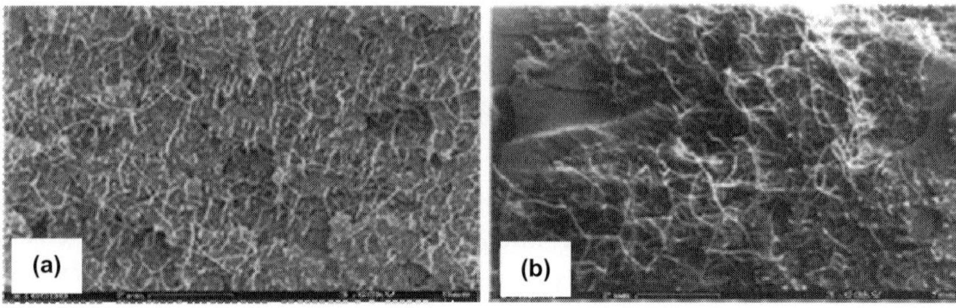

Fig. 5. SEM images of MWNTs in epoxy with surfactant $C_{12}EO_8$. (a) good dispersion (b), presence of occasional non-uniform aggregates [21].

Polymer Wrapping

As shown in Figure 6A, polymer wrapping is to envelope the CNT surface by a polymer layer so that the CNT would be compatible with the solution or the melt [20,28–31]. O'Connell *et al.* made the SWNT to be soluble in water by non-covalently associating them with linear polymers, PVP and PS as shown in Figure 6A. The function of polymer wrapping is to eliminate the hydrophobic interface between the CNT and their aqueous medium [28]. Tang *et al.* wrapped CNT with poly(phenylacetylenes) (PPA) and found very good dispersion quality of CNT in PPA as shown in Figure 6B [30]. Compared to surfactants, polymers usually have better surface adsorption on CNT surface. In addition, polymer wrapping usually does not damage the surface and keeps the pristine CNT mechanical properties unharmed. However, the bonding force between the CNT and matrix is usually non-covalent and weaker than the covalent bond which can be created by functionalizing the CNT surface as will be discussed in the next section.

Surface Functionalization

Many methods can be used to functionalize the CNT surface to graft chemical groups onto the CNT surface [15] [32–58]. These functional groups could make CNT soluble or link the CNT with the polymer solution or melt. The most well known functionalization method is the strong acid oxidation of the CNTs. The CNTs are refluxed in a concentrated acid, such as sulfuric acid or nitric acid or their mixture.

The oxidation helps purify the CNT and also exfoliates the ends of CNTs and attaches carboxylic acid group such as –COOH to the tips. The carboxylic groups make the CNTs disperse well in water, acetone and polymer solutions because the negative charge functional groups attached to the CNT surface, counteract the van der Waals forces and keep them from agglomerating. Shaffer *et al.* have achieved homogeneously separated CNT dispersion in water with the acid treated CNT [44]. Fan *et al.* [2005] found oxidized MWNT that were separated and homogeneously dispersed in Vinyl ester resin [39], as shown in Figure 2a.

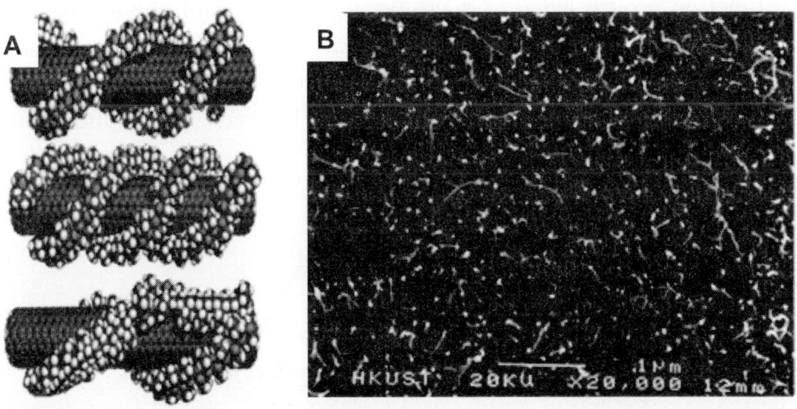

Fig. 6. (A) A cartoon depicting the wrapping arrangement of PVP on a SWNT [28]; (B). SEM image showing evidence of good dispersion of poly(phenylacetylenes) (PPA) wrapped around CNT [30].

Further functionalization could be performed on oxidized CNTs to improve bonding between the CNTs and the resin molecules. By chemical reaction, carboxylic groups on CNT could be replaced by other functional groups, favoring reaction with matrix molecules. For example, Zhu *et al.* [58] used HCl to oxidize SWNT to facilitate the terminations of opened ends of SWNTs with carboxylic acid groups rather than carboxylates. Then he fluorinated the oxidized SWNT or SWNT-COOH by heating the SWNT-COOH in C_2F to gain the fluorinated F-SWNT-COOH, which had –COOH groups attached at the tip and fluorine group on the surface, as shown in Figure 7a. The –COOH group would favor the esterification with epoxy and fluorine like to react with diamine in epoxy as shown in Figure 7b. Thus, SWNT is dispersed well in epoxy and bonding force also increases due to the reaction between functional groups and epoxy resin.

In the work by Eitan *et al.*, oxidized MWNT reacted with di-glycidyl ether of bisphenol A (DGEBA). Thus some new epoxide groups attached to carboxyl groups which can further interact with the polymer matrix [37]. Chen *et al.* used the octadecylamine (ODA) instead of DGEBA and introduced amide groups to connect with carboxyl groups [35] and the new function groups acted as a bridge, having one end bonded with the CNT surface and the other end with the polymer resin molecules, such as epoxy. Moniruzzman *et al.* also achieve good dispersion by grafting epoxy resin molecules by an etherification between carboxylic groups and epoxy [15].

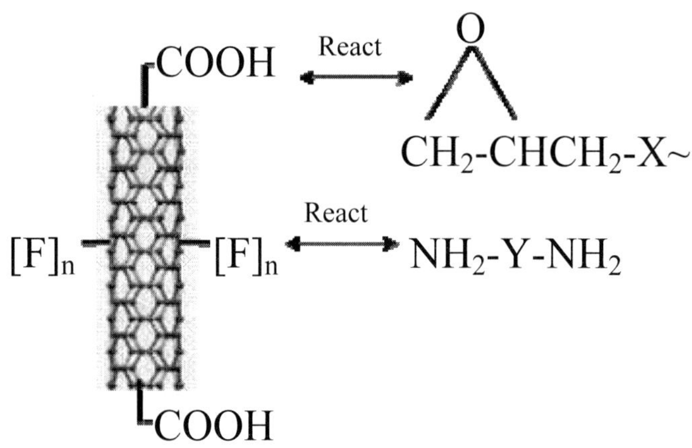

Fig. 7. Schematic of functionalized SWNT and its reaction with epoxy [58].

Wang *et al.* grafted PMMA onto the MWNT surface by a series of chemical reactions to functionalize MWNT in the following sequence, MWNT-COOH, MWNT-Cl, MWNT-OH, MWNT-Br, PMMA-MWNT. The PMMA grafted MWNT had very good compatibility with PMMA and was uniformly disperse in PMMA matrix [46]. Sabba *et al.*

functionalized SWNT by ultrasonic agitation SWNT in (NH2OH)(HCl) for 24h. The functionalization induces a positive charge on the nanotubes, which resulted in a repulsive force between the neighboring CNTs. This repulsive force was capable of overcoming the attractive van der Waals force and thus enabled the exfoliation of SWNT ropes and achieved good dispersion [43]. The advantage of most chemical functionalization is that it usually establishes covalent bonds between CNT and matrix, which is stronger and improves the efficiency of the load transfer from the matrix to the CNT. However, the chemical reaction can also cause some defects on the CNT surface and deteriorate the CNT mechanical properties.

2.4. *Characterization of dispersion of CNT in the suspension*

Although many methods have attempted to improve CNT dispersion, there is no established standard to characterize and quantitatively describe the quality of dispersion. Most researchers have claimed that their method improved the dispersion quality, by showing a SEM or TEM picture and qualitatively accessing the dispersion state by visual inspection. Moreover, there is no standard as to what magnification one should observe under to claim there are no aggregates. For example if one observes under optical microscope and does not observe CNT aggregates, does that imply that there are no aggregates present or is the resolution not able to spot them? It is also possible that one may conclude that CNTs have dispersed well if one does not observe CNT aggregates under Scanning Electron Microscope (SEM). This may not be true as some CNT aggregates may be hidden under the matrix. Transmission Electron Microscope (TEM) may uncover these but they only show very localized CNT dispersion image in the composite. Here, we review various methods that have been used to characterize CNT dispersion.

Solvent method

This method is common and straightforward. If CNT can dissolve in a liquid by a physical displacement method such as ultrasonic agitation, and if the suspension remains stable and the nanotubes do not separate or precipitate over a prolonged period of time that is of the order of days, it is concluded that the CNT have dispersed well in the liquid. Islam *et al.* checked several surfactants and found SWNT could dissolve in water with surfactant SDS: $CH_3(CH_2)_{11}OSO_3Na$, TX100: $C_8H_{17}C_6H_4(OCH_2CH_2)_nOH$, and NaDDBS ($C_{12}H_{25}C_6H_4SO_3Na$). However, after 5 days, NaDDBS-SWNT suspensions appeared homogeneous whereas SDS-SWNT and TX100-SWNT suspensions had coagulated nanotubes at the surface and the bottom of the vials, as shown in Figure 8 [19]. Thus NaDDBS proved to be a good surfactant. The solvent method is limited to very low viscosity suspending fluids such as water, acetone and Tetrahydrofuran (THF). It is common to check the dispersion quality in biological aqueous media and solution sonication method. In case of highly viscous fluids, the solvent method is not very effective. For example, we found that MWNT could dissolve in Epoxy by solution sonication method and stay without any precipitation for a long time, even though the dispersion state was in form of heterogeneous aggregates.

Fig. 8. Effect of Surfactant on SWNT/Water dispersion (from left to right: SDS, TX100 and NaDDBS). NaDDBS was most effective in dispersing SWNT in water as the SWNTs remain dissolved in the liquid [19].

Optical Microscope

Optical Microscope is another quick method to check CNT dispersion. Moniruzzaman *et al.* claimed that because there were no obvious agglomerates in the optical microscope picture, the SWNT dispersed

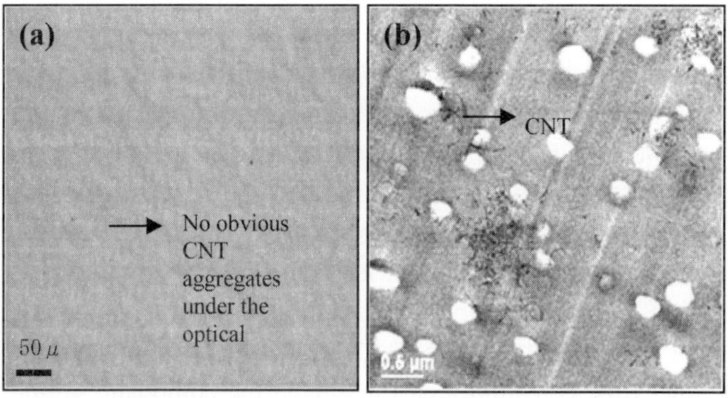

CNT aggregates under the TEM

Fig. 9. In (a) at 1000X, no obvious CNT aggregates are visible, and claim is made of good dispersion; however, in (b), at 30,0000X, TEM image shows non-uniform CNT aggregates dispersion.

uniformly in epoxy [15]. However, optical microscope is limited by its magnification, which could only prove that CNT dispersion was uniform at micro scale as shown in Figure 9a. We found that under less than 1000x optical microscope, the dispersion of MWNT in Epoxy was uniform, while at 30,000x, some MWNTs were still in the form of small aggregates, about 500 nm in diameter as shown in Figure 9b. Presence of aggregates suggest fair to poor dispersion.

Transmission Electron Microscope (TEM)

TEM is a useful method to check the CNT dispersion quality. Solid CNT composite is sliced into thin sections of 90nm thickness and observed under a TEM. The electrons can penetrate the section and generate signals which can visualize all the CNTs inside the composite slice, as shown in Figures 2, 3 and 8a. The advantage of TEM is that it shows all details inside the volume of the selected sample. However, as TEM can show a very small section, one can gauge only the state of CNT dispersion in that region only. This small slice may not represent the dispersion state overall. In addition, preparation of TEM section is time consuming and TEM is a very expensive microscope and is not that readily available.

Scanning Electron Microscope (SEM)

SEM is more commonly available and is used to characterize the CNT dispersion quality. Electrons emit from an electronic gun and hit the surface of the sample. The signals generated from the sample surface are collected and help visualize the sample surface at the nano scale. Thus SEM is used to investigate the CNT dispersion state on the fractured surface of a composite sample. As shown in Figure 6B, CNTs pulled out on the surface of PPA clearly exhibits the good dispersion. Compared to TEM, preparation and operation of SEM is more convenient and less expensive. The only time consuming part is to coat the composite sample surface with a layer of gold before observation if the composite surface is not conductive. SEM can scan a much large sample than the TEM, which is why it is more commonly used to observe and characterize CNT dispersion in polymer matrix. However, SEM can only observe the detail on the surface, not under the surface. That's why we need to fracture the CNT/Polymer composites to pull out most CNTs for clear observation. Sometimes, if CNTs are short, for example some oxidized CNT [59], most CNTs might not appear on the surface and then SEM image cannot ascertain the dispersion state of the CNT in the composite. We believed that if SEM scan showed uniform dispersion and TEM also showed localized separate and homogeneously dispersed CNT at the same time, one could conclude with some confidence that CNT are dispersed well in the suspending matrix.

Raman Imaging

Raman Imaging is one way to quantitatively characterize CNT dispersion although it's not commonly employed. This approach takes advantage of the resonance enhancement effect of Raman scattering from SWNTs relative to that of the polymer matrix [60]. Du *et al.*, used a laser to light a thin SWNT/PMMA film and Raman spectrum was collected. The approximation that Raman intensity is directly proportional to the number of SWNTs is used to quantify the dispersion. The Raman intensity map represents the state of nanotube distribution in the sample on a scale of tens of microns. When the nanotube bundles are uniformly

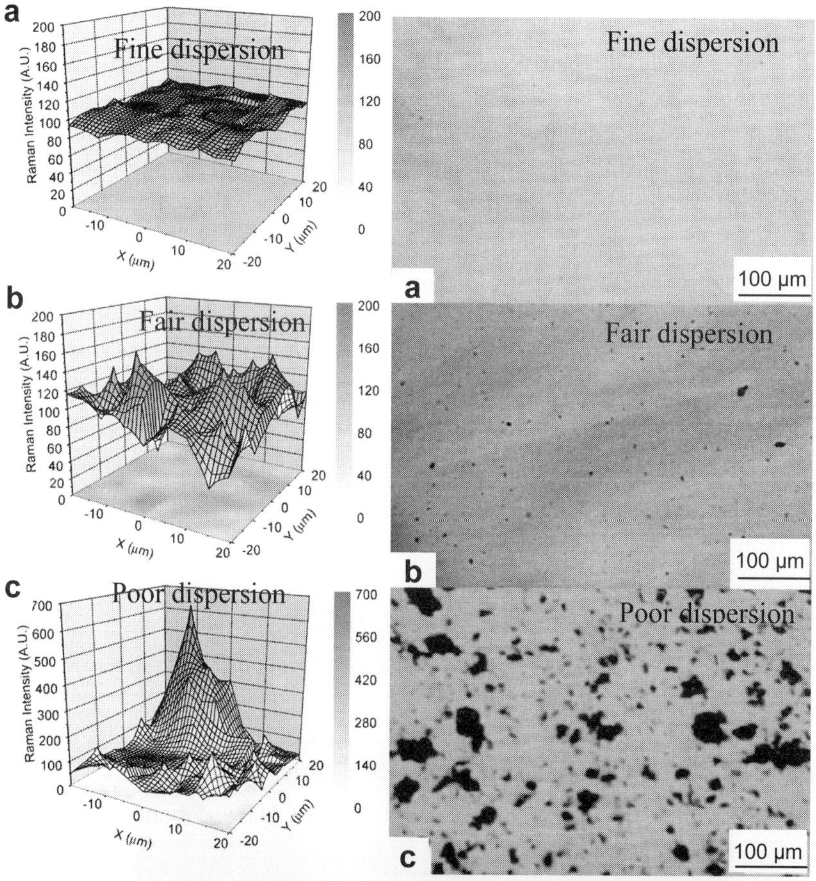

Fig. 10. Raman Images (Left) and Optical Micrographs (Right) of (a) fine (b) fair and (c) poor SWNT dispersion in Epoxy [60].

distributed and the surface is smooth, the Raman map is featureless [60]. As shown in Figure 10, a comparison of fine, fair and poor SWNT dispersion quality were shown by Raman Image and Optical Microscope.

Capillary method

A simple capillary method was proposed to qualitatively characterize dispersion [39]. It takes advantage of micro gaps between single glass fibers in a glass fiber bundle. The gaps form a series of parallel capillaries with diameters ranging from 5 μm to 30 nm. If the polymer solvent has low viscosity of the order of 10 to 1500 cP and compatible with the glass fiber surface, the capillary pressure will pull the suspension up between the fiber bundles as soon as one end of the bundle is placed in the suspension. If the CNTs are homogeneously dispersed in the polymer in the form of separate CNTs or micro aggregates, the CNT/Polymer suspension will rise up in the glass fiber bundle. On the other hand, if the CNTs are poorly dispersed and exist as large aggregates in the suspending polymer, the suspension won't be able to rise to a high level in the glass fiber bundle because the CNT aggregates will block the micro capillaries. Figure 11 showed the height reached in glass fiber bundle by MWNT/Vinyl ester suspensions of (a) Non-uniform MWNT aggregates dispersion (b) Uniform smaller MWNT aggregates dispersion and (c) Homogeneously separated MWNT dispersion. The higher the MWNT suspension rises in the glass fiber tow, the better the MWNT dispersion state [39].

Fig. 11. Qualitative characterization of the MWNT dispersion: The quality of MWNT dispersion is proportional to the height h, the level to which the suspension rises in the fiber tow. (Insert: SEM image of the MWNT on the glass fiber in c) [39].

3. Interfacial Bonding

It is believed that when subjected to tension or shear, if the CNTs have a strong bond with the polymer matrix, the mechanical load will be effectively transferred to the CNT. Thus the strong mechanical properties of CNT can fully benefit the composites.

3.1. *Methodology*

There are almost no physical approaches that can be attempted to increase the interfacial bonding between the CNT and the polymer matrix. However, one approach to disperse CNT well is to exfoliate CNT aggregates as much as possible. The hypothesis is that if the CNT can exfoliate, the contact area between the surface of the CNT and matrix will be larger and the friction on the CNT surface will help in effectiveness of the load transfer. Thus, dispersion plays a key role in enhancement of interfacial bonding.

Most methods to increase interfacial bonding are chemical approaches. Actually, there is no clear boundary between the chemical approaches to improve interfacial bonding and to improve dispersion as discussed above. Surfactants, polymer wrapping and function groups not only make CNT surface compatible with polymer matrix, but also link the CNT surface to the polymer matrix and increase the bond force. Jia *et al.* used the *in situ* polymerization method to make MWNT/PMMA and found that AIBN, which was used to polymerize PMMA, reacted with CNT by openning their π bond and linking them with the PMMA matrix. It was found that the MWNTs were more dispersed and wrapped by PMMA layers, demonstrating strong bonding force between MWNT and PMMA matrix [61]. Zhu *et al.* functionalized SWNT via a dicarboxylic acid peroxide treatment to attach alkylcarbonyl groups onto SWNT sidewalls. The functionalized SWNT could act as a curing agent for epoxy and in addition they covalently integrated into epoxy matrix and became part of the cross linked structure rather than a separate filler as shown in Figure 12. Breakage of SWNTs on the fracture surface verified the strong interfacial bonding between SWNT and epoxy matrix [52].

Many chemical methods have been explored to improve the boding and success has been achieved to a limited degree [47–53,61,62] . But it is believed that creating a strong covalent bond between CNT and the matrix molecules would be more effective than non-covalent bonds to

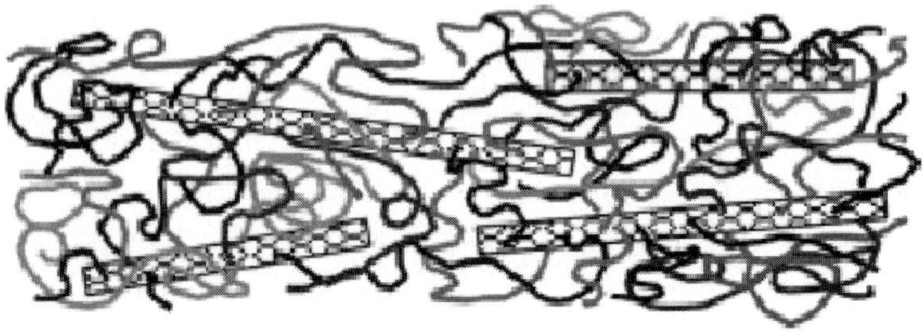

Fig. 12. Integration of the nanotubes into the polymer cross linked structure. [52].

improve bonding for effective load transfer. However, as discussed previously, creating covalent bond may involve aggressive chemical treatment and will damage the CNT surface. Thus one needs to balance the pros and cons when functionalizing CNT surfaces to improve bonding characteristics to enhance final composite mechanical properties.

3.2. *Characterization*

Unlike the CNT dispersion which can be directly observed under SEM or TEM, there is almost no standardized method available to measure the bonding force at the nano scale. Due to the nano scale of CNTs, it's extremely difficult to grasp and pull a single CNT out from a composite. However, Cooper *et al.* created the *in situ* test by detaching a single MWNT and SWNT from the Epoxy matrix, observing deformation under TEM and tracking the force, as shown in Figure 13A & B [63]. They reported the shear strength of MWNT was 91 MPa and SWNT was 366 MPa. Similarly, Barber *et al.* achieved this by using AFM. They attached one single CNT on the tip of an AFM, pushed the CNT into a heated soft polymer and then pulled it out after it cooled down and solidified as shown in Figure 13C. By recording the pulling force as the CNT was being pulled out, Barber *et al.* calculated the shear stress to be about 47 MPa, [64].

Direct measurement of these forces is difficult. So some researchers have proposed indirect ways to characterize the interfacial bonding force between CNT and the matrix. TEM is a commonly used method to gauge

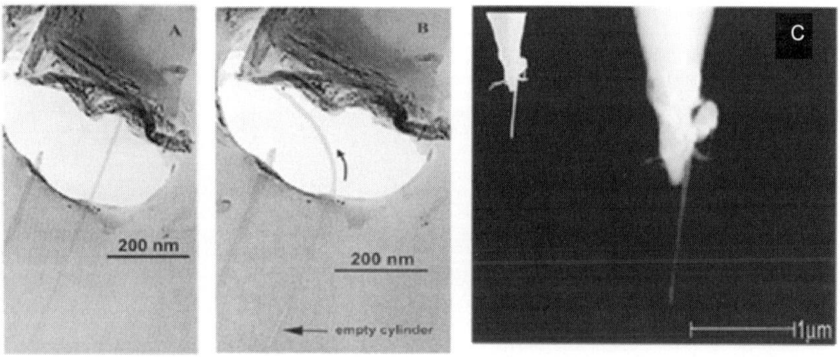

Fig. 13. (A&B) TEM images of a MWNT crossing a hole in an epoxy resin matrix [63] and (C) Use of AFM to measure the force on a single MWNT [64].

interfacial bonding force between CNT and matrix. By observing the CNT composite fracture, one can estimate the bonding force. As show in Figure 14, Gojny *et al.* hypothesized that if the single CNT did not break when subjected to a tension force, and came out completely intact, the bonding between the CNT and polymer matrix was weak [65]. On the other hand, if CNT breaks due to load transfer, it is indicative of strong bonding. Under TEM, Lourie *et al.* observed CNT fracture due to compression during composites processing, which indicated a strong bond between CNT and the epoxy matrix [66], as shown in Figure 15. Qian *et al.* made the in situ TEM observation of load transfer and deformation in CNT/Polystyrene and the breakage of individual MWNT, which also suggested strong bonding force, as shown in Figure 16 [67]. But Ding *et al.* thought that because the yield strength of the CNT was very high, even in the case of strong bonding between the CNT and matrix, it's still possible that CNT could be pulled out completely intact and still signal strong bond strength. They found that in their CNT/Polycarbonate composite, before the breakage of CNT and CNT bond with the matrix, the polymer around the CNT cracked before the CNT was pulled out with fragments of polymer matrix attached on the CNT surface [68]. This to them implied very strong bonding between the CNT and the polycarbonate.

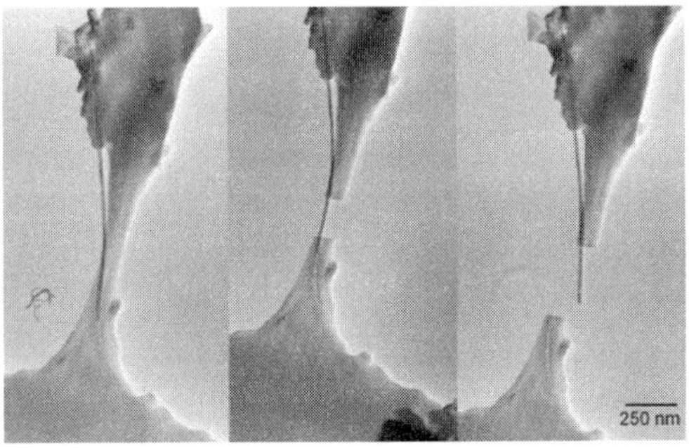

Fig. 14. A weak interaction between the epoxy matrix and the nanotube leads to easy pull-out [65].

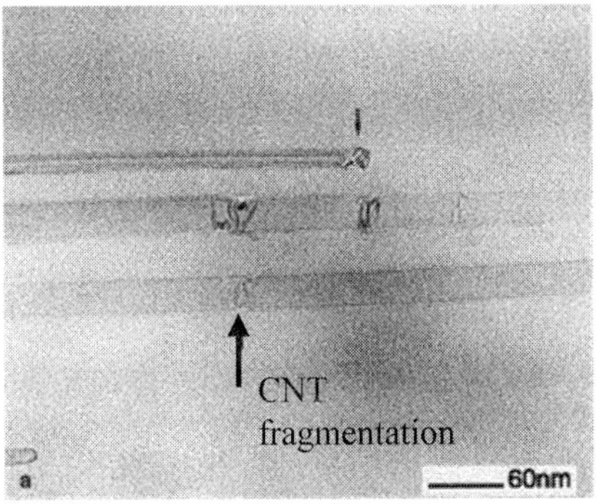

Fig. 15. CNT fragmentation due to compression indicates strong bonding between the CNT and the epoxy matrix [66].

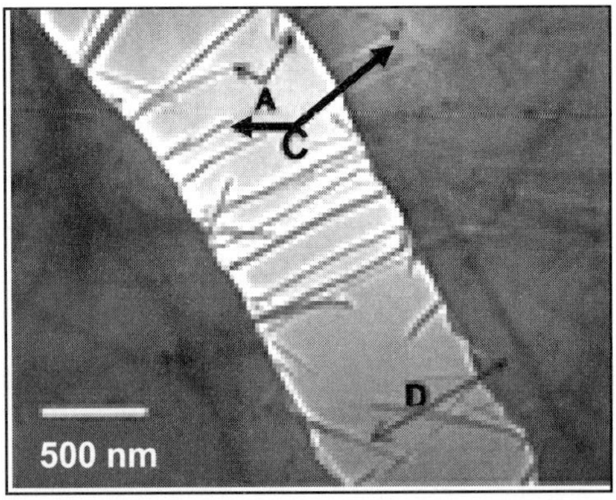

Fig. 16. MWNT breakage due to tension load at location C in Polystyrene [67].

Besides observing breakage of CNT, Gojny *et al.* believed that good wetting or good polymer wrapping around a single CNT by the polymer molecules also implied good bonding between the CNT and the polymer matrix [65] as shown in Figure 17. This was also agreed by similar observation by others [61,69,70]. Compared with SWNT, interfacial bonding between MWNT and the polymer matrix is weaker due to its multi layered structure. Under TEM, sliding of inner nanotube from MWNTs or telescopic pull-outs behavior was observed, indicating weak bonding between MWNT layers [65,71], as shown in Figure 18.

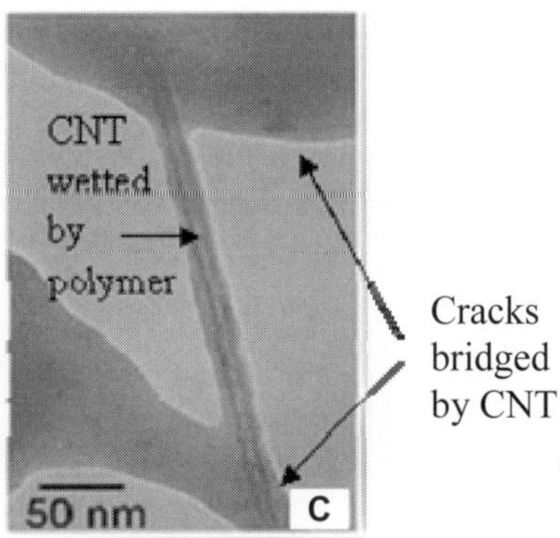

Fig. 17. The single MWNT bridged the micro-cracks and wets epoxy creating good bonding [65].

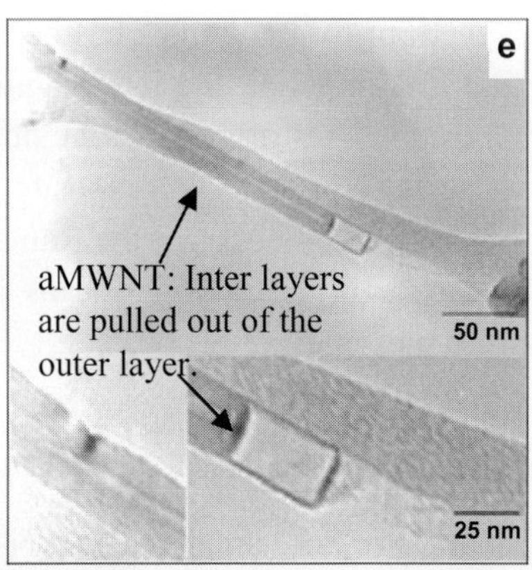

Fig. 18. The telescopic pull-outs from MWNT: The inter layers cannot sustain load because of the weak force between layers [71].

Observation of the fracture surface of CNT composite under SEM could also be used to characterize the interfacial bonding force between the CNT and the polymer matrix. If the CNT were pull out and was long and it left an imprint on the fractured surface, the bonding force was characterized as weak. If CNTs were found on the fractured surface and pulled out with a very short length, it would signify a strong bond between CNT and the matrix [58,72,73]. Under SEM, Figure 19A showed the imprint due to poor bonding [72] and breakage of MWNT due to strong bonding in Figure 19B [66].

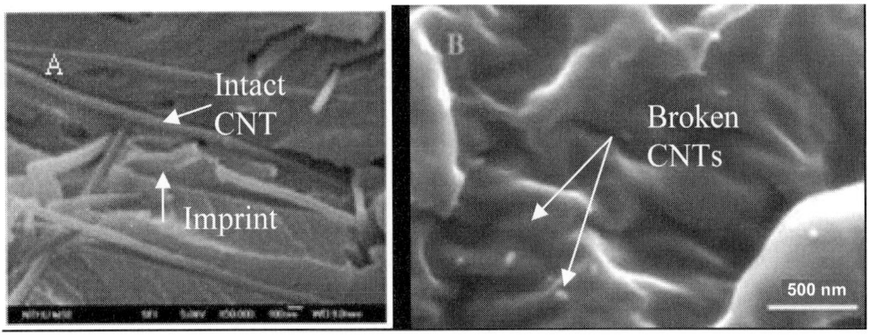

Fig. 19. (A) CNT imprint on the fracture surface due to weak bonding [72]; (B) broken CNT on the fracture surface indicates strong bonding [66].

Raman spectrum could also be used to characterize the interfacial bonding. The Raman spectrum for nanotubes has been well documented. When a strain is applied to a material, the interatomic distances change, and thus the vibrational frequencies of some of the normal modes change causing a Raman peak shift. When CNT composite was subjected to a load deformation, it was assumed that larger the Raman peak shift, larger is the strain carried by the nanotubes, which means good bonding force between CNT and the polymer matrix [74].

4. CNT Orientation State

Similar to short fiber reinforced composites, CNT orientation state determines the anisotropic functionality of CNT composites. Hence, CNT orientation state plays a crucial role in defining the internal micro structure which will influence the composite physical and mechanical properties. In this part, we focus on two issues: (1) Characterization of CNT orientation state in a composite (2) Alignment of CNTs in composites.

4.1. *CNT orientation characterization*

Unlike short fiber suspensions, CNT cannot be seen by the naked eye or effectively under an optical microscope. There is no direct and straightforward way to characterize CNT orientation state in a composite. We will discuss the few methods that have been used to characterize the orientation state of the carbon nanotubes.

TEM and SEM are two popular tools to observe CNT directly. However, because SEM can only see the surface, it is not very useful in characterizing the CNT orientation state. TEM is reliable because it can show the characterized CNT orientation. Figure 20 showed the TEM image of aligned CNTs caused by stretching in the matrix [75]. Complete alignment is easy to characterize as all or most CNT are oriented in a particular direction. However, if the CNTs are oriented randomly or exhibit partial CNTs alignment, one can quantify the CNT orientation state, which will discussed later.

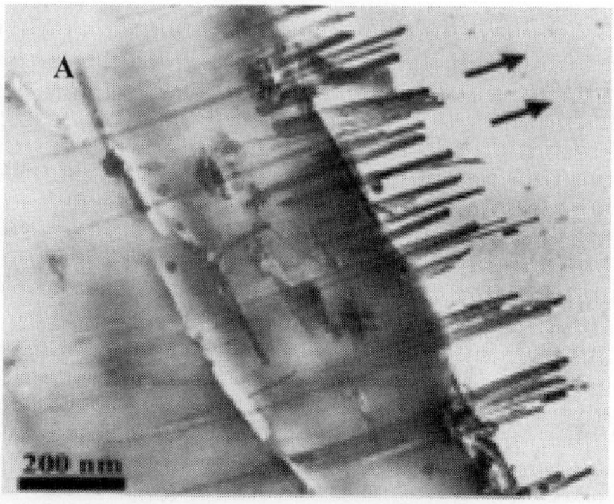

Fig. 20. TEM image of aligned Carbon Nanotubes in Polyhydroxyaminoether (PHAE) [75].

TEM is a direct observation method to characterize CNT orientation state. As direct observation method is tedious and time consuming, indirect ways to characterize CNT orientation have been attempted. Raman Spectrum is one of them. Wood *et al.* found that slope of Raman wavenumber vs. strain was very different between load axis direction (–467/cm.strain) and transverse direction (–214/cm.strain) because of a substantial alignment of CNT in the load axis direction (Figure 21). Thus Raman Spectrum could be used to characterize CNT orientation state as well [76].

Wide Angle X-ray diffraction (WAXD) is also a powerful tool for characterization of the degree of crystallinity of CNTs aligned in a polymer matrix. Kima *et al.* applied WAXD to check the CNT alignment from the WAXD photographs of CNT composite sample, as shown in Figure 22. The thinner white arrow indicates the Debye–Scherrer ring of the (002) plane of the CNTs. Although the Debye–Scherrer ring is of a circular form, the whiteness of the ring is more concentrated in the equatorial direction for Figure 22(a) and (b), indicating that the CNTs were aligned in the X-direction [77].

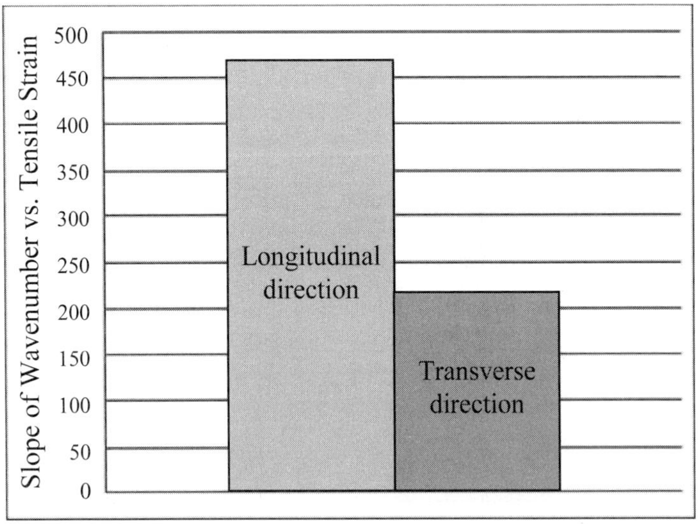

Fig. 21. Different slope of Raman Wavenumber vs. Tensile Strain in longitudinal and transverse direction because most CNT are aligned in the longitudinal direction [76].

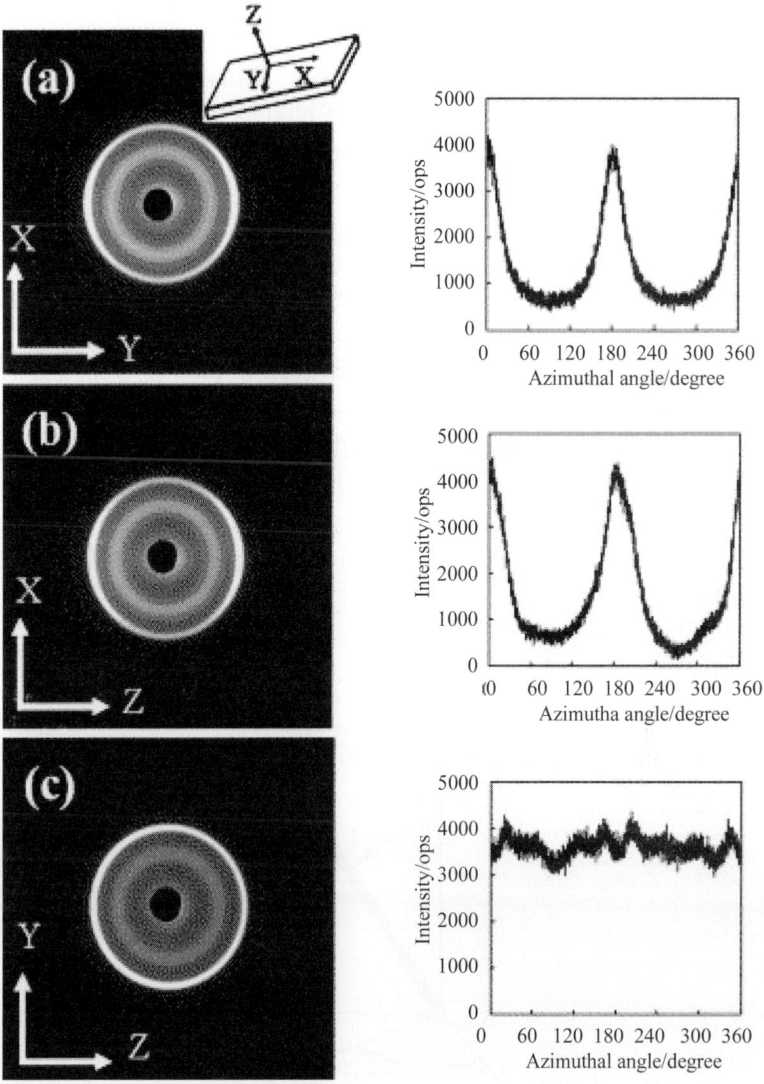

Fig. 22. X-ray photographs and azimuthal X-ray intensity profiles for 30 wt.%-CNTs filled rubber composites; X-ray incident directions are (a) Z, (b) Y, and (c) X: the whiteness of the ring is more concentrated in the equatorial direction for (a) and (b), indicating that the CNTs were aligned in the X-direction [77].

Small-angle X-ray scattering (SAXS) can also be used to characterize CNT orientation, if the CNT does not have good crystallinity (made in HiPco, Inc.). From the two-dimensional X-ray pattern, Du *et al.* integrated along the radial direction and plotted the intensity versus the azimuthal angle. These data were then fit with a simple Lorentzian function, the full width at half-maximum (FWHM) which quantitatively described the degree of nanotube alignment in the composites [78].

To further characterize the orientation quantatively, one can first obtain an SEM or TEM image of the CNTs. Next, one must calculate the average direction of each nanotube in the region of interest. Each CNT is then represented by a unit vector P, as shown in Figure 23. In a selected region, one may find many CNTs which are oriented in different directions as shown in Figure 25. The CNT orientation in Figure 25 is not as straight forward to characterize as in Figure 20 as the nanotubes are oriented in different directions to a different degree of orientation. Hence, it will be important to use a description that will embody many different directions.

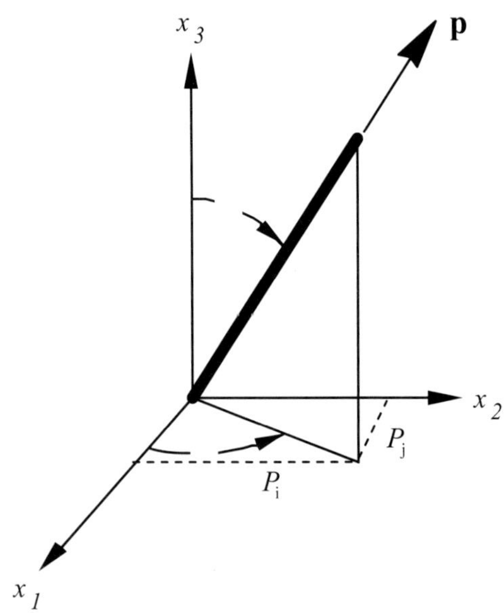

Fig. 23. The orientation of a single CNT can be described by two angles, θ and ϕ, or by a unit vector **p** directed along the particle axis.

One approach as followed by, Shaffer *et al.* is to apply fast Fourier transformation on the SEM micrograph (Figure 24) to obtain a power spectrum. In the resulting power spectrum, the angular density distribution is indirectly represented by the angular distribution of CNTs. Integrating the intensity radially, but omitting the solid which can be formed into a variety of macroscopic shapes and readily handled, a director could be derived which was defined as the angle at which this CNT distribution is maximum (Figure 24) [44].

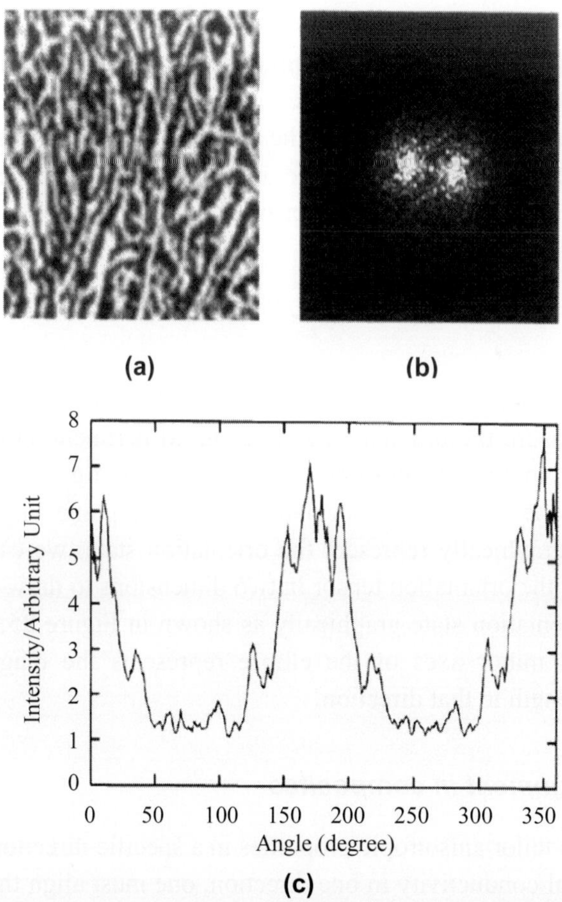

Fig. 24. Steps in the Fast Fourier image analysis technique for determining the ordering parameter: (a) randomly selected SEM image (b) its power spectrum; and (c) the angular intensity of the power spectrum [44].

We use the direct method of measuring the orientation of each and every nanotube in the TEM picture and describing the orientation state at that location is by using the second order orientation tensor description [59]. In this method we first assign a vector for each nanotube along the nanotube length direction. The second order orientation tensor is defined as follows

$$\langle a \rangle = \langle p_i p_j \rangle = \oint p_i p_j \varphi(P) dP \tag{1}$$

Here P_i and P_j refer to the components of the unit vector P as shown in Figure 23 and $\varphi(P)$ is the distribution function, describing the probability of a nanotube to be oriented in the direction p. Experimentally, one would measure the direction of each and every unit vector along each and every nanotube and express the average value of the components of the second order tensor a_{ij} as follows

$$a_{ij} = \frac{1}{N} \sum_{k=1}^{N} p_i^k p_j^k \tag{2}$$

Here N is the number of nanotubes in the region. Thus as shown in Figure 25 we can measure the orientation state from all pertinent TEM images and express it in two dimensions as shown in the figure or in three dimension if one is able to calculate the out of plane angle of the nanotube. To graphically represent the orientation state, we can use the components of the orientation tensor in two dimensions to draw an ellipse to describe orientation state graphically as shown in Figure 25, in which the major and minor axes of the ellipse represents the magnitude of orientation strength in that direction.

4.2. *CNT alignment in composites*

If one wants to tailor anisotropic properties in a specific direction, such as strong electrical conductivity in one direction, one must align the CNT in that specified direction. Several methods have been explored to align CNTs in composites. In the following, we will review how CNT are aligned by these methods.

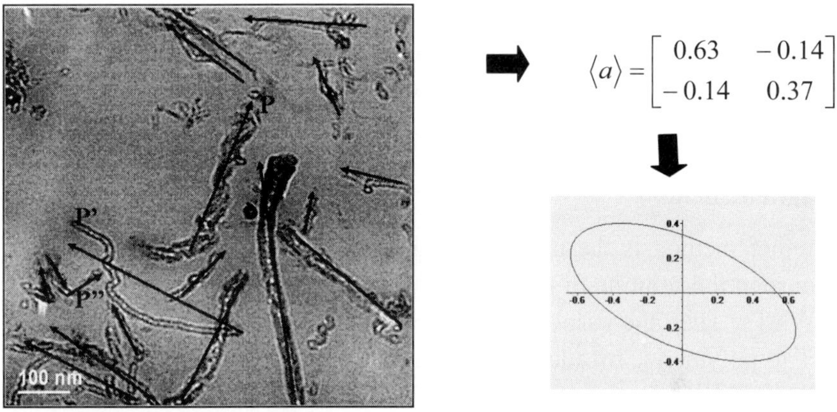

Fig. 25. Characterization of MWNT orientation state from a TEM image using second order orientation tensor and use of ellipse to represent the strength and direction of orientation: ratio of major axis over minor axis represents the degree of alignment in the direction of the major axis [59].

Mechanical stretching

The shear flow has been known to orient short fibers in the direction of the shear or stretching [79]. Jin *et al.* fabricated MWNT/Polyhydroxyaminoether (PHAE) composites by casting a suspension of carbon nanotubes in a solution of a thermoplastic polymer and chloroform. The composite was then stretched uniaxially at 100°C and the MWNTs alignment inside the composite was found to remain in the direction of the elongation after removal of the load at room temperature as shown in Figure 20 [75].

Extruder

Extruder is one common way to align CNT in highly viscous thermoplastic resins, such as PS and PE. The screw in the extruder can generate very high shear flow which can help dispersion of the nanotubes in the melt. When this highly viscous melt is drawn from the thin film die, the stretching will align the CNTs in drawing direction. Thostenson *et al.* not only achieved good dispersion of MWNT in PS, but also aligned the MWNTs in PS when the hot soft composite was drawn from the small die of an extruder, as shown in Figure 26. They also found that the storage modulus of CNT/PS composite was five times higher in the aligned direction as compared to the composite in which the CNT did not have a preferred orientation [12].

Magnetic field

Another method is the use of strong magnetic field. Kimura *et al.* first sonicated the solution of unsaturated polyester and styrene monomer with MWNTs. Then the suspension was put into a constant magnetic field of 10T to polymerize by adding the radical initiator. As shown in the Figure 27, most MWNTs were found aligned in the magnetic field direction [80]. The author also found that the storage modulus of the aligned MWNT/PE composites was higher than the random MWNT/PE composite. However, this method requires a very strong magnetic force and is not readily available.

Electric AC Field

Bando *et al.* used an electric field with a direct current and successfully aligned SWNT in polyimide and characterized it by Polarized Raman spectroscopy [81]. Liu *et al.* tried to use AC field to align MWNT in dielectric solvent and found that aligned MWNT increased capacitance and conductance of the suspension [82]. Figure 28 shows the aligned MWNT bundles.

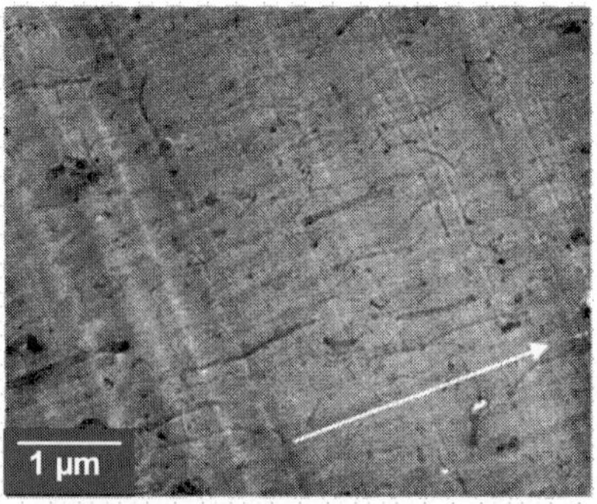

Fig. 26. TEM image of well dispersed and aligned MWNT due to high shear rates experienced by the material in the in extruder [12].

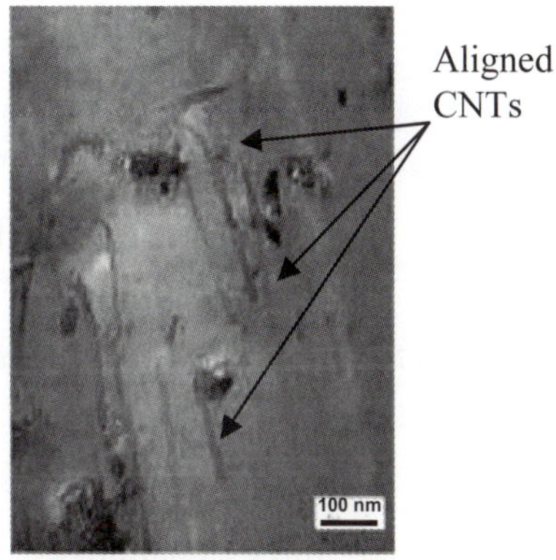

Fig. 27. TEM image of MWNT aligned by high magnetic field (10T) in Polyethylene [80].

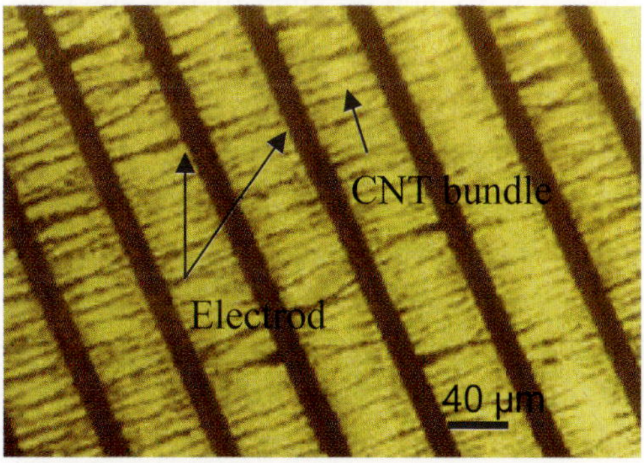

Fig. 28. Optical image of aligned MWNT bundles in AC field [82].

Shear flow field in micro channel

Similar to extruder, during composite processing, shear flow could be created that will align CNTs to a certain degree in the direction of shear. Fan *et al*. obtain good alignment of MWNTs when the MWNT suspension was forced to flow through nano channels within glass fiber bundles. Because the gaps between single glass fibers in a bundle was of the order of 20 nm, high shear flow is generated in this gap that was able to align the MWNT as shown in Figure 29 [59].

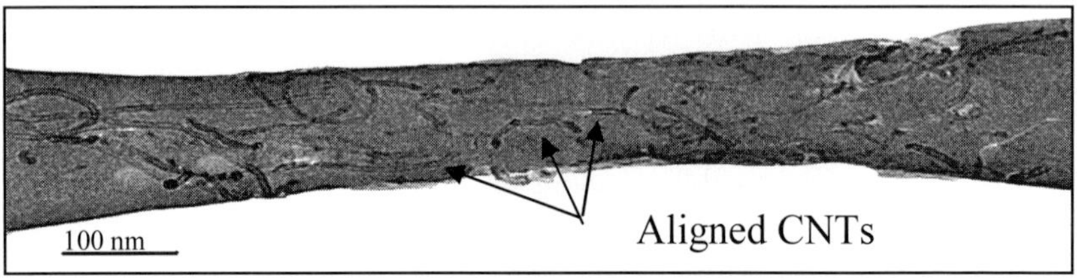

Fig. 29. TEM image of aligned MWNTs due to shear in a nano channel [59].

Conclusions and Outlook

In this chapter, we reviewed three challenges that arise when manufacturing CNT/Polymer composites: dispersion, interfacial bonding and orientation of the nanotubes. Good CNT dispersion, strong interfacial bonding force between CNT and polymer matrix and tailoring of the CNT orientation state are crucial in enhancing the CNT based functional polymer composites.

Physical and chemical approaches to disperse have been attempted by various researchers. To date, for high temperature and very viscous thermoplastic polymers, such as PS, homogeneous and separate CNT dispersion has been achieved by high shear rates experienced in an extruder. For moderately viscous thermoset polymers, such as epoxy, there are no standard methods to exfoliate and disperse CNT in the matrix. Combinations of solution sonification and chemistry treatments on CNT have produced good CNT dispersion to a certain degree. However, no technique that can scale this process for mass production is available. Chemical modification of CNT surfaces is important for dispersion and formation of strong bonds with the suspending matrix Most methods to date use a qualitative measure to characterize CNT dispersion state and there are still no quantitative methods to gauge CNT dispersion quality. Practice has converged on observation of nanotubes under SEM and TEM to characterize CNT dispersion quality.

Interfacial bonding force between CNT and polymer matrix is more difficult to gauge as no standard methods are available. Researchers have used atomic force microscopy to measure the force required to separate an embedded nanotube from a polymer matrix. They also found that the polymer properties in the viscinity of the nanotube were also higher compared to the bulk polymer properties. Surface treatment is the path that is being followed to improve bonding between the CNT and the surrounding matrix. One promising approach seems to be grafting organic groups or polymer molecules on the CNT surface and CNT tips that are compatible with the polymer bulk matrix. Covalent molecular bonding creates the strongest interfacial bonding and effective load transfer. However, during chemical treatment, potential damage to the CNT surface can have an adverse effect on the bond strength.

One could align the nanotubes using the stretching and shearing experienced during flow. For highly viscous thermoplastic polymers containing CNT, alignment could be achieved by use of an extruder with most of the nanotubes aligned in the direction of shear or stretching. There are other methods practiced to a smaller extent such as use of an electrical field or a magnetic field which could help achieve the goal of preferential alignment. The properties of the CNT reinforced poymer do depend on the orientation of the CNT in addition to dispersion, aspect ratio and interfacial bonding. Thus, one could tailor the orientation to obtain the desired properties.

We predict that the 21st century will focus more and more on "nano scale" engineering in which the role of nanoparticles such as nanoclays and nanotubes will continue to evolve. Nanotubes have exhibited unique bulk properties but the surface of the nanotube is not usually ideal for the desired application. The challenge is can we improve and make the nanotube surface more compatible with the host material without degrading its bulk properties? Can we introduce an atomic layer coating on the surface of the nanotube to yield more desirable surface properties? Can we also produce these materials on a bulk scale at an affordable price? Can we process them effectively with different polymer matrices? These challenges must be addressed if nanotubes are to find their way into the applications of the future.

Acknowledgements

The authors gratefully acknowledge the partial support provided by the National Science Foundation under Grant DMI- 0115127.

References

1. Iijima, S., *Helical Microtubules of Graphitic Carbon.* Nature, 1991. **354**(6348): p. 56–58.
2. Lordi, V. and N. Yao, *Molecular mechanics of binding in carbon-nanotube-polymer composites.* Journal of Materials Research, 2000. **15**(12): p. 2770–2779.
3. Krishnan, A., *et al., Young's modulus of single-walled nanotubes.* Physical Review B, 1998. **58**(20): p. 14013–14019.
4. Yu, M.F., *et al., Tensile loading of ropes of single wall carbon nanotubes and their mechanical properties.* Physical Review Letters, 2000. **84**(24): p. 5552–5555.
5. Berber, S., Y.K. Kwon, and D. Tomanek, *Unusually high thermal conductivity of carbon nanotubes.* Physical Review Letters, 2000. **84**(20): p. 4613–4616.
6. Frank, S., *et al., Carbon nanotube quantum resistors.* Science, 1998. **280**(5370): p. 1744–1746.
7. Lau, K.T., *Interfacial bonding characteristics of nanotube/polymer composites.* Chemical Physics Letters, 2003. **370**(3–4): p. 399–405.
8. Lau, K.T., S.Q. Shi, and H.M. Cheng, *Micro-mechanical properties and morphological observation on fracture surfaces of carbon nanotube composites pre-treated at different temperatures.* Composites Science and Technology, 2003. **63**(8): p. 1161–1164.
9. Lin, Y., *et al., Advances toward bioapplications of carbon nanotubes.* Journal of Materials Chemistry, 2004. **14**(4): p. 527–541.
10. Haggenmueller, R., *et al., Aligned single-wall carbon nanotubes in composites by melt processing methods.* Chemical Physics Letters, 2000. **330**(3–4): p. 219–225.
11. McNally, T., *et al., Polyethylene multiwalled carbon nanotube composites.* Polymer, 2005. **46**(19): p. 8222–8232.
12. Thostenson, E.T. and T.W. Chou, *Aligned multi-walled carbon nanotube-reinforced composites: processing and mechanical characterization.* Journal of Physics D-Applied Physics, 2002. **35**(16): p. L77–L80.
13. Potschke, P., *et al., Orientation of multiwalled carbon nanotubes in composites with polycarbonate by melt spinning.* Polymer, 2005. **46**(23): p. 10355–10363.
14. Potschke, P., A.R. Bhattacharyya, and A. Janke, *Melt mixing of polycarbonate with multiwalled carbon nanotubes: microscopic studies on the state of dispersion.* European Polymer Journal, 2004. **40**(1): p. 137–148.
15. Moniruzzaman, M., *et al., Increased flexural modulus and strength in SWNT/epoxy composites by a new fabrication method.* Polymer, 2006. **47**(1): p. 293–298.
16. Sandler, J., *et al., Development of a dispersion process for carbon nanotubes in an epoxy matrix and the resulting electrical properties.* Polymer, 1999. **40**(21): p. 5967–5971.
17. Safadi, B., R. Andrews, and E.A. Grulke, *Multiwalled carbon nanotube polymer composites: Synthesis and characterization of thin films.* Journal of Applied Polymer Science, 2002. **84**(14): p. 2660–2669.
18. Vaccarini, L., *et al., Reinforcement of an epoxy resin by single walled nanotubes.* AIP Conf. Proc., 2000. **544**: p. 521.
19. Islam, M.F., *et al., High weight fraction surfactant solubilization of single-wall carbon nanotubes in water.* Nano Letters, 2003. **3**(2): p. 269–273.

20. Cui, S., *et al.*, *Characterization of multiwall carbon nanotubes and influence of surfactant in the nanocomposite processing.* Carbon, 2003. **41**(4): p. 797–809.
21. Gong, X.Y., *et al.*, *Surfactant-assisted processing of carbon nanotube/polymer composites.* Chemistry of Materials, 2000. **12**(4): p. 1049–1052.
22. Long, Y.Z., *et al.*, *Synthesis and electrical properties of carbon nanotube polyaniline composites.* Applied Physics Letters, 2004. **85**(10): p. 1796–1798.
23. Matarredona, O., *et al.*, *Dispersion of single-walled carbon nanotubes in aqueous solutions of the anionic surfactant NaDDBS.* Journal of Physical Chemistry B, 2003. **107**(48): p. 13357–13367.
24. Sinani, V.A., *et al.*, *Aqueous dispersions of single-wall and multiwall carbon nanotubes with designed amphiphilic polycations.* Journal of the American Chemical Society, 2005. **127**(10): p. 3463–3472.
25. Velasco-Santos, C., *et al.*, *Dynamical-mechanical and thermal analysis of carbon nanotube-methyl-ethyl methacrylate nanocomposites.* Journal of Physics D-Applied Physics, 2003. **36**(12): p. 1423–1428.
26. Zhang, X.T., J. Zhang, and Z.F. Liu, *Conducting polymer/carbon nanotube composite films made by in situ electropolymerization using an ionic surfactant as the supporting electrolyte.* Carbon, 2005. **43**(10): p. 2186–2191.
27. Xu, Y.S., G. Ray, and B. Abdel-Magid, *Thermal behavior of single-walled carbon nanotube polymer-matrix composites.* Composites Part a-Applied Science and Manufacturing, 2006. **37**(1): p. 114–121.
28. O'Connell, M.J., *et al.*, *Reversible water-solubilization of single-walled carbon nanotubes by polymer wrapping.* Chemical Physics Letters, 2001. **342**(3–4): p. 265–271.
29. Star, A., *et al.*, *Noncovalent side-wall functionalization of single-walled carbon nanotubes.* Macromolecules, 2003. **36**(3): p. 553–560.
30. Tang, B.Z. and H.Y. Xu, *Preparation, alignment, and optical properties of soluble poly(phenylacetylene)-wrapped carbon nanotubes.* Macromolecules, 1999. **32**(8): p. 2569–2576.
31. Liu, P., *Modifications of carbon nanotubes with polymers.* European Polymer Journal, 2005. **41**(11): p. 2693–2703.
32. Arnold, M.S., *et al.*, *Encapsulation of carbon nanotubes by self-assembling peptide amphiphiles.* Langmuir, 2005. **21**(10): p. 4705–4709.
33. Banerjee, S. and S.S. Wong, *Rational sidewall functionalization and purification of single-walled carbon nanotubes by solution-phase ozonolysis.* Journal of Physical Chemistry B, 2002. **106**(47): p. 12144–12151.
34. Berguiga, L., *et al.*, *Carbon nanotube silica glass composites in thin films by the sol-gel technique.* Optical Materials, 2006. **28**(3): p. 167–171.
35. Chen, J., *et al.*, *Solution properties of single-walled carbon nanotubes.* Science, 1998. **282**(5386): p. 95–98.
36. yke, C.A. and J.M. Tour, *Overcoming the insolubility of carbon nanotubes through high degrees of sidewall functionalization.* Chemistry-a European Journal, 2004. **10**(4): p. 813–817.

37. Eitan, A., *et al.*, *Surface modification of multiwalled carbon nanotubes: Toward the tailoring of the interface in polymer composites.* Chemistry of Materials, 2003. **15**(16): p. 3198–3201.
38. Esumi, K., *et al.*, *Chemical treatment of carbon nanotubes.* Carbon, 1996. **34**(2): p. 279–281.
39. Fan, Z.H., K.T. Hsiao, and S.G. Advani, *Experimental investigation of dispersion during flow of multi-walled carbon nanotube/polymer suspension in fibrous porous media.* Carbon, 2004. **42**(4): p. 871–876.
40. Hill, D., *et al.*, *Functionalization of carbon nanotubes with derivatized polyimide.* Macromolecules, 2005. **38**(18): p. 7670–7675.
41. Philip, B., *et al.*, *Polyaniline/carbon nanotube composites: starting with phenylamino functionalized carbon nanotubes.* Polymer Bulletin, 2005. **53**(2): p. 127–138.
42. Qu, L.W., *et al.*, *Polyimide-functionalized carbon nanotubes: Synthesis and dispersion in nanocomposite films.* Macromolecules, 2004. **37**(16): p. 6055–6060.
43. Sabba, Y. and E.L. Thomas, *High–concentration dispersion of single-wall carbon nanotubes (vol 37, pg 4815, 2004).* Macromolecules, 2004. **37**(17): p. 6662–6662.
44. Shaffer, M.S.P., X. Fan, and A.H. Windle, *Dispersion and packing of carbon nanotubes.* Carbon, 1998. **36**(11): p. 1603–1612.
45. Velasco-Santos, C., *et al.*, *Chemical functionalization of carbon nanotubes through an organosilane.* Nanotechnology, 2002. **13**(4): p. 495–498.
46. Wang, M., K.P. Pramoda, and S.H. Goh, *Enhancement of the mechanical properties of poly (styrene-co-acrylonitrile) with poly(methyl methacrylate)-grafted multiwalled carbon nanotubes.* Polymer, 2005. **46**(25): p. 11510–11516.
47. Chen, J., *et al.*, *A versatile, molecular engineering approach to simultaneously enhanced, multifunctional carbon-nanotube-polymer composites.* Advanced Functional Materials, 2006. **16**(1): p. 114–119.
48. Gao, J.B., *et al.*, *Continuous spinning of a single-walled carbon nanotube-nylon composite fiber.* Journal of the American Chemical Society, 2005. **127**(11): p. 3847–3854.
49. Liu, L.Q., *et al.*, *Mechanical properties of functionalized single-walled carbon-nanotube/poly(vinyl alcohol) nanocomposites.* Advanced Functional Materials, 2005. **15**(6): p. 975–980.
50. Paiva, M.C., *et al.*, *Mechanical and morphological characterization of polymer-carbon nanocomposites from functionalized carbon nanotubes.* Carbon, 2004. **42**(14): p. 2849–2854.
51. Ramanathan, T., H. Liu, and L.C. Brinson, *Functionalized SWNT/polymer nanocomposites for dramatic property improvement.* Journal of Polymer Science Part B-Polymer Physics, 2005. **43**(17): p. 2269–2279.
52. Zhu, J., *et al.*, *Reinforcing epoxy polymer composites through covalent integration of functionalized nanotubes.* Advanced Functional Materials, 2004. **14**(7): p. 643–648.
53. Zhu, J., *et al.*, *Processing and properties of polymer composites reinforced by functionalized SWNTs.* Pricm 5: The Fifth Pacific Rim International Conference on Advanced Materials and Processing, Pts 1–5, 2005. **475–479**: p. 1059–1062.

54. Breton, Y., *et al., Mechanical properties of multiwall carbon nanotubes/epoxy composites: influence of network morphology.* Carbon, 2004. **42**(5–6): p. 1027–1030.
55. Dyke, C.A. and J.M. Tour, *Covalent functionalization of single-walled carbon nanotubes for materials applications.* Journal of Physical Chemistry A, 2004. **108**(51): p. 11151–11159.
56. Shi, X.F., *et al., Rheological behaviour and mechanical characterization of injectable poly(propylene fumarate)/single-walled carbon nanotube composites for bone tissue engineering.* Nanotechnology, 2005. **16**(7): p. S531–S538.
57. Lin, Y., *et al., Polymeric carbon nanocomposites from carbon nanotubes functionalized with matrix polymer.* Macromolecules, 2003. **36**(19): p. 7199–7204.
58. Zhu, J., *et al., Improving the dispersion and integration of single-walled carbon nanotubes in epoxy composites through functionalization.* Nano Letters, 2003. **3**(8): p. 1107–1113.
59. Fan, Z. and S.G. Advani, *Characterization of Orientation State of Carbon Nanotubes in Shear Flows.* Polymer, 2005. **46**(14): p. 5232–5240.
60. Du, F.M., *et al., Nanotube networks in polymer nanocomposites: Rheology and electrical conductivity.* Macromolecules, 2004. **37**(24): p. 9048–9055.
61. Jia, Z.J., *et al., Study on poly(methyl methacrylate)/carbon nanotube composites.* Materials Science and Engineering a-Structural Materials Properties Microstructure and Processing, 1999. **271**(1–2): p. 395–400.
62. Park, J.U., *et al., Effective in-situ preparation and characteristics of polystyrene-grafted carbon nanotube composites.* Korea-Australia Rheology Journal, 2005. **17**(2): p. 41–45.
63. Cooper, C.A., *et al., Detachment of nanotubes from a polymer matrix.* Applied Physics Letters, 2002. **81**(20): p. 3873–3875.
64. Barber, A.H., S.R. Cohen, and H.D. Wagner, *Measurement of carbon nanotube-polymer interfacial strength.* Applied Physics Letters, 2003. **82**(23): p. 4140–4142.
65. Gojny, F.H., *et al., Surface modified multi-walled carbon nanotubes in CNT/epoxy-composites.* Chemical Physics Letters, 2003. **370**(5–6): p. 820–824.
66. Lourie, O. and H.D. Wagner, *Evidence of stress transfer and formation of fracture clusters in carbon nanotube-based composites.* Composites Science and Technology, 1999. **59**(6): p. 975–977.
67. Qian, D., *et al., Load transfer and deformation mechanisms in carbon nanotube-polystyrene composites.* Applied Physics Letters, 2000. **76**(20): p. 2868–2870.
68. Ding, W., *et al., Direct observation of polymer sheathing in carbon nanotube-polycarbonate composites.* Nano Letters, 2003. **3**(11): p. 1593–1597.
69. Chen, W., *et al., Enhanced mechanical properties and morphological characterizations of poly(vinyl alcohol)-carbon nanotube composite films.* Applied Surface Science, 2005. **252**(5): p. 1404–1409.
70. Velasco-Santos, C., *et al., Improvement of thermal and mechanical properties of carbon nanotube composites through chemical functionalization.* Chemistry of Materials, 2003. **15**(23): p. 4470–4475.
71. Yu, M.F., *et al., Strength and breaking mechanism of multiwalled carbon nanotubes under tensile load.* Science, 2000. **287**(5453): p. 637–640.

72. Yeh, M.K., N.H. Tai, and J.H. Liu, *Mechanical behavior of phenolic-based composites reinforced with multi-walled carbon nanotubes.* Carbon, 2006. **44**(1): p. 1–9.
73. Leelapornpisit, W., *et al.*, *Effect of carbon nanotubes on the crystallization and properties of polypropylene.* Journal of Polymer Science Part B-Polymer Physics, 2005. **43**(18): p. 2445–2453.
74. Schadler, L.S., S.C. Giannaris, and P.M. Ajayan, *Load transfer in carbon nanotube epoxy composites.* Applied Physics Letters, 1998. **73**(26): p. 3842–3844.
75. Jin, L., C. Bower, and O. Zhou, *Alignment of carbon nanotubes in a polymer matrix by mechanical stretching.* Applied Physics Letters, 1998. **73**(9): p. 1197–1199.
76. Wood, J.R., Q. Zhao, and H.D. Wagner, *Orientation of carbon nanotubes in polymers and its detection by Raman spectroscopy.* Composites Part a-Applied Science and Manufacturing, 2001. **32**(3–4): p. 391–399.
77. Kim, Y.A., *et al.*, *Fabrication of aligned carbon nanotube-filled rubber composite.* Scripta Materialia, 2006. **54**(1): p. 31–35.
78. Du, F.M., J.E. Fischer, and K.I. Winey, *Coagulation method for preparing single-walled carbon nanotube/poly(methyl methacrylate) composites and their modulus, electrical conductivity, and thermal stability.* Journal of Polymer Science Part B-Polymer Physics, 2003. **41**(24): p. 3333–3338.
79. Advani, S.G. and M. Sozer, *Process Modeling in Composite Manufacturing.* 2002, NY: Marcel Dekker, Inc,.
80. Kimura, T., *et al.*, *Polymer composites of carbon nanotubes aligned by a magnetic field.* Advanced Materials, 2002. **14**(19): p. 1380–1383.
81. Banda, S., *et al.*, *Electric field alignment of single wall carbon nanotubes (SWNT) in polymers.* Proceedings of SPIE - The International Society for Optical Engineering, 2005. **5761**: p. 12–22.
82. Liu, X.M., *et al.*, *Electric-field oriented carbon nanotubes in different dielectric solvents.* Current Applied Physics, 2004. **4**(2–4): p. 125–128.

CHAPTER 3

SWNT Buckypaper Nanocomposites: High Nanotube Loading and Tailoring Nanostructures

Zhiyong (Richard) Liang*, Ben Wang and Chuck Zhang

Florida Advanced Center for Composite Technologies (FAC^2T)
Department of Industrial & Manufacturing Engineering
FAMU-FSU College of Engineering, Florida State University
Tallahassee, FL 32310-6046, USA
**liang@eng.fsu.edu*

A novel technical approach was developed to fabricate bulk polymeric matrix nanocomposites with both randomly oriented and quasi-unidirectional single-walled nanotubes (SWNTs) for producing high performance structural materials and devices. Instead of directly mixing nanotubes in a resin matrix, a special resin infiltration method was developed to impregnate buckypapers, which are thin preformed sheets consisting of well-controlled and dispersed porous SWNT networks. Buckypaper preparation involves a multiple-step process of dispersion and filtration of nanotube suspensions. Magnetically aligned SWNT buckypapers were produced by filtering well-dispersed nanotube suspensions in high strength magnetic fields to achieve nanotube alignment. Multiple-layers of resin-impregnated buckypapers were stacked together and cured to fabricate solid bulk composite samples. The nanotube loading in the nanocomposites ranged from 20 w/w% to 60 w/w%. Both the randomly oriented and aligned buckypapers and nanocomposites consisted of well-dispersed nanotube networks, which are critical for developing high performance nanostructured materials. The storage moduli of the quasi-unidirectional nanocomposites were as high as 45 GPa, which is almost equal to that of current quasi-isotropic IM7 carbon fiber composites. Theoretical models developed to predict the influence of SWNT dispersion, orientation and contents on the mechanical properties indicate that the formation of nanotube ropes significantly affects mechanical performance.

1. Introduction

Unlike conventional fibrous reinforcements, SWNTs are nanoscale dimensional materials with exceptionally large surface areas (1,000~1,350 m^2/g), which result in strong interactions between the nanotubes due to the van der Waals forces, causing the SWNTs to form into ropes or bundles. The SWNT ropes also have the tendency to aggregate together and demonstrate intensive molecular interactions with resin matrix molecules. Nanocomposites produced by conventional methods of directly mixing the nanotubes and polymers in casting or utilizing injection processes do not yield acceptable uniform dispersion and high loading due to the aggregation tendency of SWNTs, high resin viscosity and the inability to control the formation of the SWNT networks or nanostructures during composite consolidation. Consequently, a major challenge of nanotube composite processing is to produce a nanotube network in polymeric matrix with the desired uniform dispersion, alignment and high loading. The functionalizing of nanotubes to enhance nanotube dispersion and interfacial bonding is another vital step for developing high performance materials.

Typically, SWNT reinforced composites are fabricated by directly mixing SWNTs into polymers [1–6], followed by employing casting or injection molding techniques. However, SWNTs have a strong tendency to form bundles by aggregating together because of their large surface area and their strong van der Waals interaction. Their nanoscale dimension, stable chemical characteristics and smooth surface make efficiently dispersing SWNTs into a polymer matrix and controlling the final nanostructure of the composites during processing difficult, if not impossible, especially for high SWNT loaded composites. Some researchers chemically modified the nanotube surface to improve the nanotubes dispersion in polymer matrix [7,8]. However, these chemical modifications may destroy the pristine structure of the SWNTs and decrease maximum modulus and strength of the SWNTs [9]. Therefore, other dispersion methods that would avoid nanotube structure damage are required. Park *et al.* tried an *in situ* polymerization method to effectively disperse as-pristine SWNT bundles into polyimide with the aid of sonication [10]. Besides the dispersion issue, the rapidly increasing viscosity of the polymer/nanotube mixture during processing makes fabricating high SWNT loading in composites more difficult. The reported SWNT loading of the composites was usually less than 10~15% by weight [11–12]. A more comprehensive review of fabrication and properties of carbon nanotube-reinforced composites can be seen in Reference 11 and 13–15.

To fabricate bulk nanotube-reinforced nanocomposites with good nanotube dispersion, alignment and high nanotube loading, a novel buckypaper/resin infiltration process was developed. In this process, SWNTs are first dispersed into water with the aid of surfactant and sonicated to form a stable suspension. The SWNT suspension is filtrated to form buckypapers, which are thin sheets composed of uniform SWNT rope networks. A diluted low viscosity resin solution is added to infiltrate the buckypaper to impregnate the SWNT rope networks. A compressive molding process is used to cure and make the final composite from the multilayered resin-impregnated buckypapers. Nanotube alignment of the composites can be achieved by magnetically aligning the SWNTs in the buckypapers in a high magnetic field. This chapter describes this unique composite manufacturing process and the resultant composite properties

2. Raw Materials

The SWNTs used in this study were produced by Carbon Nanotechnologies Inc. (CNI) under the brand name BuckyPearls™. According to the technical sheet provide by CNI, the individual nanotube is about 0.8–1.2 nm in diameter and 100–1000 nm long. The BuckyPearls are purified SWNTs, as shown in Figure 1. The main impurities are metal catalyst particles (10~12 w/w%) and amorphous carbon. The bulk density of the BuckyPearls is 0.4 g/cm^3. The TEM image of the SWNTs is given in Figure 2. The SWNTs were directly used without further purification.

Shell Chemical produced the epoxy resin system used in this study. EPON 862 is a low-viscosity, liquid epoxy resin manufactured from epichlorohydrin and Bisphenol-F (DGEBPF). This resin contains no diluents or modifiers. When cross-linked with appropriate curing agents,

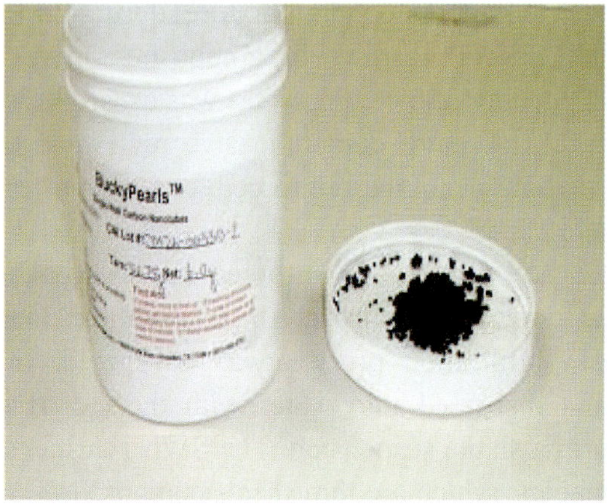

Fig. 1. BuckyPearls from CNI .

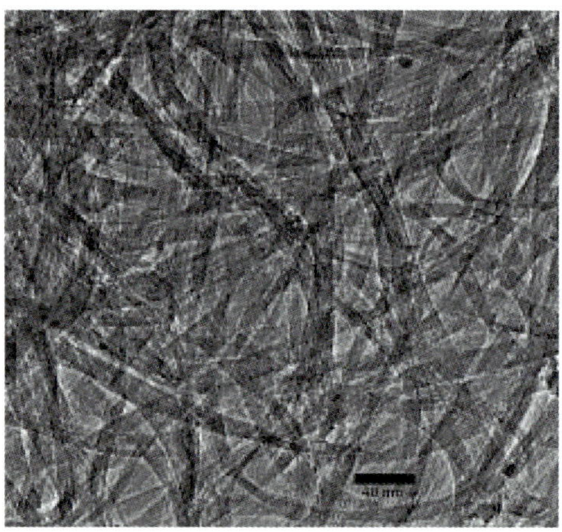

Fig. 2. Clear SWNT ropes in BuckyPearls.

superior mechanical, adhesive, electrical and chemical resistance properties can be obtained. The epoxide equivalent weight of EPON 862 is 166–167. The viscosity of EPON 862 at 25°C is 2700cp. EPI Cure W was used as the DETDA hardener or curing agent. The resin and curing agent molecules are shown in Figure 3.

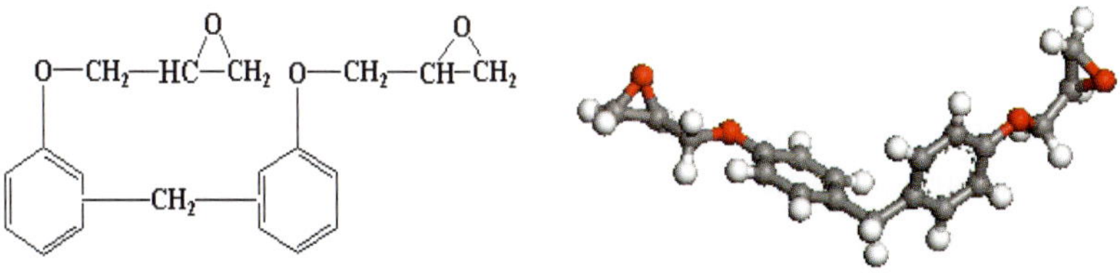

(a) Chemical structure and molecular model of EPON 862

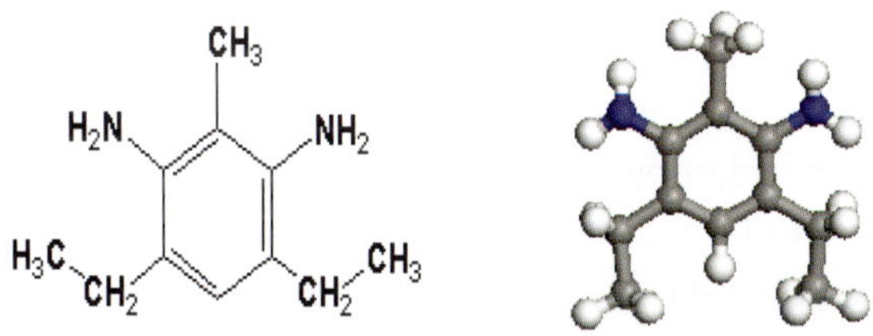

(b) Chemical structure and molecular model of DETDA

Fig. 3. Molecular models of an EPON 862 epoxy resin and curing agent (DETDA).

3. Fabrication of Randomly Oriented and Magnetically Aligned SWNT Buckypapers

Buckypapers are thin (10~50 μm) membranes of nanotube networks produced by a multiple-step process of nanotube dispersion and suspension filtration. Since aqueous suspension and a number of different water-based surfactants can be used, relatively good nanotube dispersion can be expected in the buckypapers. The formed nanotube networks in the buckypapers can be retained during the composite fabrication process and transferred into the final solid nanocomposites. In other words, the buckypaper can provided a preformed nanotube reinforcement structure for the final nanocomposites.

3.1. *Randomly orientated SWNT buckypapers*

Buckypapers of randomly oriented SWNTs are produced by filtrating SWNT suspensions. The concentration of the aqueous SWNT suspension is 10~100mg SWNTs/L. A selected surfactant is used to enhance dispersion and stabilize the suspension. The major steps of buckypaper fabrications includes:

- Grinding SWNTs with a small amount water (~10 ml) using a mortar and pestle, until the mixture becomes paste;
- Adding the surfactant and water to the paste and transferring the paste into 250 ml deioned water;
- Sonicating under full power for 30 minutes using the sonicator (Misonix Model 3000);
- Transferring the suspension into 500ml deioned water and sonicating another 30 minutes;
- Adding an additional 500 ml deioned water and sonicating 30 minutes to form a stable ink-like suspension;
- Filtrating resultant suspension through a nylon filter membrane (0.45 μm pore size) with the aid of a vacuum;
- Washing the buckypaper separately with water and iso-propanal to thoroughly remove the surfactant;
- Peeling the buckypaper from the nylon membrane; and
- Drying overnight at room temperature and then drying in a vacuum oven at 80°C.

The filtration process takes 8–12 hours due to the small pore size of the filter membranes. The resultant buckypapers are shown in Figure 4. The produced buckypapers are thin black membranes less than 50 μm thick. Since the buckypaper are strong and flexible, they can be handled similar to traditional glass fiber mats.

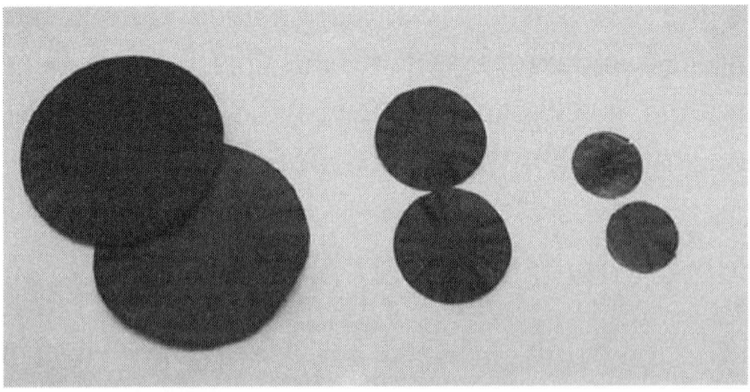

90 mm 47 mm

Fig. 4. Small randomly oriented buckypapers.

The nanostructures of the produced random buckypaper were characterized using Scanning Electron Microscope (Joel 6400F, Joel Co.) and Atomic Force Microscope (Dimension 3000, Digital Instrument Co.). SEM and AFM images of the randomly oriented SWNT buckypapers are shown in Figure 5 and 6, respectively. The buckypaper showed porous network structures composed of "continuous" SWNT ropes. These continuous ropes were the result of nanotube self-assembly by van der Waals force during buckypaper filtration. The rope size and porous structure were uniform, which indicates very good dispersion of the nanotubes in the suspension. The rope diameter was 15–60 nm. The images shows the nanoscale porous structures of the buckypapers with pore sizes around 100–200 nm, which are much smaller than those found in the traditional glass fiber and carbon fiber fabrics or mats. This indicates that the wetting and resin infiltration of buckypaper-based composites will occur at nanoscale and significantly create intense molecular interactions.

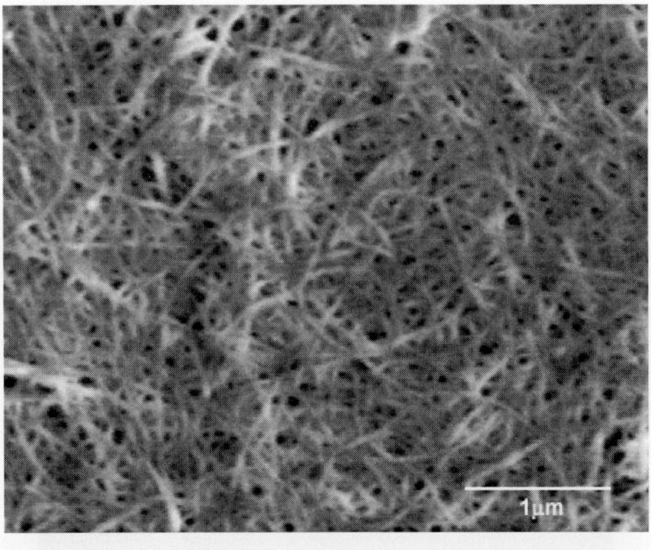

Fig. 5. SEM image of the randomly oriented buckypaper nanostructure.

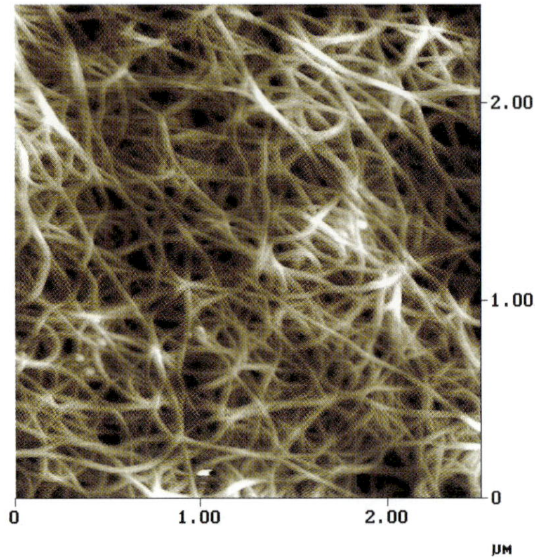

Fig, 6. AFM image of the randomly oriented buckypaper.

Large randomly oriented SWNT buckypapers were also produced using custom-made filter systems. Figure 7 shows a 9 in × 9 in randomly oriented buckypaper produced during the research.

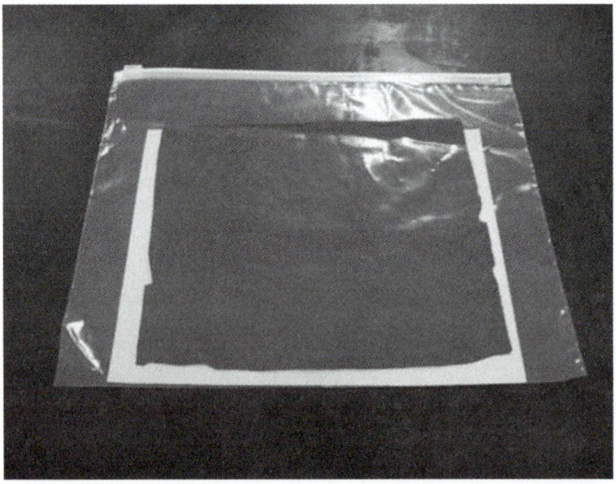

Fig. 7. 9 in × 9 in randomly oriented SWNT buckypaper.

3.2. Magnetically aligned SWNT buckypapers

Controlled SWNT alignment is critical for both structural and multifunctional applications of SWNTs-reinforced nanocomposites. Randomly oriented buckypapers and composites have very good nanotube dispersion; however, nanotube alignment does not occur during manufacturing. The literature review indicates that the molar susceptibilities for a (10, 10) SWNT (armchair) is $+85.4 \times 10^{-6}$ emu (mol C)$^{-1}$ when parallel to B (magnetic vector), and -21.0×10^{-6} emu (mol C)$^{-1}$ when perpendicular to B. Theoretical predictions show that a (10, 10) SWNT 300 nm long can be aligned in a magnetic field of 10–25 Tesla (T) when suspended in a very low viscous solution. Dr. Richard Smalley's team experimentally demonstrated the effectiveness of aligning SWNTs in buckypapers within high magnetic fields of greater than 8 T [16–18].

The research team further modified the filter design and processing parameters to produce high quality and relatively large buckypapers for making nanocomposites with aligned SWNTs. The team used the high magnetic fields at the National High Magnetic Field Laboratory (NHMFL) in Tallahassee, Florida to carry out the nanotube alignment experiments. Large magnetically aligned buckypapers were fabricated using custom-made cylinder filters, which can make buckypapers of approximately 38 cm (15 in) long and 10 cm (3.94 in) wide, resulting in an area of 387 cm^2 (60 in^2).

Non-magnetic materials were required for all parts of the custom-designed filter since it was used inside the high strength magnet of 5~25 T. Any magnetic materials, such as steel, would damage to the filter apparatus as well as the magnet; therefore, aluminum was used for making the filter. The design requirements for the filter were constrained by the size of the magnet bore and area covered by the magnetic field in Cell 4 at the NHMFL. The magnet bore diameter was 190 mm. The depth of the bore was 1,165.2 mm and the magnetic field was centered at 582.6 mm. Allowing for clearance, the maximum diameter of the filter assembly was less than 172 mm (6.75 in). The total length of the filter could vary but the membrane remained within 76 mm (~3.0 in) from the magnetic center, which maintained good uniformity in magnetic strength.

The total filter length was 152 mm (~6.0 in). The custom-made filter is shown in Figure 8. Three parts make up the external filter assembly: End Cap A, End Cap B and the External Cylinder. Five parts make up the internal filter assembly: Internal Cap C, Internal Cap D, Flange C, Flange D and the rolled perforated tube, as shown in Figure 8. When these components are assembled, they form the structure upon which the Millipore filter membrane of 0.45 μm pore size was attached to the surface of the preformed nanotube surface. A 5/16″ 18 threaded rod bolted the internal filter assembly together. End Cap B was then bolted to the External Cylinder to form a watertight chamber.

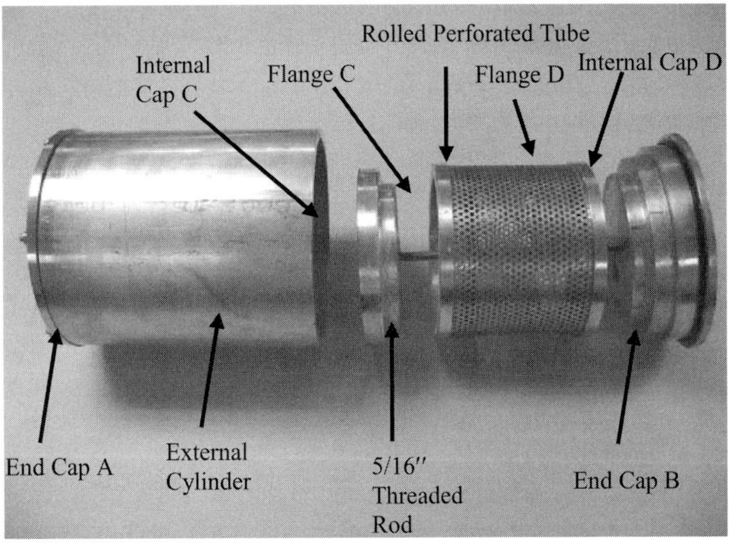

Fig. 8. Custom-made cylindrical filter assembly.

The nylon membrane of 0.45 μm pore size, made by the Millipore Corporation, comes at a standard width of 15 cm and 300 cm long (Product ID: INYU00010). Once the flanges C and D were attached to the rolled and perforated nanotube, the length of the inner (internal) filter assembly was approximately 13.5 cm. The membrane was cut so that the length of the membrane was 5 cm more than the circumference internal filter assembly. The filter membrane was wrapped onto the inner filter assembly and a layer of double sided permanent tape was applied on the inner side of the membrane. The membrane was sealed by using West System fast cure epoxy glue. The outer edges of the membrane were attached to the flanges C and D using double-sided permanent tape. A silicone gasket was placed on internal cap C. Silicone glue was uniformly applied onto the outer periphery of the inner filter assembly and then onto internal cap C. A silicone gasket was placed onto the internal cap D and silicone glue was applied onto the gasket. The internal cap D was then placed onto the inner filter assembly. The end cap D was placed on flange D and tighten using a nut and bolt. The inner assembly, external cylinder, the end cap A and B were then attached using nuts and bolts. The filter was then ready to use. The assembled filter used in the magnetic field is shown in Figure 9.

Filter membrane on the surface of the perforated tube

Complete assembly of the internal parts

Filter assembly

Final filter assembly with suspension supply and outlet tubes

Fig. 9. Various phases of filter assemble.

The assembled filter was used to fabricate large magnetically aligned buckypapers. During the experiments the peristaltic pump was found to cause the suspension to frequently breakdown due to the peristaltic motion. To ensure a stable suspension could pass through the filter, an in-situ sonication system was developed for subsequent experiments. In this way, the suspension was resonicated prior to entering the filtration system, ensuring good nanotube dispersion and a stable suspension. Figure 10 shows the experimental setup for fabricating the large aligned buckypapers at the NHMFL. After fabrication, the produced buckypapers were cleaned and dried.

Fig. 10. Modified experimental setup for producing large magnetically aligned nanotube buckypapers.

The world's largest magnetically aligned buckypapers were successfully produced in the research, as shown in Figure 11. The produced buckypapers have a working area more than 387 cm^2 (60 in^2) with a thickness of 15~30 μm. Good nanotube alignment was observed in the produced buckypapers as shown Figure 12. The successful development of large aligned buckypaper fabrication approach provides the possibility for producing bulk nanocomposite samples with in-plane controlled tube orientation.

Fig. 11. World's largest magnetically aligned buckypaper (17.3 Tesla magnetic field).

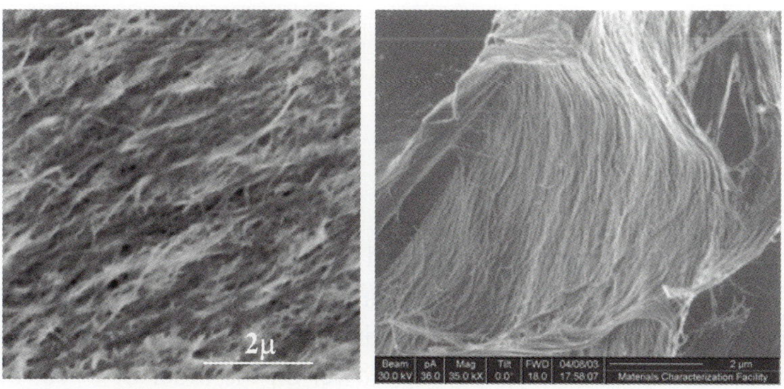

Fig. 12. Significant SWNT alignment in the buckypaper produced by filtering SWNT suspension within a 17.3 Tesla magnetic field.

4. Resin Infiltration of the Buckypapers

4.1. *Through-thickness permeability of the buckypapers*

Since the produced nanotube buckypapers have nanoscale porous structures, it is important to understand their permeability for developing resin infiltration processes. The through thickness (z-direction) permeability of the produced buckypaper was measured to estimate the resin infiltration time along the thickness direction. Water was the fluid in the permeability test. The experiment was carried out with a filtration system. A buckypaper was sealed inside the filter. The buckypaper acted as a filter membrane in this test. Water was forced through the buckypaper along the thickness direction with the aid of a vacuum. The water flow rate, buckypaper thickness, buckypaper surface area and vacuum pressure were recorded to calculate the permeability. Under this experimental condition, the permeability in the z-direction of a buckypaper of 10–50μm thick and the nanoscale pore structure were measured. The z-direction permeability K_z was calculated using the following equation derived from Darcy's Law:

$$K_z = \frac{Q\eta L}{AP} \tag{1}$$

where: Q is the flow rate;

η is the viscosity of water;
L is the thickness of buckypaper;
P is the vacuum pressure; and
A is the surface area of buckypaper.

The permeability test results are listed on Table 1. The average value of the buckypaper's z-direction permeability was about 2×10^{-19} m^2. Compared to normal glass fiber reinforcement fabrics or mats, the permeability of the buckypaper was lower on the order of 8–10. Small in-plane permeability of a buckypaper can also be expected due to its small pore sizes. Therefore, allowing the resin to flow and infusing the buckypaper in the in-plane direction, as for conventional RTM and VARTM processes, would be extremely difficult due to the thinness (20–50 μ) of the produced buckypaper, as well as their very low permeability.

Table 1. Kz of the randomly oriented buckypapers.

Buckypaper Thickness (μm)	25.4	31.2	35.7
z-Permeability (m^2)	3.36×10^{-19}	1.11×10^{-19}	1.44×10^{-19}
Average z-Permeability (m^2)		1.97×10^{-19}	

To allow the resin system to completely impregnate the entire thickness direction of the buckypaper, the minimum infiltration time can be calculated by Equation 2, which ips also derived from Darcy's Law.

$$t = \frac{\eta L^2}{K_z P} \tag{2}$$

where: K_z is the z-direction permeability of buckypaper;

η is the viscosity of resin;
L is the thickness of buckypaper; and
P is the vacuum pressure for resin infusion.

The viscosity of EPON 862/EPI Cure W (DETDA) system was 2700 cp at room temperature, which means that under a full vacuum, complete impregnation through a 40 μm thickness buckypaper at room temperature will take more than 60 hours. During such a long period, the EPON 862/EPI Cure W system will partially gel, which will increase the viscosity and require an even longer time to impregnate, making the process unfeasible. Therefore, decreasing the viscosity of EPON 862/EPI Cure W system was necessary. Acetone was used to dilute the resin system and decrease the viscosity. In our experiments, the content of acetone was properly controlled such that enough resin was left within the final nanocomposites after the evaporation of the acetone. The viscosity of the diluted resin solution was 25 cp at room temperature. A 5-hour infiltration time was chosen to ensure the buckypaper was completely impregnated, which was indicated by observing the liquid resin layer appearing on the opposite surface of the buckypaper during resin infiltration.

4.2. *Resin infiltration of the buckypapers*

The resin infiltration was performed under the same experimental setup as the permeability test. Instead of water, epoxy resin was used as the fluid. The EPON 862 and EPI Cure W were mixed at a weight ratio of 100:26.4. The mixed resin and curing agent were diluted with acetone to decrease the viscosity. To properly impregnate the SWNTs, the diluted resin system was allowed to flow through the buckypaper along its thickness direction for 5 hours. Figure 13 illustrates the buckypaper before and after resin infiltration, showing that the resin completely infiltrated though and impregnated the buckypaper. After infiltration, the impregnated buckypaper was soaked overnight in the EPON 862 resin system. The acetone was vaporized in the vacuum oven, and made ready for use.

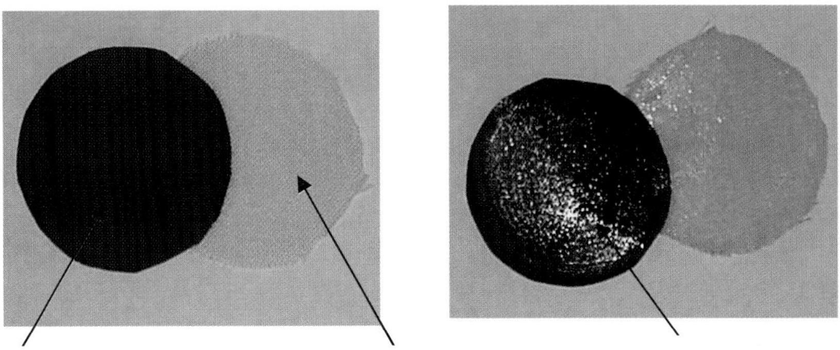

Fig. 13. Buckypapers before and after resin infiltration.

5. Nanostructures of SWNT Buckypaper/epoxy Nanocomposites

5.1. *Nanostructures of random buckypaper nanocomposites*

SWNT buckypaper/epoxy resin nanocomposites were prepared using the compression molding technique. The hot press machine used in this research was the Carver Model #3925. Several resin-impregnated buckypapers were stacked to fabricate relatively thick (0.2–0.5 mm) bulk composite samples. The curing process followed the instructions from the Shell Chemical: cure at 177°C for 2.5 hours under pressure (10 kg/cm^2); cool to room temperature; post cure in an oven for another 2 hours at 177°C and cool to room temperature inside the oven.

Table 2 lists descriptions of the manufactured randomly oriented buckypaper-reinforced composites. The composites were 47 mm in diameter. In conventional processing techniques of directly mixing resin with SWNTs, even less than 1 wt% loading of SWNTs mixed in resin will result in extremely high viscosity. High viscosity causes dispersion problems and makes manufacturing the composites exceptionally difficult, if not impossible. Using the proposed method, the high viscosity issue is completely avoided and the SWNT loading of the resulting composites can be as high as 39 wt%.

Table 2. Randomly oriented buckypaper nanocomposites.

Sample No.	No. of Buckypaper Layers	Composites Thickness (mm)	SWNT Content (wt%)
A	5	0.198	28.1
B	3	0.114	31.3
C	3	0.220	37.7
D	3	0.246	39.1

Figure 14 shows the SEM image of the fracture surface of the SWNT composite samples with 31.3 w/w% nanotube loading. These samples were fractured under tensile force. Figure 14(a) shows a cross-section of the nanocomposite sample, indicating that the epoxy resin completely impregnated throughout the cross section of the buckypaper-reinforced composites. The fracture surface was coarse, as shown in Figure 14(b), which may indicate that the composite had good toughness because the SWNTs are flexible, which not only increases the stiffness but also the toughness. Figure 14(c) and (d) show that the SWNT ropes were distributed across the surface and have a continuous and evenly distributed network. Good nanotube/resin wetting was also observed. The striking observation was the morphology of SWNT ropes, which were surrounded by cured epoxy resins, as shown in Figure 15 (AFM images). These could be contrasted from the SWNT ropes in the buckypaper before resin infiltration, as shown in Figures 5 and 6. The porous structure of the original nanotube network of the buckypapers was replaced by the epoxy resin.

(a) Good resin/buckypaper impregnation

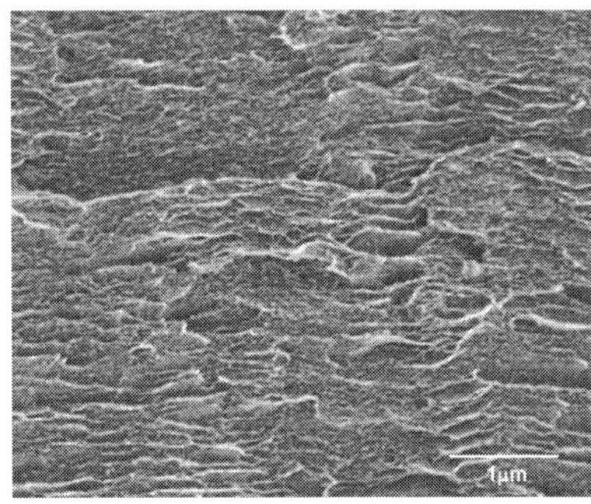

(b) Rough fracture surface

(c) SWNT network in the composites

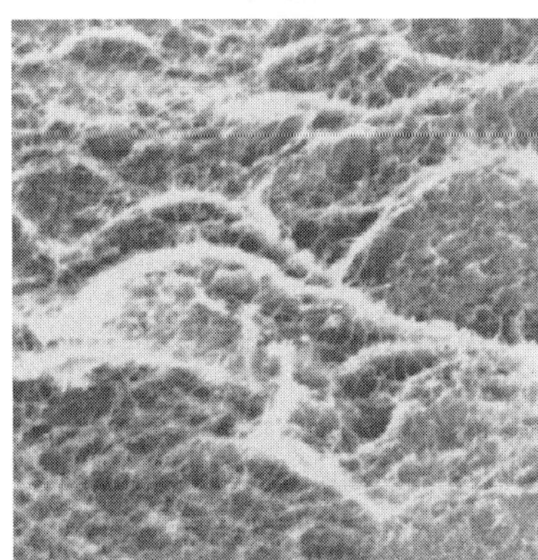

(d) SWNT network in the composites (×30K)

Fig. 14. SEM images of the fracture surface of buckypaper composites.

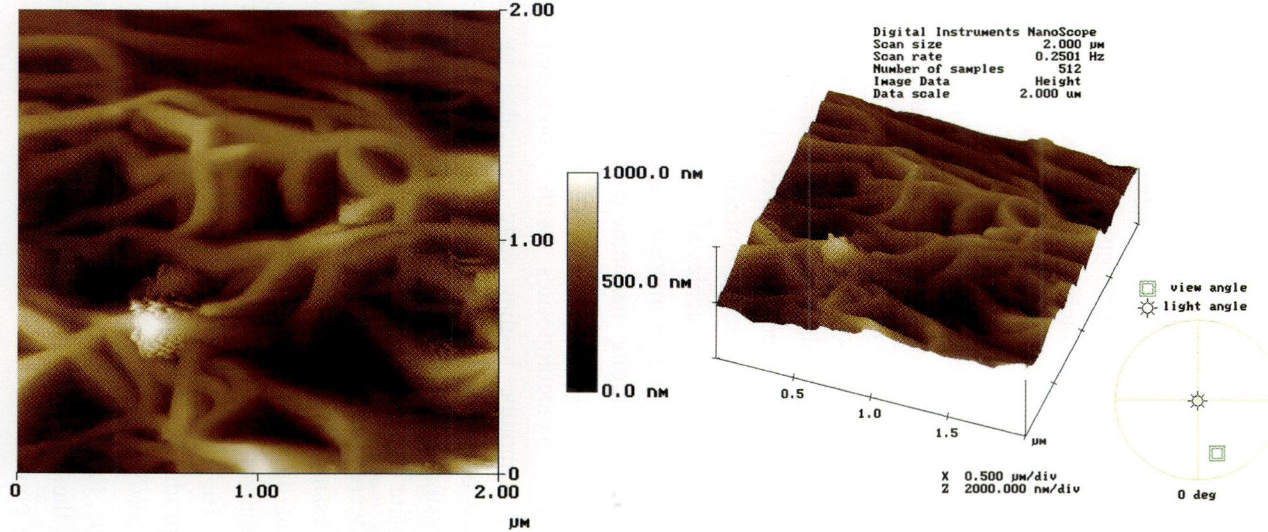

Fig. 15. AFM images of nanotube/resin wetting in buckypaper composites.

5.2. Nanostructures of magnetically aligned buckypaper nanocomposites

Due to their nanoscale dimension, large surface area and strong molecular interactions, SWNTs have the tendency to form ropes that aggregate together making direct manipulation of large quantity of nanotubes extremely difficult for achieving desired alignment during nanocomposite processing [19–26]. Several efforts have been attempted to control nanotube alignment in composites. Ajayan and his colleagues observed nanotube alignment during cutting polymer/nanotube samples [20]. Stretch-induced alignment, which involves mechanically stretching the SWNTs/PMMA composites at an elevated temperature of 90°C, has also been attempted [25,26]. Nanotube alignment has been reportedly achieved through spinning or spooling a mixture of nanotubes/resin into fibers or strip forms, which is called flow-induced alignment. However, these methods are considered ineffective since only local and relatively low degree of orientation can be achieved. Possible reasons for the inability of aligning nanotubes in composites are high resin viscosity, nanotube aggregation and strong nanotube/polymer resin molecular interactions after resin/tube mixing.

In this research, nanocomposites with controlled nanotube alignment and high nanotube loading were fabricated by using the magnetically aligned buckypapers/resin infiltration approach. By using the same resin infusion approach and curing parameters as outlined in Section 4, composite samples of six-layers of magnetically aligned buckypaper/epoxy were successfully produced, as shown in Figure 16. In the composite processing, all six aligned buckypapers were stacked along the same alignment direction to produce quasi-unidirectional SWNT-reinforced nanocomposites. The curvature of the composite samples, as shown in Figure 16, also indicates good nanotube alignment. The parameters of the produced aligned buckypaper composites are provided in Table 3. Since SWNTs are more closely packed in the magnetically aligned buckypapers, the nanotube content of the composites is relatively high compared to the randomly oriented buckypaper composites. The composite samples could have diameters of 25–90 mm and 40–200 μm thick depending on the number of layers of impregnated buckypapers used.

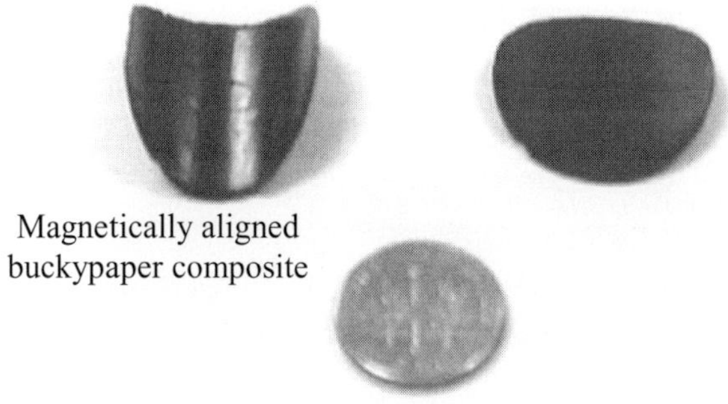

Fig. 16. Curvature of magnetically aligned buckypaper composite.

Table 3. Parameters of magnetically aligned buckypaper nanocomposites.

Sample	Magnetic Field* (T)	SWNT Content in Nanocomposites (w/w%)
Aligned Nanotube/Epoxy 1	17.3	47.3
Aligned Nanotube/Epoxy 2	17.3	63.7
Aligned Nanotube/Epoxy 3	17.3	54.5

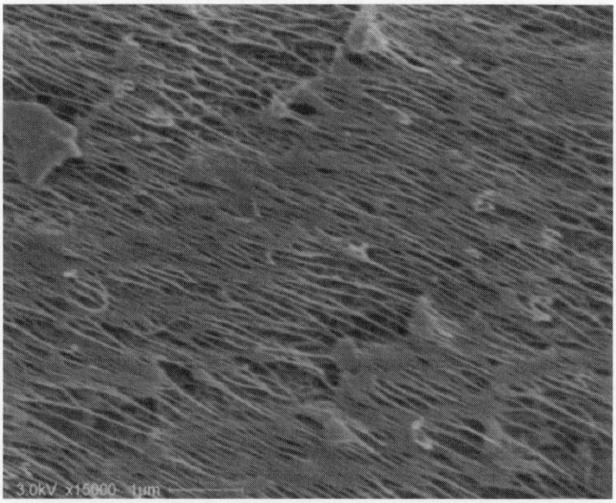

Fig. 17. Nanotube alignment in magnetically aligned buckypaper/epoxy resin.

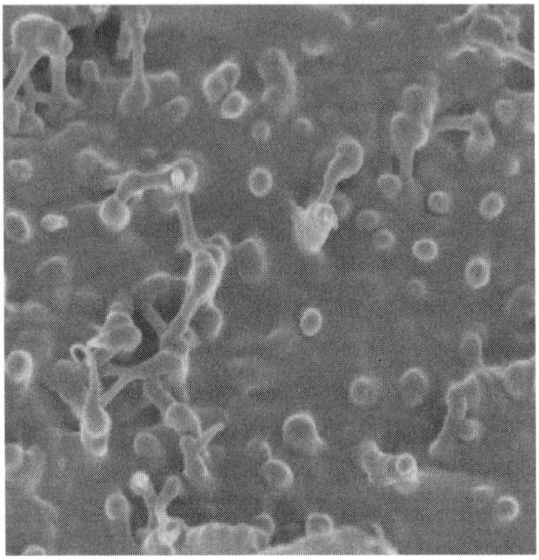

Fig. 18. Cross-section of magnetically aligned buckypaper/epoxy composite (X50, 000).

Figures 17 and 18 show the cross-sections of the aligned buckypaper composites, which indicates that the nanotube alignment in the magnetically aligned buckypaper was successfully transferred into the bulk solid composite samples. The SWNTs and resin matrix can be seen to have achieved good impregnation throughout the cross-section in the nanocomposites. The results indicate that the proposed processing approach is effective for manufacturing nanocomposites with high nanotube loading and controlled alignment nanocomposites.

6. Dynamic Property of SWNT Buckypaper/epoxy Nanocomposites

The dynamic mechanical properties and damping behaviors of the nanocomposites were tested with a dynamic mechanical analyzer (TA Instrument Co. DMA2980). The test was conducted in the tensile (film) mode. The sample was heated to 350°C at 5°C/min at a frequency of 1 Hz.

6.1. *Randomly oriented buckypaper/epoxy nanocomposites*

The storage modulus results of the randomly oriented buckypaper nanocomposites listed in Table 2 are shown in Figure 19 and Table 4. The storage moduli of the resulting nanocomposites dramatically increased by 349% to 492%, compared to the neat resin modulus. However, the storage moduli of the composites were significantly less than predicted by the rule of mixtures for randomly orientation individual SWNT-reinforced composites. The poor load transfer between the nanotubes in the ropes and between the nanotubes and resin matrix could be the major reason. The formation of large nanotube ropes in the composites would significantly decrease the load transfer in the composites. For example, Salvetat *et al.* experimentally showed that when the SWNT rope diameters increased from 3 to 20 nm, the elastic and shear modulus of SWNT ropes could be decreased from 900 GPa to around 100 GPa due to possible nanotube slippage in the ropes [27]. Therefore, the actual nanotube rope-reinforced composites may

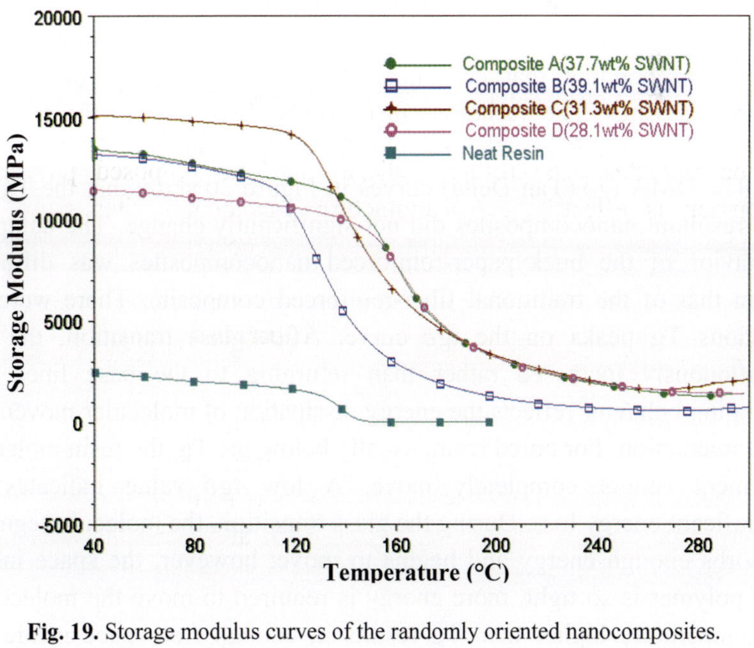

Fig. 19. Storage modulus curves of the randomly oriented nanocomposites.

Table 4. Storage modulus results of the randomly oriented buckypaper composites.

Sample No.	SWNT Loading (wt%)	Storage Modulus (GPa)	Increase (%)
Neat Resin	0	2.55	0
A	28.1	11.45	349
B	31.3	15.10	492
C	37.7	13.49	429
D	39.1	13.24	419

demonstrate poor mechanical properties. Realizing individual nanotube dispersion in the buckypapers and composites for better load transfer is an ongoing research effort. Another notable phenomenon observed from the DMA results was that the nanotube loading did not influence the storage modulus. This is possibly due to the nanostructures and mechanical properties of the nanotube-reinforced composite are highly dependent on nanotube dispersion, which directly influences the molecular interactions of tube/tube and tube/resin in the materials. Such molecular interactions will play a critical role in load transfer and interfacial bonding that determine the mechanical properties of the nanocomposites. The variations in nanotube dispersion in the resultant composites could be the major reason for this phenomenon.

The DMA tgδ (Tan Delta) curves in Figure 20 show that the Tgs of the resultant nanocomposites did not significantly change. The damping behavior of the buckypaper-reinforced nanocomposites was different from that of the traditional fiber-reinforced composite. There were no obvious Tg peaks on the tgδ curve. After glass transition, the tgδ continuously increased rather than returning to the base line. The damping behavior reflects the energy dissipation of molecular movement and interaction. For cured resin, usually below the Tg, the resin molecule segment cannot completely move. A low tgδ value indicates no significant energy loss. During the glass transition, the molecule segment absorbs enough energy and begins to move; however, the space inside the polymer is so tight, more energy is required to move the molecules. A temperature higher than Tg results in enough free space inside the

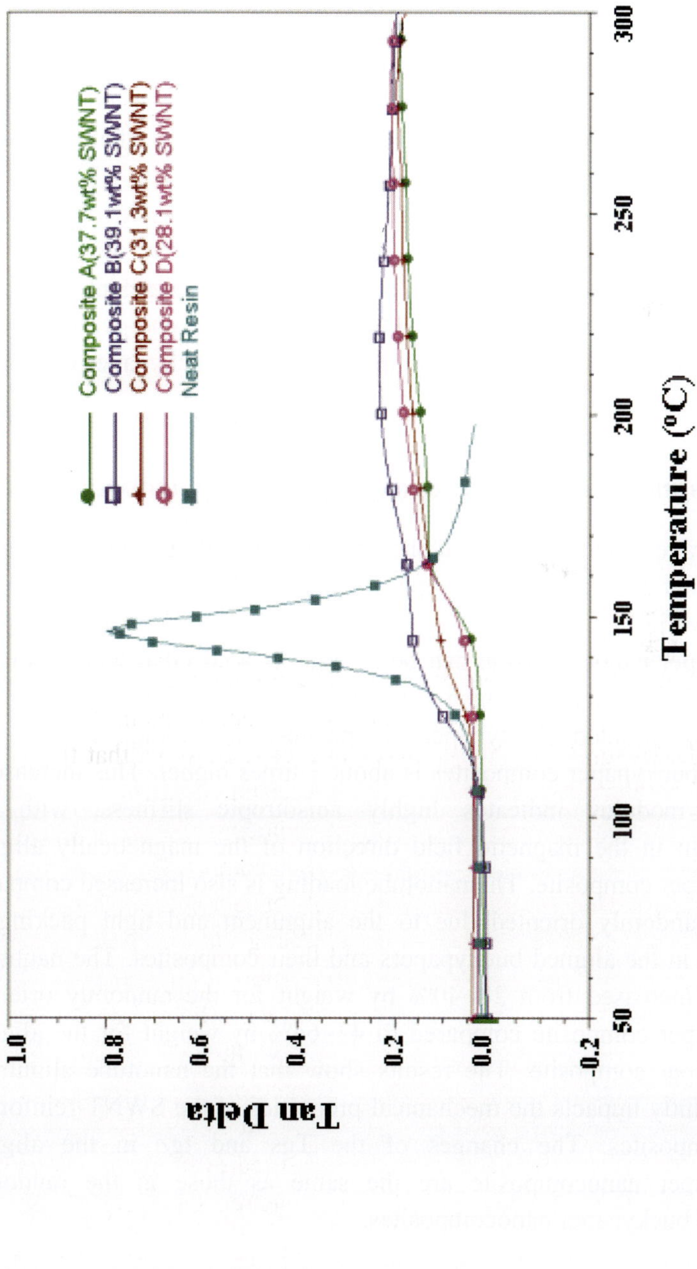

Fig. 20. tgδ curves of the randomly oriented nanocomposites.

resin to allow the molecule segments move freely, causing the tgδ to again decrease.

For traditional carbon fiber and glass fiber reinforced composites, both fiber and porous structures are at microscale and only a very small portion of the resin molecules are able to directly interact with the reinforcement. The damping behavior of the carbon fiber and glass fiber reinforced composites is almost the same as that of the neat resin. However, for the buckypaper-reinforced composite, the SWNTs have extremely large surface area and are at the same scale of the resin molecules. Interactions between nanotubes and resin molecules are expected to be intense. Therefore, the resin molecule segment movement may intensely interact with the SWNTs and consume energy, which may be the major reason for the high tgδ of the buckypaper-reinforced composites.

6.2. *Magnetically aligned buckypaper/epoxy nanocomposites*

The storage modulus results along the nanotube alignment direction of the magnetically aligned buckypaper nanocomposites listed in Table 3 are shown in Figure 21 and Table 5. The storage modulus of the aligned buckypaper nanocomposite can be as high as 45.0 GPa, which is more than 16.5 times higher compared to neat epoxy resin. Compared with the randomly oriented buckypaper composites, the storage modulus of aligned buckypaper composites is about 3 times higher. This increase in storage modulus indicates highly anisotropic stiffness, with the maximum in the magnetic field direction of the magnetically aligned buckypaper composite. The nanotube loading is also increased compared to the randomly oriented due to the alignment and tight packing of SWNTs in the aligned buckypapers and their composites. The nanotube loading increased from 28–40% by weight for the randomly oriented buckypaper composite compared to 47–64% by weight for the aligned buckypaper composite. The results show that the nanotube alignment significantly impacts the mechanical properties of the SWNT-reinforced nanocomposites. The changes of the Tgs and tgδ in the aligned buckypaper nanocomposite are the same as those in the randomly oriented buckypaper nanocomposites.

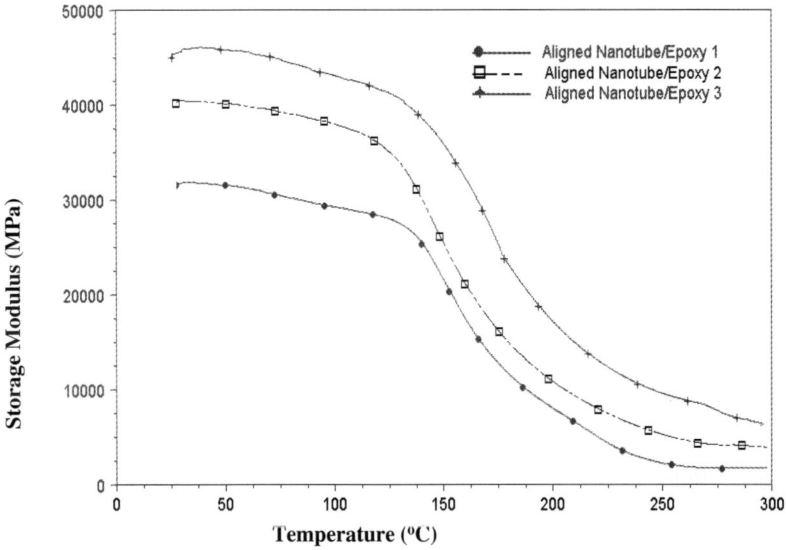

Fig. 21. Storage modulus curves of the magnetically aligned nanocomposites.

Table 5. Storage modulus results of the magnetically aligned buckypaper composites.

Sample	Magnetic Field (T)	SWNT Loading (wt%)	Storage Modulus (GPa)	Increase (percentage)
Neat Resin	0	0	2.55	0
Aligned Nanotube/Epoxy 1	17.3	47.3	31.0	1,116
Aligned Nanotube/Epoxy 2	17.3	63.7	45.0	1,665
Aligned Nanotube/Epoxy 3	17.3	54.5	40.5	1,488

7. Effect of SWNT Rope Formation on Mechanical Properties of SWNT-reinforced Nanocomposites

Although significant increases in the storage moduli of the buckypaper nanocomposites were described in Section 6, the mechanical properties are still far below the anticipated values based on exceptional properties of individual SWNTs [11,13–15]. One of the major possible reasons is that SWNTs formed ropes in the buckypapers and their composites as shown in Section 3 and 5. Unlike conventional fibrous reinforcements, SWNTs have strong interactions between each other due to van der Waals forces, causing the SWNTs to form ropes or bundles. SWNT ropes also have the tendency to aggregate together due to their molecular interactions. Therefore, to understand the effect of the formation of SWNT ropes is important for further improving the mechanical properties of the buckypaper/epoxy nanocomposites.

In this section, the property predictions of SWNT-reinforced composites are examined to theoretically reveal the effects of nanotube rope formation, loading and orientation on the composite properties based on our experimental observations. Since the SWNTs were shorter than the length of the composites in the experiments, they were discontinuous in the composites. Therefore, these nanotube composites can be considered as short reinforcement composites. In this case, the composite properties strongly depended on the aspect ratio of the nanotubes, the elastic properties of nanotube and matrix, and the interfacial shear between the nanotube and matrix. Both randomly oriented and aligned discontinuous reinforcement models of composites were used to predict the tensile moduli of individually dispersed SWNTs and SWNT rope reinforced nanocomposites. The parameters used in the models were determined based on experimental observations.

7.1. *Calculation models*

For unidirectional short fiber composites, elastic properties, such as tensile modulus, can be calculated by Halpin-Tsai equations [28]. The Halpin-Tsai method is a simple yet effective model or a set of empirical relationships that enable the properties of a composite material to be expressed in terms of the properties of the matrix and reinforcing phases, together with their proportions and microstructure geometry. These equations were validated by experimental measurements. To theoretically predict the mechanical properties of nanotube-reinforced composites, these equations were assumed to be valid for predicting nanotube-reinforced nanocomposites since nanotubes can also be considered as short-reinforcements in the resin matrix. Good wet-out and interfacial bonding were assumed to exist between the nanotubes and the matrix. Based on experimental observations of the produced buckypaper and nanocomposites, the nanotubes in the composites were also assumed to be straight. The composite moduli for aligned discontinuous SWNT/epoxy in the longitudinal and transverse direction were calculated from the following equations:

$$E_{11} = \frac{1 + 2(l_{NT} / d_{NT})\eta_L V_{NT}}{1 - \eta_L V_{NT}} E_{EP} \tag{3}$$

$$E_{22} = \frac{1 + 2\eta_T V_{NT}}{1 - \eta_T V_{NT}} E_{EP} \tag{4}$$

$$\eta_L \quad \frac{(E_{NT} / E_{EP}) \quad 1}{(E_{NT} / E_{EP}) \quad 2(l_{NT} / d_{NT})} \tag{5}$$

$$\eta_T = \frac{(E_{NT} / E_{EP}) - 1}{(E_{NT} / E_{EP}) + 2} \tag{6}$$

where,

E_{11} = Composite modulus of aligned discontinuous SWNT/epoxy in longitudinal direction
E_{22} = Composite modulus of aligned discontinuous SWNT/epoxy in transverse direction
l_{NT} = Length of the nanotubes
d_{NT} = Diameter of the nanotubes
V_{NT} = Volume fraction of the nanotubes
E_{EP} = Elastic modulus of the epoxy resin
E_{NT} = Elastic modulus of the nanotubes

For the moduli of a randomly oriented discontinuous SWNT/epoxy nanocomposite, E_{CR}, planar isotropy was assumed. Using the equations provided by Halpin and Pagano, the in-plane tensile modulus of a composite containing two-dimensional randomly orientated nanotubes were calculated,

$$E_{CR} = \left[\frac{3}{8} E_{11} + \frac{5}{8} E_{22} \right] \tag{7}$$

where E_{11} and E_{22} are the moduli of the aligned discontinuous SWNT/epoxy in longitudinal direction and transverse direction of the nanocomposites with same nanotube content, respectively, which were calculated from Equations 3 to 6.

7.2. *Model parameters of individually dispersed nanocomposites*

Figure 22 shows the molecular model of the nanocomposites of individually dispersed and unidirectional aligned SWNTs. The following assumptions were made to calculate the composite modulus of randomly oriented and aligned discontinuous SWNT/EPON 862 resin composites:

- Only individual nanotubes were present in the SWNT/epoxy composite;
- The SWNTs were assumed to be individually and uniformly dispersed in the resin matrix;
- The presence of only one type of SWNTs, (7,7) SWNTs, was considered in the nanocomposites and these nanotubes had a diameter of 1 nm, which was almost the same as that of the SWNTs used in experiments.

Based on the molecule structure of (7,7) SWNT, E_{NT} and the effective radius R_{NT} of the nanotube were calculated [29],

$$E_{NT} = \frac{4\sqrt{3}\pi bY\Lambda\nu}{3b^2\Lambda^2 + 2\sqrt{3}\pi\Lambda\nu + \pi^2\nu^2} \tag{8}$$

$$R_{NT} = \frac{b}{2\pi}\sqrt{3}\Lambda + \frac{\nu}{2} \tag{9}$$

$$\Lambda = \sqrt{(n^2 + m^2 + m^2 + nm^2)} \tag{10}$$

where,

n and *m* are chiral vector integer pairs of SWNT;
b is the C-C bond length in SWNT;
ν is equilibrium separation distance of SWNTs and resin matrix; and
Y is the in-plane Young's modulus of the graphene sheet.

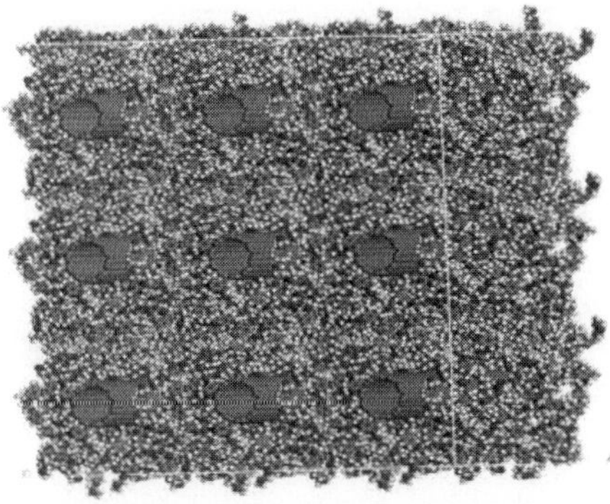

Fig.22. Molecular model of aligned individual SWNT-reinforced nanocomposites.

Using the above equation, the effective radius of (7,7) SWNT was 0.645 nm and effective diameter was 1.29 nm. The modulus of (7,7) SWNT was 801.37 GPa. The elastic modulus of EPON 862 resin was 2.72 GPa. The length of individual tube used in our experiments was ~100–1000 nm.

7.3. *Properties of individually dispersed SWNT-reinforced nanocomposites*

Using the above models and parameters, E_{CR} and E_{11} were calculated for both randomly oriented and aligned nanotube composites with 40% SWNT volume content. These properties with respect to aspect ratio are shown in Figures 23 and 24. The results show that the aspect ratio plays a significant role in changing the modulus of the composite. Results also indicate that in both randomly oriented and aligned cases, the maximum properties that can be achieved when the SWNT aspect ratio is larger than 1,000.

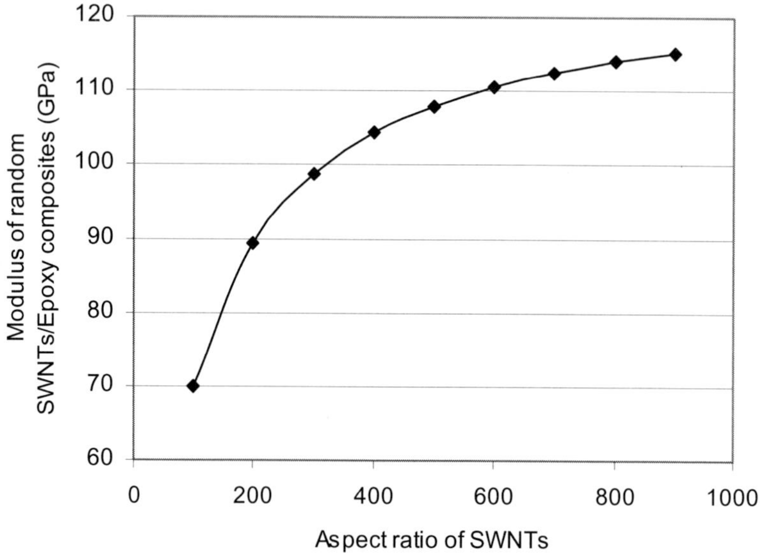

Fig. 23. E_{CR} vs. tube aspect ratio of randomly oriented individually dispersed SWNT/epoxy nanocomposites.

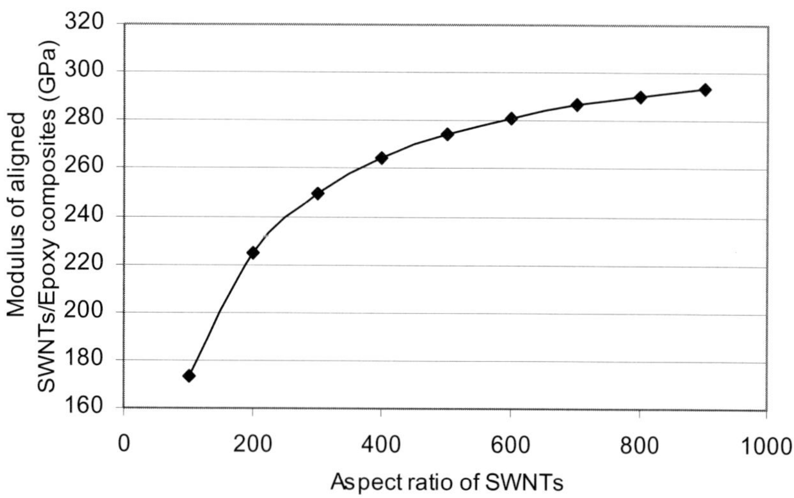

Fig. 24. E_{11} vs. tube aspect ratio of aligned individually dispersed SWNT/epoxy nanocomposites.

Figures 23 and 24 show E_{CR} and E_{11} for both randomly oriented and aligned nanotube composites with an increase in nanotube volume. Here, the length of SWNTs is assumed as 1 μm to calculate upper-boundary properties of the nanocomposites in the experiments. The results show that the nanotube content in the composite plays a significant role in changing the modulus of the composite.

Figures 25 and 26 also show a linear increase in modulus for randomly oriented and aligned discontinuous SWNT/epoxy composites with respect to nanotube content. As shown in Figure 25, for randomly oriented nanocomposites, the modulus is 177 GPa when the nanotube volume fraction is 0.6, which is close to a unidirectional carbon fiber composite with the same fiber volume fraction. As shown in Figure 26, for aligned nanocomposites, the composite modulus could be as high as 448 GPa when the nanotube volume fraction is 0.6, which is about 2.5 times higher than that of the randomly oriented discontinuous SWNT/epoxy nanocomposite. These results indicate the importance of nanotube alignment for developing high performance SWNT nanocomposites.

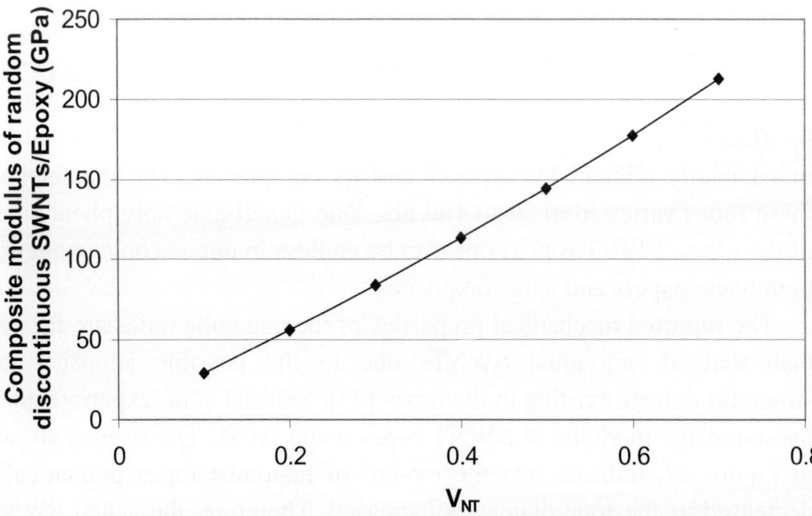

Fig. 25. E_{CR} vs. nanotube volume fraction (V_{NT}) of random individually dispersed SWNT nanocomposites

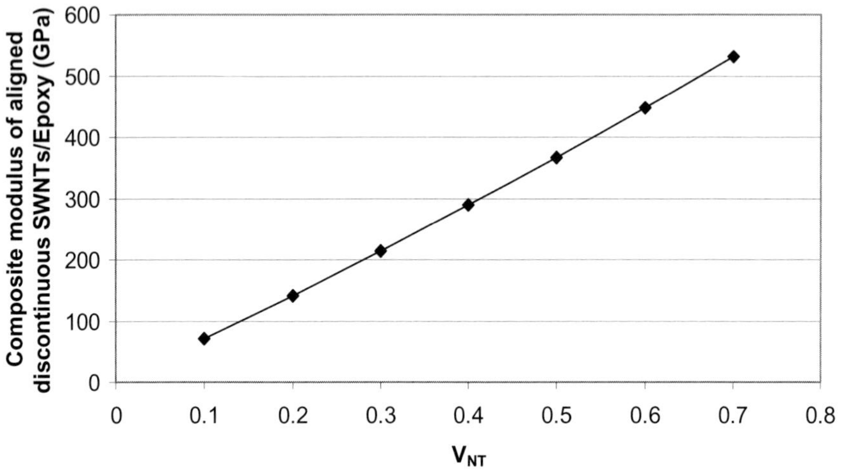

Fig. 26. E_{CR} vs. nanotube volume fraction (V_{NT}) of aligned individually dispersed SWNT nanocomposites

7.4. *Properties of SWNT rope-reinforced nanocomposites*

The predicted properties of individually dispersed SWNT-reinforced nanocomposites are exceptionally high compared to current experimental results and available data in literature. SWNTs usually form ropes or bundles in the buckypapers and composites, as shown in Figures 2, 5 and 6. Figures 12, 17 and 18 show the oriented nanotube ropes in the magnetically aligned buckypaper and its composites. The diameter of these ropes varies from 20 to 100 nm. Due to self-assembly phenomena of the ropes, SWNT ropes appear to be endless in microscopic images of both buckypapers and nanocomposites.

The reported mechanical properties of the nanotube ropes are far less than that of individual SWNTs due to the possible slippage and structural defects existing in the ropes [27]. Salvetat *et al.* experimentally measured the modulus of SWNT ropes using AFM. The results, shown in Figure 27, indicate that the moduli of nanotube ropes dramatically decreased as the rope diameters increased. Therefore, the actual SWNT rope properties should be able to predict the mechanical properties of the SWNT buckypaper-reinforced nanocomposites.

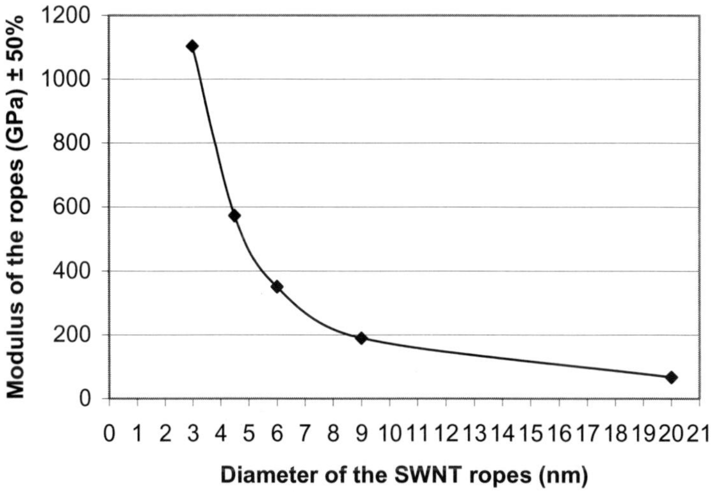

Fig. 27. Modulus decreases with increase of SWNT rope diameters [27].

Based on the observations of SWNT rope geometry in the buckypaper and composites, the diameter of the ropes was assumed to be 20 nm, and from Figure 27, the modulus for a 20 nm rope diameter was 67 GPa ± 50%. These values were used to recalculate the nanocomposite properties using the equations. The predicted results of both the randomly oriented and aligned nanotube SWNT rope-reinforced nanocomposites are provided in Figures 28 to 31. The predicted results were compared with experimental results for validation.

Figures 28 and 29 show that the composite moduli of the randomly oriented and aligned SWNT ropes/epoxy increased with respect to the increase in aspect ratio, where the rope volume content was 40%. The results show that the composite modulus remained nearly constant for both randomly oriented and aligned discontinuous SWNT ropes/epoxy composite after the rope aspect ratio reached 400. Hence, for subsequent calculations, the length of the nanotube ropes was chosen to be 8 μm since the ropes appear endless in the available microscope images for the buckypapers and their composites. However, the rope length is believed to be significantly less than the size of nanocomposite samples (25–47 mm long). Based on SEM analysis, 8 μm long ropes were

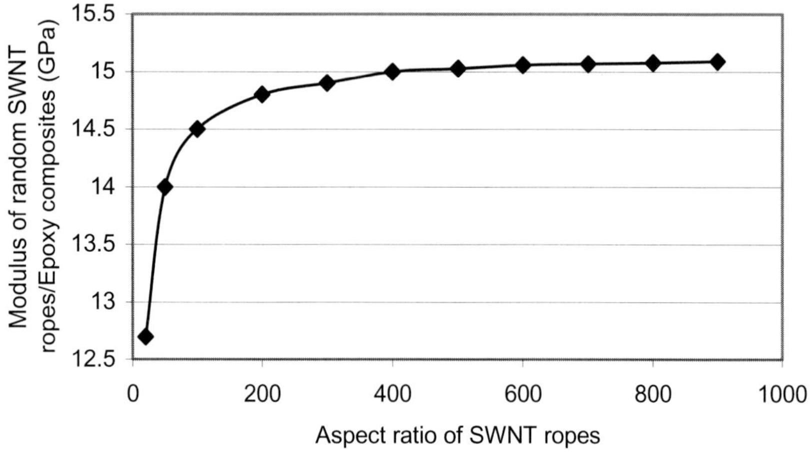

Fig. 28. E_{CR} vs. nanotube rope aspect ratio of random SWNT rope/epoxy nanocomposites

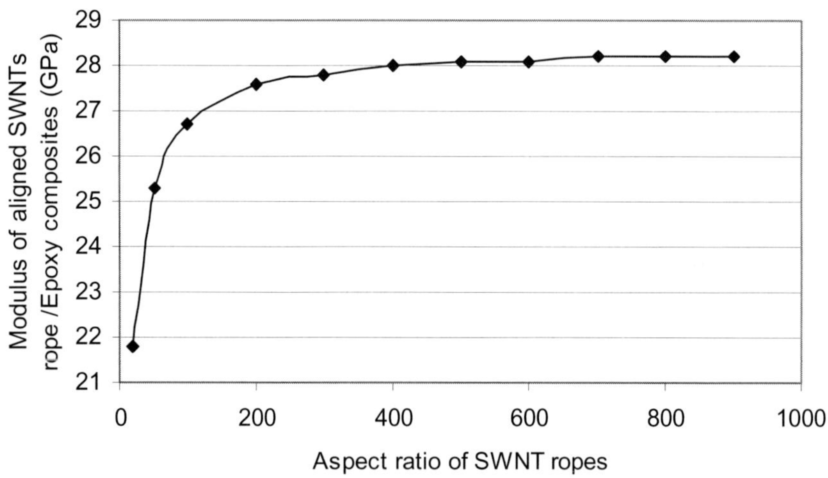

Fig. 29. E_{11} vs. tube rope aspect ratio of aligned SWNT rope/epoxy nanocomposites.

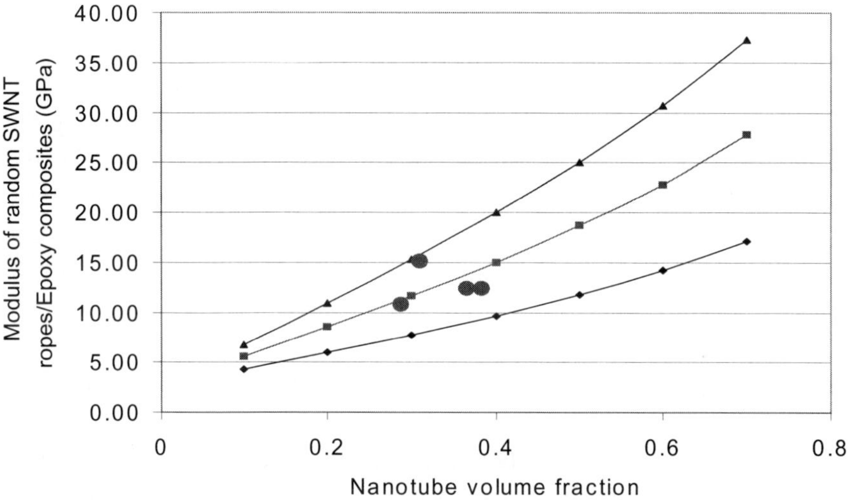

Fig. 30. E_{CR} vs. nanotube volume fraction of randomly oriented SWNT rope nanocomposites.

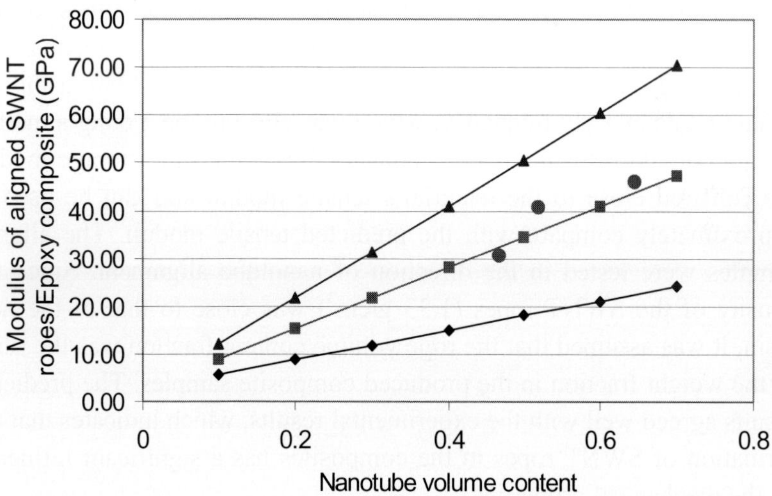

Fig. 31. E_{11} vs. nanotube volume fraction of aligned SWNT rope nanocomposites.

assumed for predicting the upper-boundary properties of the buckypaper composites. Figures 30 and 31 show an almost linear increase in modulus of randomly oriented and aligned SWNT ropes/epoxy composites with respect to an increase in nanotube content. Furthermore, considering the large variation of the measured properties of SWNT ropes [27], the upper-boundary (upper line), middle (middle line) and lower-boundary (bottom line) composite properties are also shown in the Figures when using moduli of 110.5, 67.0 and 33.5 GPa, respectively, for the SWNT ropes in the prediction.

Unlike the direct mixing of nanotubes in the resin, the buckypaper/resin infiltration process can produce nanocomposites with uniform nanotube dispersion and high loading. More importantly, the desired in-plane nanotube alignment in the nanocomposites was achieved by using the aligned buckypapers made in high magnetic fields. The experimental results of the storage moduli of both randomly oriented and aligned buckypaper nanocomposites listed in Tables 4 and 5 are also shown as round dots in Figures 30 and 31, respectively. The dots in Figures 30 and 31 represent the actual experimental values of the randomly oriented and aligned buckypaper/epoxy nanocomposite moduli. These storage moduli were tested with a DMA 2980, using the tensile mode at 1 Hz frequency, which was close to the Young's moduli testing condition for the nanocomposites. Therefore, the storage moduli are consired close to the material's tensile moduli and can be used to approximately compare with the predicted tensile moduli. The aligned samples were tested in the direction of nanotube alignment. Since the density of the SWNT ropes (1.33 g/cm^3) was close to that of the neat resin, it was assumed that the rope volume content fraction was the same as the weight fraction in the produced composite samples. The predicted results agreed well with the experimental results, which indicates that the formation of SWNT ropes in the composites has a significant influence on the mechanical properties.

8. Conclusions

SWNTs possess remarkable mechanical, electrical and thermal properties, which exceed those of any existing composite reinforcement materials, such as IM7 carbon fibers. However, using nanotubes as reinforcements to produce quality composites is still a very challenging issue due to their nanoscale dimension, unique chemical characteristics and intensive molecular interactions of the nanotubes and resin molecules during composite processing.

A novel technical approach of buckypaper/resin infiltration was developed to fabricate bulk polymeric nanocomposites with controlled nanostructures, desired alignment and high nanotube loading. Good nanotube dispersion and uniform pore structure of the produced buckypaper were demonstrated. Significant nanotube alignment in the magnetically aligned buckypapers was observed. The results also show that the new developed processing approach can effectively fabricate nanotube composites with controlled nanotube network structures. The storage modulus of the magnetically aligned nanocomposites was as high as 45 GPa. The produced nanocomposites can have nanotube loading from 20 w/w% to 60 w/w%. Both good dispersed nanotube networks and good alignment have been observed in the resultant randomly oriented and aligned nanocomposites. The world's largest magnetically aligned buckypaper with a working area of 60 in^2 was successfully manufactured, which provides the possibility for making bulk nanocomposites or devices with desired nanotube orientation. The influences of the formation of nanotube ropes in SWNT-reinforced nanocomposites were revealed by using composite models of short fibers and observed nanostructures of the produced buckypaper and nanocomposites. The research results well demonstrate the feasibility of using buckypaper/resin filtration process to make nanocomposites with controlled nanostructure and high nanotube loading, which are critical for developing high performance composite materials. The amount of nanostructure and property data provided from this research will enhance our understanding of nanocomposite processing and properties.

9. Future Work

The feasibility of using the novel SWNT buckypaper/resin infiltration process to fabricate nanocomposites with improved dispersion, alignment and high nanotube loading has been demonstrated in the research.

However, the mechanical properties of the resultant nanocomposites are still far below the anticipated values. Further research is needed to reduce SWNT rope size, functionalize nanotubes to improve interfacial bonding, optimize magnetic alignment process to reduce cost and scale-up buckypaper production as well as create a comprehensive database for structural design of nanocomposites. Electrical and thermal properties of buckypapers and their composites also must be explored for developing multifunctional structural applications. Several follow-on research projects are continuing to solve these issues to further reveal properties and application potentials of SWNT buckypaper nanocomposites.

The preformed SWNT buckypapers are easy to handle and use during conventional composite processes, including hand lay-up, vacuum bagging and RTM/VaRTM processes. The success of developing SWNT buckypaper nanocomposites will significantly accelerate applications of nanotube materials for high performance structural and multifunctional composite.

Acknowledgements

This work is supported by MN and ML Directorates of the Air Force Research Laboratory (Grant #: F08630-01-1-0010), AFOSR (Grant #: FA9550-04-1-0349 and FA9550-05-1-0271), Cornerstone Program of FSU Research Foundation, and NSF I/UCRC Program (Award #: 0224612). The authors also would like to thank the support of Dr. Leslie Kramer and Mr. Percy Funchess of Lockheed Martin Missiles and Fire Control-Orlando. The authors would also like to thank the support of the National High Magnetic Field Laboratory (NHMFL).

We also would like to acknowledge the other contributors for this research at the FAMU-FSU College of Engineering and Florida Advanced Center for Composite Technologies.

Dr. Jihua (Jan) Gou , Dr. Zhi (Michael) Wang, Mr. Gonnet Philippe, Mr. Ravi Shankar, Mr. Olivier Marietta-Tondin, Ms. Liao (Ashley) Yu-Hsuan, Ms. Oni Sadiki, Mr. Clinton Baker, Mr. Kevin Barefield, Mr. Blake Cullum

References

1. Satish Kumar, Thuy D. Dang, Fred E. Arnold, Arup R. Bhattacharyya, Byung G. Min, Xiefei Zhang, Richard A. Vaia, Cheol Park, W. Wade Adams, Robert H. Hauge, Richard E. Smalley, Sivarajan Ramesh and Peter A. Willis, "Synthesis, structure and properties of PBO/SWNT composites," *Physics*, 2002, 35, 9039.
2. M.J. Biercuk, M.C. Llaguno, M. Radosavljevic, J.K. Hyun and A.T. Johnson and J.E. Fischer, "Carbon nanotube composites for thermal management," *Applied Physics Letters*, 2002, 20, 15.
3. Marc in het Panhuis, Amitesh Maiti, Alan B. Dalton, Albert van den Noort, Jonathan N. Coleman, Brendan McCarthy and Werner J. Blau, "Selective interaction in a polymer-single-walled carbon nanotube composites," *Applied Physics Letters*, 2003, 107, 478.
4. Kin Liao and Sean Li, "Interfacial characteristics of a carbon Nanotube-polystyrene composite system," *Applied Physics Letters*, 2001, 79, 4225.
5. J. Sandler, M.S.P. Shaffer, T. Prasse, W. Bauhofer, K. Schulte and A.H. Windle, "Development of a dispersion process for carbon nanotubes in an epoxy matrix and the resulting electrical properties," *Applied Physics Lettes*, 1999, 40, 5967.
6. B.C. Rosen and L. Jin, "Single-walled carbon nanotube-polymer composites: electrical, optical and structural investigation," *Synthetic Metals*, 2002, 127, 59.
7. Jiang Zhu, Jong Dae Kim, Haiqing Peng, John L. Margrave, Valery N. Khabashesku and Enrique V. Barrera, "Improving the dispersion and integration of single-walled carbon nanotubes in epoxy composites through Functionalization," *Nano. Letters*, 2003, 3, 1107.
8. E.T. Mickelson, C.B. Human, A.G. Rinzler, R.E. Smalley, R.H. Hauge and J.L. Margrave, "Fluorination of single-wall carbon nanotubes," *Chemical Physics Letters*, 1998, 296, 188.
9. A. Garg and S.B. Sinnott, "Effect of chemical functionalization on the mechanical properties of carbon nanotubes," *Chemical Physics Letters*, 1998, 295, 273.
10. C. Park, Z. Ounaies, K.A.Watson, R.E. Crooks, J. Smith Jr., S.E. Lowther, J.W. Connell, E.J. Siochi, J.S. Harrison and T.L. St. Clair, "Dispersion of single wall carbon nanotubes by in situ polymerization under sonication," *Chemical Physics Letters*, 2002, 364, 303.
11. E.T. Thostenson, Z.F. Ren and T.W. Chou, "Advanced in the science and technology of carbon nanotube and their composites", *Composites Science and Technology*, 2001, 61, 1899.
12. R. Saito, G. Dresselhaus and M.S. Dresselhaus. *Physical properties of carbon nanotubes*. 1st ed. London: Imperial College Press, 1999.
13. E.T. Thostenson, C. Li and T.W. Chou, "Nanocomposites in context", *Composites Science and Technology*, 2005, 65, 491.
14. K.T. Lau and D. Hui, "The revolutionary creation of new advanced materials-carbon nanotubes composites", *Composites: Part B*, 2002, 33, 267.
15. X.L. Xia, Y.W. Ma and X. Zhou, "Dispersion and alignment of carbon nanotubes in polymer matrix: A review", *Materials Science and Engineering R*, 2005, 49, 89.

16. A. Walters, M.J. Casavant, X.C. Qin, C.B. Huffman, P.J. Boul, L.M. Ericson, E.H. Haroz, M.J. O'Connell, K. Smith, D.T. Colbert and R.E. Smalley, "In-plane-aligned membranes of carbon nanotubes," *Chemical Physics Letters*, 2001, 338,14.
17. J.E. Fischer, W. Zhou, J. Vavro, M.J. Casavant, D.E Walters and R.E. Smalley, "Magnetically aligned single wall carbon nanotubes filmes: Preferred orientation and anisotropic transport properties," *Journal of Applied Physics*, 2003, 93, 2157.
18. J. Hone, M.C. Llaguno, N.M. Nemes, A.T. Johnson, D.A. Walters, M.J. Casavant, J. Schmidt, and R.E. Smalley, "Electrical and thermal transport properties of magnetically aligned single wall carbon nanotube films," *Applied Physics Letters* 2000, 77, 666.
19. Xiao-Lin Xiea, Yiu-Wing Maia and Xing-Ping Zhou, "Dispersion and alignment of carbon nanotubes in polymer matrix: A review," *Materials Science and Engineering R,* 2005, 49, 89.
20. Ajayan, P.M., Stephan, O., Colliex, C. and Trauth, D., "Aligned Carbon Nanotube Arrays Formed by Cutting a Polymer Resin-nanotube Composite," *Science*, 1994, 265, 1212.
21. Jin, L., Bower, C. and Zhou, O., "Alignment of Carbon Nanotubes in a Polymer Matrix by Mechanical Stretching," Applied Physics Letters, 1998, 73, 1197.
22. Mei Zhang, Shaoli Fang, Anvar A. Zakhidov, Sergey B. Lee, Ali E. Aliev, Christopher D. Williams, Ken R. Atkinson and Ray H. Baughman, "Strong, transparent, multifunctional, carbon Nanotube sheets," *Science*, 2005, 309, 1215.
23. Thostenson, E.T. and Chou, T.W., "Aligned multi-walled carbon nanotube-reinforced composites: processing and mechanical characterization," *Journal of Physics D-Applied Physic,* 2002, 35, L77.
24. Haggenmueller, R., Gommans, H.H., Rinzler, A.G., Fischer, J.E. and Winey, K.I., "Aligned Single-wall Carbon Nanotubes in Composites by Melt Processing Methods," *Chemical Physics Letters*, 2000, 330, 219.
25. Andrews, R., Jacques, D., Rao, A.M., Rantell. T., Derbyshire, F., Chen, Y., Chen, J. and Haddon, R.C., "Nanotube composite carbon fibers," *Applied Physics Letters*, 1999, 75, 1329.
26. Ko, F., Gogotsi, Y., Ali, A., Naguib, N., Ye. H.H., Yang, G.L., Li, C. and Willis, P., "Electrospinning of continuons carbon nanotube-filled nanofiber yarns," *Advanced Materials*, 2003, 15, 1161.
27. J.-P. Salvetat, G.A.D. Briggs, J-M. Bonard, R.R. Bacsa, A.J. Kulik, T. Stöckli, N.A. Burnham and L. Forró, Elastic and Shear Modulus of Single-Walled Carbon Nanotube Ropes. *Physical Review Letters*, 1999, 82, 944.
28. P.K Mallick, *Handbook of fiber reinforced composites*, Second Edition, Marcel Dekker Inc., 1998.
29. R.B. Pipes, S-J. Frankland, P. Hubert and E. Saether, "Self-Consistent Properties of the SWCN and Hexagonal Arrays as Composite Reinforcements," *Composites Science and Technology*, 2003, 63, 1349.

CHAPTER 4

Processing and Mechanical Properties Characterization of Hybrid Thermoset Polymer Composites with Micro-Fiber and Carbon Nano-Fiber Reinforcements

Kuang-Ting Hsiao

Department of Mechanical Engineering
University of South Alabama
Mobile, AL 36688, USA
kthsiao@jaguar1.usouthal.edu

1. Introduction

In this chapter, the use of carbon nanotubes (CNTs) and carbon nanofibers (CNFs) in micro-fiber reinforced thermoset polymer matrix composites will be reviewed. Micro-fiber reinforced polymer composite materials have been widely used in many structural applications such as aerospace, automotive, civil and marine structures due to their excellent strength-to-weight ratio, chemical and weather resistances, tailor-able mechanical, thermal and electrical properties. The conventional micro-fiber reinforced polymer composite usually consists of polymer (either thermoset of thermoplastic resin) as the matrix system and micro-fibers with diameter of 5–30 μm as the reinforcement system. To obtain the best strength-to-weight ratio, continuous micro-fibers and thermoset resin are usually selected to form the composite material. Such thermoset polymer composites can have strength and Young's modulus comparable to the commonly used structural metal materials such as steel, titanium, and aluminum but with much lighter weight.

The general approach to manufacture the conventional micro-fiber reinforced thermoset polymer composites includes several steps. The first step is to have liquid resin to impregnate the continuous fiber system, which can be in the form of tows, mats, or preform. The second step is to hold the liquid resin saturated fiber system in a mold and apply heat to cure the resin. The curing process is an exothermal process that cross-links the polymer molecules into a rigid network thus converts the liquid resin into solid. After the curing process, the micro-fibers are bonded together by the polymer matrix so that the mechanical load can be effectively transferred and shared from one fiber to another. The system consisting of micro-fibers and polymer matrix forms the "micro-fiber reinforced thermoset polymer composite material."

Different from all the conventional micro-fiber reinforced thermoset polymer composites that are reinforced by the micro-fiber system, a new type of polymer composite which is reinforced by both the micro-fibers and recently discovered nano-fibers will be discussed in this chapter. As the gap between two micro-fibers in a composite should be in micrometer range, theoretically, it is possible to add another kind of fiber with diameter in nano-meter range and create a hybrid nano-/micro-fibers reinforced thermoset polymer composite material. In this chapter, we will review different approaches to manufacturing such hybrid nano-/ micro- fibers reinforced thermoset polymer composite material and then compare their mechanical properties to identify the advantages and disadvantages of this new composite material. The challenges and outlook for further developing this kind of hybrid nano-/micro- fibers reinforced thermoset polymer composite material will be discussed in the end of this chapter. In next section, we will briefly review carbon nanotubes, carbon nanofibers, and their mechanical properties as blended with different kinds of polymers. This background information will be employed to analyze the processes for manufacturing the hybrid nano-/ micro- fibers reinforced thermoset polymer composites and the mechanical properties improvement.

2. Background

2.1. *Carbon Nanotube (CNT) and Carbon Nanofiber (CNF)*

Carbon nanotubes (CNTs) were first discovered by Iijima in 1991 [1] in the form of multiwalled concentrical tubular structure. The multiwalled carbon nanotube (MWNT) can be considered as several layers of graphite sheets rolled into a cylinder with diameter ranges from 2 nm to 50 nm and with length of several microns. As a result, the multiwalled carbon nanotubes have excellent mechanical, electrical, and thermal properties as graphite. Another type of carbon nanotube is the single-walled carbon nanotube (SWNT) which only has one layer of graphite in the tubular structure [2] and with diameter typically ranges from 1 nm to 2 nm. The lattices of the carbon nanotube are usually hexagonal but sometimes can be pentagons to fit between the neighboring hexagonal lattices especially at a curved surface. When a graphite sheet is rolled into a tube, the chiral angle between the direction of hexagonal lattices and graphite sheet closing direction can be different [3]. This difference in chiral angle can vary the tube's properties such as the electrical conductivity.

Carbon nanotubes can be synthesized by using electric arc discharge method [1,4], chemical vapor deposition method (CVD) [4–7], and laser ablation [8–13]. The electrical arc discharge method and the laser ablation method use high energy level and catalyst under high temperature to transform the graphite into carbon nanotubes. These two methods usually produce CNTs with different chiralities and diameters. The laser ablation has been considered as a possible method for mass producing single-walled carbon nanotubes (SWNTs). On the other hand, the chemical vapor deposition (CVD) method has been successfully used for producing multiwalled carbon nanotubes (MWNTs). It was reported by Schlitter *et al.* [14] that this method can be modified to achieve uniform diameter, chirality and high purity SWNTs. They used C_{60} and Ni vapor mixture along with a moving mask with an array of holes of 300 nm diameter to grow SWNTs on a molybdenum substrate plate. With this improved CVD method, it has been found that the diameter, chirality and purity of CNTs can be well-controlled. More detailed discussion regarding the synthesis of CNTs can be found in other chapters of this book.

Similar to carbon nanotubes, carbon nanofibers (CNFs) are formed by graphite as well. However, the graphite structures of a CNF usually does not form into a continuous tube like that of a CNT. Instead, the graphite sheet may form different geometries such as the platelet structure, the

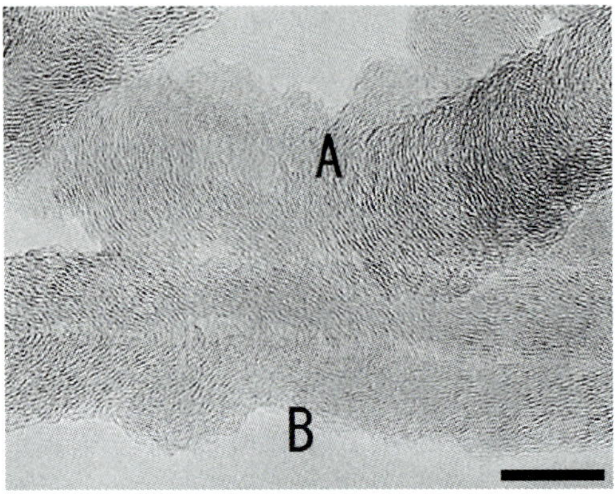

Fig. 1. HRTEM image of two CNF structures: (A) platelet structure and (B) herringbone structure. Note the scale bar represents 10 nm. (Picture is taken from [15]).

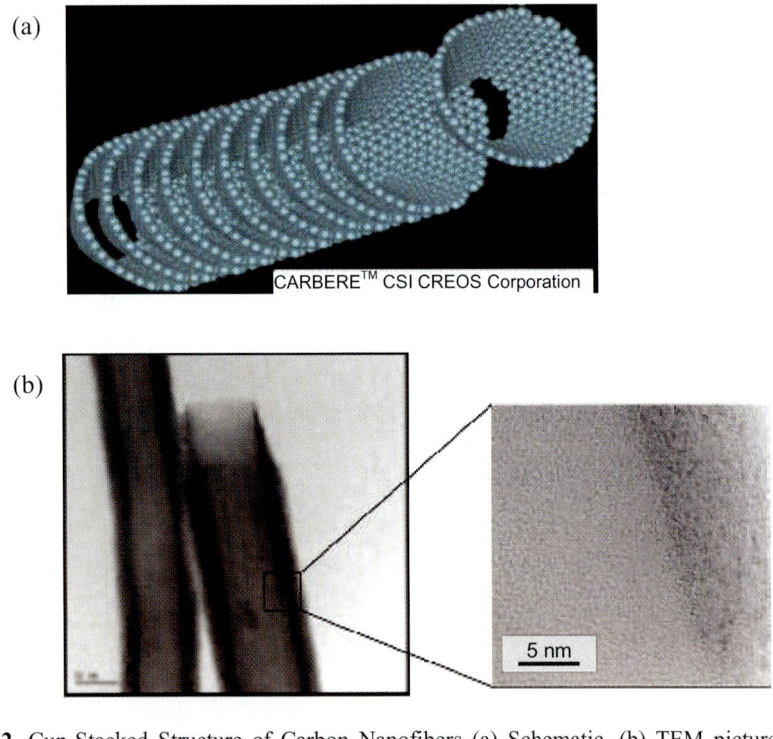

Fig. 2. Cup-Stacked Structure of Carbon Nanofibers (a) Schematic, (b) TEM picture. (Pictures are taken from [26]).

herringbone structure (see Figure 1), and the cup-stacked structure (see Figure 2). The distance between two graphitic planes in a cup-stacked structure is 0.334 nm [16].

Carbon nanofibers can be produced by different methods. Catalytic chemical vapor deposition method [17] is widely used to grow CNFs. Different from the CVD method used to grow carbon nanotubes, the CNFs are grown by the chemical vapor deposition method at lower temperature (500–700°C) [18]. Zheng *et al.* [18] studied the effect of CVD temperature to CNF growth. The alumina-supported Ni catalyst was used in CVD experiments. The inflows of C_2H_2, H_2, and N_2 were set at the flow rates of 30, 150, and 100 sccm, respectively. Figure 3 shows different geometries of CNFs produced under different CVD temperatures. In addition to temperature, the growth of CNFs can be affected by the flow rates of acetylene and hydrogen as well [19]. Other methods to synthesize CNFs can be hot filament assisted sputtering [20], polymer blend technique [21], co-catalyst deoxidization [22], template assisted method [23], and floating catalyst method [24]. Sometimes, the as-grown CNFs will be followed by heat treatment [15,25], which can deform the open edges of the graphitic layers (see Figure 4) thus change the material properties.

Due to the discontinuous structure of CNFs, carbon nanofibers are regarded as less perfect than carbon nanotubes. However, a CNF has a rough surface (see Figure 1, Figure 2, and Figure 4) so it should have stronger interfacial bond with polymer than that of a CNT-polymer system [26]. Nevertheless, carbon nanofiber still possesses the excellent mechanical, thermal, and electrical properties inherited from graphite and

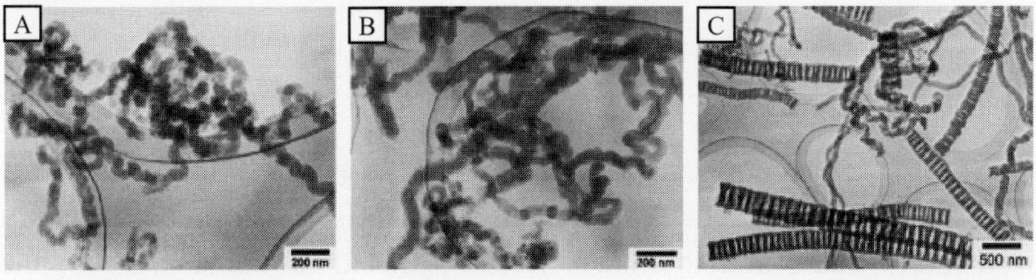

Fig. 3. TEM pictures of CNFs synthesized at (A) 550°C, (B) 600°C, and (C) 700°C. (Pictures are taken from [18]).

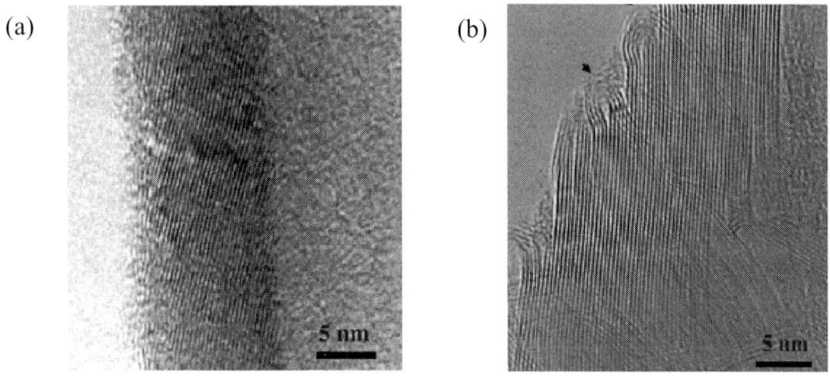

Fig. 4. The influence of heat treatment on the edge of graphitic layers of a CNF. (a) As-grown CNF. (b) CNF after 3000°C heat treatment. (Pictures are taken from [25]).

Table 1. Comparison of CNT, CNF, E-glass fiber, carbon fiber, and commonly used thermoset resins.

Description	Carbon Nanotube	Carbon Nanofiber [29]	E-glass Fiber	Carbon Fiber	Thermoset Polyester	Thermoset Vinyl Ester	Thermoset Epoxy
Ultimate Strength (GPa)	200–30 [27,4]	7	3.5	3.8–6.0	0.033–0.070	0.070–0.620	0.062–0.172
Tensile Modulus (GPa)	~1000 [28,4]	600	72.4	228–440	3–10	24–27	12–17
Diameter (micron)	0.005–0.050	0.060–0.2	~10	~7	–	–	–
Length, (micron)	>100	~100	continuous	continuous	–	–	–
Density, (g/cm^3)	1.8	1.8	2.6	1.81	1.3–2	1.6–1.95	1.79–2.1

could be an excellent reinforcement for polymer matrix composites. A mechanical property comparison between CNT, CNF, and commonly used micro-fibers and thermoset resins is listed in Table 1. From the comparison, it is clearly shown that CNT and CNF are excellent candidates for the reinforcement in polymer matrix composites.

2.2. Carbon nanotube and carbon nanofiber modified polymer matrix

The use of CNT and CNF in reinforcing polymer system has received extensive attention and been investigated by many researchers [30–37]. It was reported that one can increase the tensile modulus and tensile strength about 42% and 25%, respectively, by adding 1 wt% MWNTs with diameter of 30 nm and length of 50 μm in polystyrene matrix [30]. In addition, the TEM picture (see Figure 5) indicates that CNTs can bridge the crack surface of the polystyrene matrix.

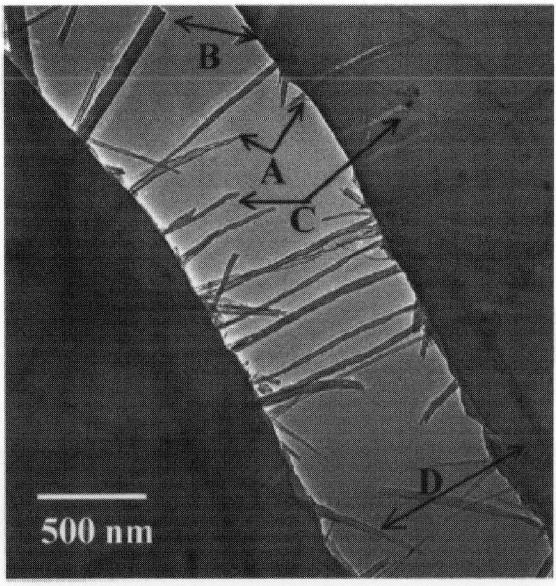

Fig. 5. TEM picture shows CNTs can bridge the crack of polystyrene matrix. (Picture is taken from [30]).

CNTs can also be used to reinforce and toughen the polymer matrix. Gojny *et al.* [35] used amino-functionalized Double-Walled Carbon Nanotubes (DWCNT-NH_2) with diameter of 2.8 nm and length of several micrometers to reinforce epoxy. The tensile and fractural properties of the DWCNT-epoxy composites, the pure epoxy control samples and the carbon black-epoxy composites were measured and compared. Table 2 summarizes the measured tensile and fractural properties of different specimens. It was found that the Young's modulus consistently increases with the amount of carbon black or DWCNTs added into the epoxy matrix. The composites with amino-functionalized Double-Walled Carbon Nanotubes (DWCNT-NH_2) showed more improvement in the Young's modulus due to the polar amino-groups. A Halpin Tsai's equation modified for a randomly orientated short fiber composite was used to calculate the maximum obtainable Young's modulus as:

$$\begin{aligned} E_c = {} & \frac{3}{8}\left(1+2\left(\frac{l}{d}\right)\left(\frac{E_{NT}/E_m - d/4t}{E_{NT}/E_m + l/2t}\right)V_{NT}\right) \\ & \times\left(1-\left(\frac{E_{NT}/E_m - d/4t}{E_{NT}/E_m + l/2t}\right)V_{NT}\right)^{-1} E_m \\ & +\frac{5}{8}\left(1+2\left(\frac{E_{NT}/E_m - d/4t}{E_{NT}/E_m + d/2t}\right)V_{NT}\right) \\ & \times\left(1-\left(\frac{E_{NT}/E_m - d/4t}{E_{NT}/E_m + d/2t}\right)V_{NT}\right)^{-1} E_m \end{aligned} \tag{1}$$

where E_c is the Young's modulus of the composite, l is the length of DWCNT (= 10 micrometers), d is the outer diameter of DWCNT (= 2.8 nm), E_{NT} is the Young's modulus of DWCNT (= 1 TPa), E_m is the Young's modulus of epoxy matrix (= 3.29 GPa), t is the thickness of the wall of DWCNT (= 0.34 nm), V_{NT} is the volume fraction of DWCNT (= 0.121% for 1 wt% DWCNT loading). Using Equation (1), Gojny *et al.* [35] predicted that 0.1 wt% DWCNTs could increase the composite E_c to 3.51 MPa, which agreed with the measured value listed in Table 2. However, the theory predicted E_c = 5.5 GPa for 1 wt% DWCNT-epoxy composite, which is much higher than the measured value E_c = 3.51 GPa. They explained that the difference was caused by inhomogeneous dispersion of DWCNTs and the presence of voids due to the increasing difficulty to degas the resin when higher DWCNT weight fraction was used.

Table 2. Tensile properties and toughness of Carbon Black and Double-Walled Carbon Nanotube reinforced epoxy [35].

	Epoxy Control Sample	0.1 wt% Carbon Black (diameter = 30 nm)	0.1 wt% DWCNTs	0.1 wt% DWCNT-NH2	1 wt% DWCNT-NH2
Young's Modulus (GPa)	3.29	3.30	3.35	3.50	3.51
Tensile Strength (MPa)	63.8	60.1	62.4	63.7	65.0
Fracture Strain (%)	7.3	7.2	7.3	7.6	6.8
Fracture Toughness K_{IC} (MPa*m$^{1/2}$)	0.65	0.77	0.75	0.77	0.82

From Table 2, one finds that the addition of carbon black and CNTs may sometimes reduce the tensile strength of a composite specimen. This may be due to the combined effect of nanoparticle agglomeration and the weak interface between nanoparticles and polymer. However, when the amino-functionalized DWCNTs were used, the agglomerate's size was reduced. As a result, the improvement in the tensile strength was achieved due to better load transfer between DWCNTs and resin.

This significant increase in the fracture toughness K_{IC} was obtained by adding carbon black or DWCNTs into the epoxy matrix [35] as listed in Table 2. This can be explained as DWCNTs bridging the cracks in the epoxy matrix as shown in Figure 6. The DWCNTs can bridge the crack with the distance 500–1000 times the average diameter of the DWCNTs. Some shorter DWCNTs shown in Figure 6 suggest the DWCNTs were broken and absorbed the fracture energy during the crack development.

Cooper *et al.* [34] has conducted experiments to test the mechanical properties of PMMA matrix reinforced by different types of carbon nanoparticles including SWNTs (diameter = 1.4–2 nm, length = 200–400 μm), MWNTs (diameter = 10–15 nm, length = 2–3 μm), and CNFs (diameter = 200 nm, length = 200–500 μm). They manufactured

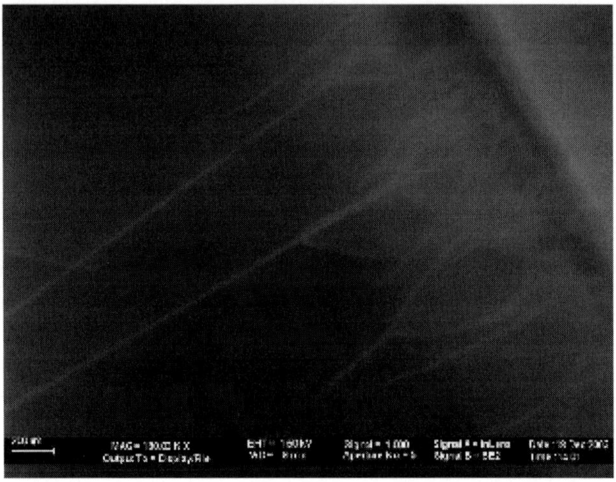

Fig. 6. DWCNT-NH_2 bridged the crack in epoxy matrix; the shorter DWCNTs were the evidence of DWCNTs being broken and pulled out to absorb the fracture energy during the crack development. (Picture is taken from [35]).

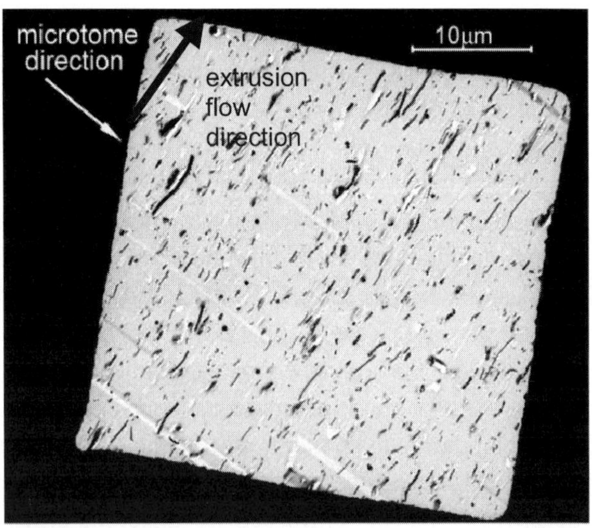

Fig. 7. TEM picture shows 10 wt% Carbon Nanofibers (CNFs) were uniformly distributed in PMMA matrix and were aligned in the extrusion flow direction. (Picture is taken from) [34].

CNT/PMMA and CNF/PMMA composite samples by using sonication, mechanical mixing and then followed by extrusion. The samples were microtomed with cutting surface parallel to the extrusion flow direction and examined under TEM as shown in Figure 7. Good dispersion and alignment control of MWNTs and CNFs in the PMMA matrix were observed under TEM (see Figure 7) and statistically analyzed as summarized in Table 3. Note that similar extrusion induced CNF alignment effect has also been reported in other independent experimental study [37]. However, in Table 3, MWNT/PMMA samples show slightly better alignment than the CNF/PMMA samples. Cooper *et al.* suspected that the larger surface area and shorter length of MWNTs caused the alignment of MWNTs slightly better than that of CNFs. They then performed tensile tests on the specimens and found that the tensile modulus was not sensitive to the addition of CNTs, but was positively influenced by the CNFs (see Figure 8). Lack of strong interfacial bond between CNTs and PMMA matrix was claimed to be responsible for the low load transfer efficiency thus caused little improvement in tensile modulus of the CNT/PMMA composite. In addition to the tensile modulus, they also tested the impact strength and found significant improvement by adding CNFs and CNTs in the PMMA matrix (see Figure 9). The significant impact strength improvement is possibly due to the high fracture strain [39–42] of the CNTs and CNFs as well as the pullout and sliding of CNTs and CNFs in the PMMA matrix.

Table 3. Statistical data of the CNFs and MWNTs alignment in PMMA matrix; the angle is with respect to the microtome direction as shown in Figure 7 [34].

Specimen	Mean (deg)	Standard Deviation (deg)	Full Width at Half Maximum (deg)
PMMA & 0.1 wt% CNFs	95	7	6
PMMA & 4 wt% CNFs	87	14	16
PMMA & 10 wt% CNFs	83	30	35
PMMA & 4 wt% MWNTs	90	9	12

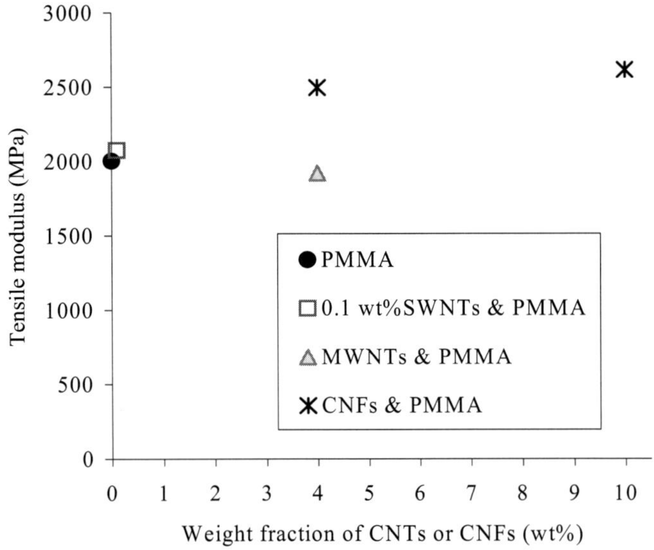

Fig. 8. Tensile modulus of CNT/PMMA composites and CNF/PMMA composites [34].

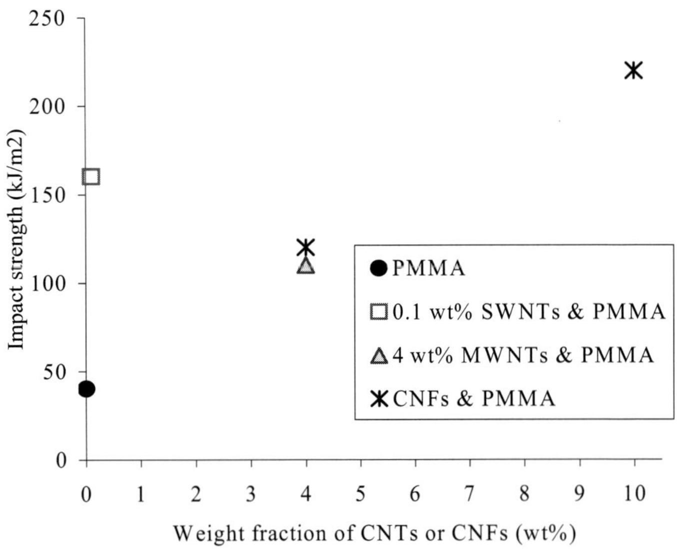

Fig. 9. Impact strength of CNT/PMMA composites and CNF/PMMA composites [34].

In addition to tensile properties, CNT is an effective filler to improve the shear strength of polymer matrix. For instance, CNTs can be used for reinforcing the adhesive for joining two composite laminates. Hsiao *et al.* [33] added MWNTs in epoxy to join two carbon fiber-epoxy laminates and tested the shear strength (ASTM-D5868-01). It was found that by adding 1 wt% and 5 wt% MWNTs into epoxy, one could improve the average shear strength of the adhesive joint about 31.2% and 45.6%, respectively (see Figure 10). By comparing the fracture surfaces of the tested specimens under SEM, they observed significant carbon fiber failure in the adherends jointed by MWNT enhanced epoxy adhesive. This shows that the addition of MWNTs in the epoxy adhesive effectively enhanced the shear strength and transferred the shear load between the two carbon fiber/epoxy composite adherends. As a result, the adherends failed before the adhesive did during the shear test. Note that similar shear strength improvement has also been reported for SWNT/epoxy adhesive by Meguid and Sun [43].

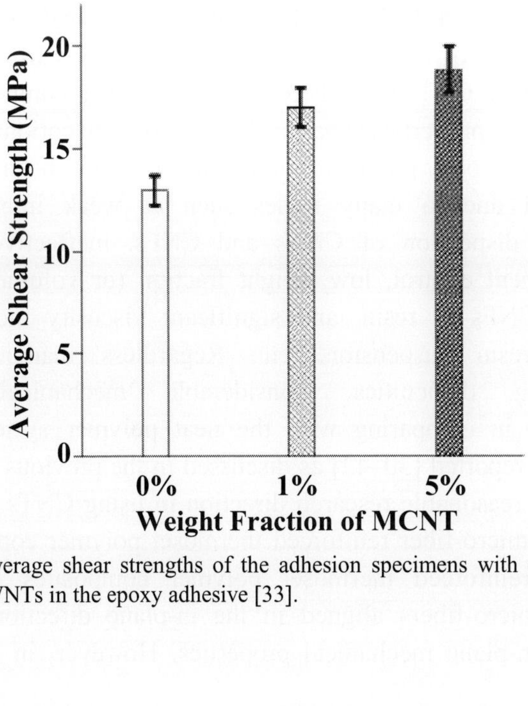

Fig. 10. The average shear strengths of the adhesion specimens with different weight fractions of MWNTs in the epoxy adhesive [33].

(a) (b)

Fig. 11. (a) SEM picture of the fracture surface of the bonding area of the 5 wt% MWNTs + epoxy case; failure at the graphite fiber of the adherends was observed. (b) SEM picture of the fracture surface of the bonding area of the epoxy-only case; failure occurred at the epoxy surface of the adherends and no significant graphite fiber fracture was observed [33].

3. Use of Carbon Nanotube and Carbon Nanofiber in Micro-fiber Reinforced Thermoset Polymer Composites

Although CNT and CNF exhibit better mechanical properties than conventional micro-fibers such as glass fibers and carbon fibers (see Table 1), a pure CNT or CNF reinforced polymer composite material with mechanical properties superior than that of conventional micro-fiber reinforced thermoset polymer composites has not been successfully manufactured due to many issues such as weak interfacial bond, non-uniform dispersion of CNTs and CNFs, ineffective CNTs and CNFs alignment control, low weight fraction (or volume fraction) of CNTs and CNFs in resin, and significant viscosity increase of the nanoparticle/resin suspension, etc. Regardless many encountered manufacturing difficulties, considerable mechanical properties improvement in comparing with the neat polymer system has been continuously reported [30–42] as discussed in the previous section. As a result, it is a reasonable research direction in using CNTs and CNFs to improve the micro-fiber reinforced thermoset polymer composites. The micro-fiber reinforced thermoset polymer composites usually have continuous micro-fibers aligned in the in-plane directions thus have remarkable in-plane mechanical properties. However, in the thickness direction, the only effective constituent to hold the micro-fibers together is the polymer matrix. If the CNTs and CNFs can substantially improve the polymer matrix, one can add them into micro-fiber reinforced thermoset polymer composites to improve the matrix-dominated mechanical properties such as delamination and inter-laminar shear properties. Next section we will review how to manufacture such hybrid nano-/micro- fibers reinforced thermoset polymer composites.

3.1. *Processes for manufacturing hybrid nano-/micro- fibers reinforced thermoset polymer composites*

To manufacture the hybrid nano-/micro- fibers reinforced thermoset polymer composites, many processing issues need to be addressed. In order to manufacture such composites, one has to: (1) successfully disperse the CNTs or CNFs in the polymer resin, (2) uniformly impregnate the micro-fiber bed (such as glass fiber preform) with the CNT or CNF modified resin, (3) prevent CNTs or CNFs being filtered by the micro-fiber bed, (4) remove excessive gas in the composite and minimize void formation. In the following subsections, we will review several manufacturing techniques reported by different researchers and discuss the results with respect to the aforementioned four processing issues.

3.1.1. *Resin Transfer Molding (RTM)*

The first manufacturing issue is to uniformly disperse CNTs or CNFs in polymer resin without forming any aggregate. Since one of the important features of using CNTs and CNFs as reinforcement is to use their large surface area, being able to uniformly disperse CNTs and CNFs in the polymer is critical. The undesired CNT or CNF agglomerates will significantly reduce the effective bonding surface area between the polymer matrix and the reinforcement system. Furthermore, the large CNT or CNF agglomerates will be filtered by the micro-fiber bed during resin impregnation if the agglomerate size is close to the pore-size of the micro-fiber bed. Fan *et al.* [44] tested three different methods for dispersing 3 wt% MWNTs in vinyl ester resin and then infused the

MWNT/vinyl ester suspension through a glass fiber preform using the resin transfer molding (RTM) technique. In a typical RTM process, the liquid resin with low viscosity infiltrates the fiber preform inside a mold cavity due to the pressure difference between the resin injection gate and the vent. After the resin completely saturates the preform, the resin starts to cure and binds the fibers together to form a solid composite. During the resin impregnation stage, the resin curing process is usually inhibited to maintain the resin viscosity low because the goal is to have the pressure driven resin to impregnate the preform completely before the resin gels. During the resin filling stage, the flow of resin and mold filling pattern can be modeled and controlled effectively based on the well-characterized preform permeability, porosity, and resin viscosity [45–50]. Once the resin fills the mold, the cure cycle starts to convert the liquid resin into solid and forms a solid composite part [51–55]. After the cure cycle, the complete cured composite part is then demolded. Note that RTM is an advanced composite manufacturing technique and is capable of producing high quality composite parts (less void and higher fiber volume fraction) with mid-to-high production volume.

In the experimental investigation of using RTM to manufacture hybrid MWNT and glass fiber reinforced thermoset polymer composites conducted by Fan *et al.* [44], a pure mechanical stirring method was first tested and showed no improvement in dispersing MWNTs in the resin and the MWNTs were almost completely filtered by the glass fiber preform. Second, a surfactant method combining the surfactant treatment (Poly-oxyethylene 8 lauryl, or $C_{12}EO_8$), sonication, and mechanical stirring [56] was tested and showed the MWNT aggregates were significantly reduced to the size that was allowed to penetrate the glass fiber preform without being filtered. Third, they tested a previously reported oxidation method [57,58] by oxidizing MWNTs in concentrated nitric acid and sulfuric acid for 30 minutes and then cleaning the oxidized MWNTs with water. The oxidized MWNTs were mixed with resin and the MWNT-resin suspension was infused into the glass fiber preform. There were no noticeable MWNT aggregates in the resin and no MWNT filtration in the preform under TEM. However, the MWNTs with original length of 10–15 μm were cut into shorter segments with

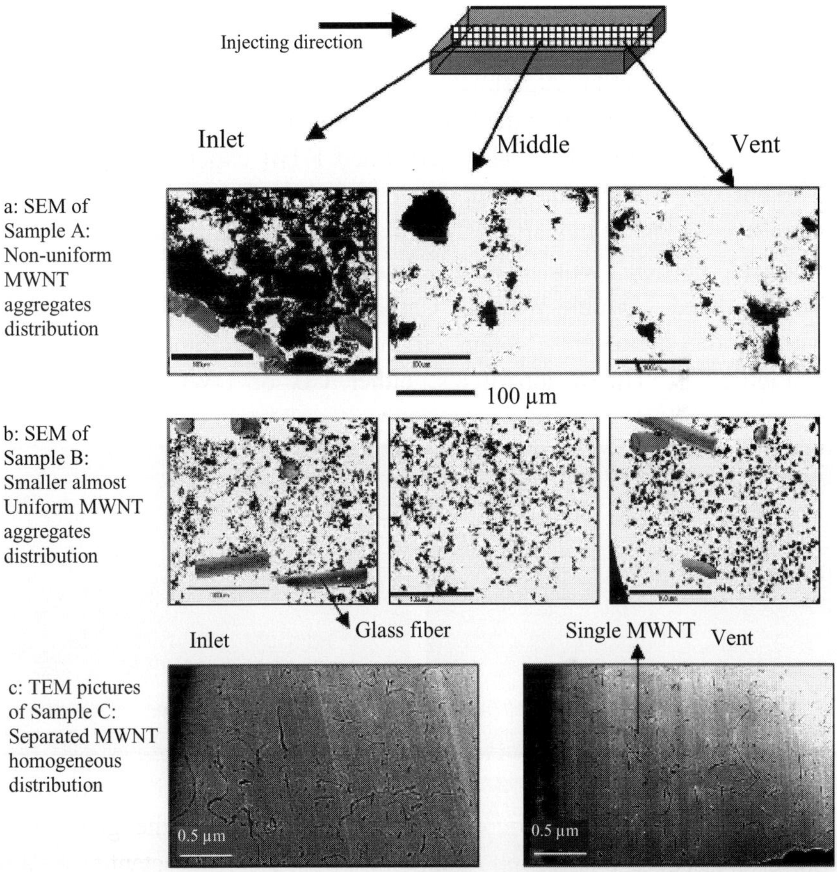

Fig. 12. TEM and SEM pictures of MWNT-resin infusing through glass fiber bed at different locations of the composite specimens: (a) sample A (mechanically stirred), (b) sample B (MWNTs were treated with surfactant), (c) sample C (MWNTs were oxidized). (Figures are taken from [44]).

length of less than 1 μm. The study showed that the MWNT/vinyl ester suspension with good dispersion state can flow through the glass fiber preform without MWNT filtration. Although the surfactant-assisted dispersion method eliminated the noticeable MWNT filtration in the preform, the MWNT aggregates with significantly reduced size could still be found in the vinyl ester. On the other hand, the oxidization method achieved extremely good dispersion result, but the MWNTs were cut into shorter segments by the acids which may reduce the MWNT reinforcing effectiveness in the polymer matrix. The CNT/resin suspension infiltration experimental results for the three different types of CNT dispersion treating methods are summarized in Figure 12.

Gojny *et al.* [59] have also performed RTM experiments by using different types of carbon nanoparticles in glass fiber reinforced epoxy composites. The nanoparticles they used were carbon black (CB, spherical particle with average diameter = 30 nm) and amino-functionalized Double-Walled Carbon Nanotubes ($DWCNT\text{-}NH_2$, average outer diameter = 2.8 nm, length = several micrometers) as shown in Figure 13. The nanoparticles, either CB or $DWCNT\text{-}NH_2$ were

Fig. 13. SEM pictures of (a) as-produced double-walled carbon nanotubes (DWCNTs) [35], (b) amino-functionalized double-walled carbon nanotubes (DWCNT-NH2), (b) Carbon Black (CB). Note that DWCNT-NH2 and CB have been used as fillers to modify the epoxy matrix of glass fiber-epoxy composites by Gojny *et al.* [59]. (Pictures are taken from [59]).

(a)

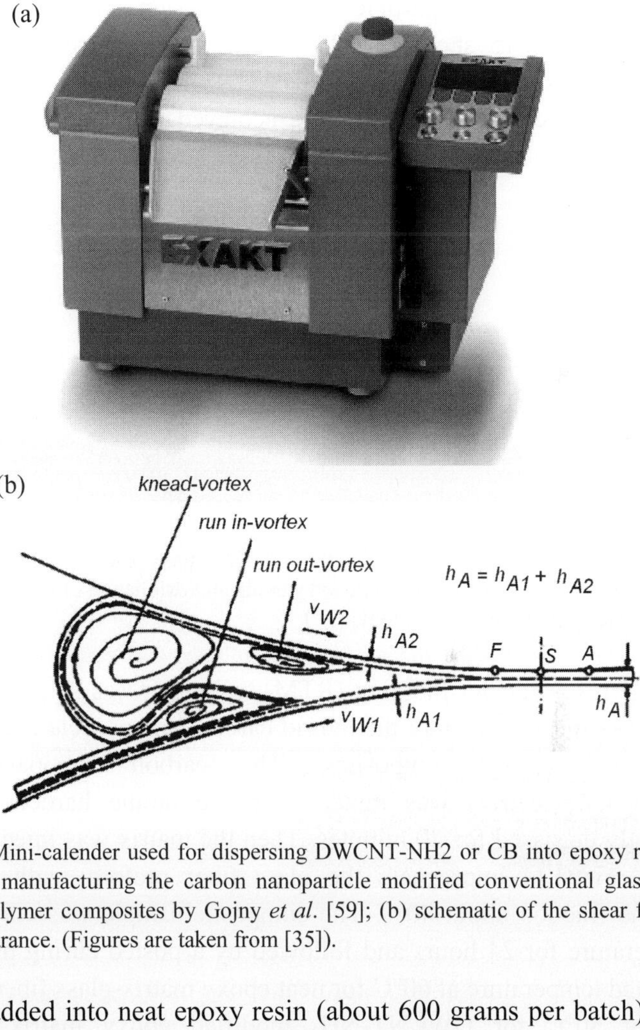

Fig. 14. (a) Mini-calender used for dispersing DWCNT-NH2 or CB into epoxy resin in the study of manufacturing the carbon nanoparticle modified conventional glass fiber reinforced polymer composites by Gojny *et al.* [59]; (b) schematic of the shear flow in the roller clearance. (Figures are taken from [35]).

manually added into neat epoxy resin (about 600 grams per batch). The nanoparticle/epoxy mixture was then fed to a three-roller mini-calender for shear mixing as schematically illustrated in Figure 14. The gap size between the ceramic rollers was 5 μm and the speeds for the first, the second, and the third rollers were set to be 20 rpm, 60 rpm, and 180 rpm, respectively. The dwell time for each batch in the mini-calender was about 2 minutes. The nanoparticle/epoxy suspension was collected for further use in manufacturing the final composites. This kind of calendering shear-mixing technique can be very effective to produce large batches of nanoparticle/epoxy suspension with good dispersion quality. Figure 15 shows the TEM picture of a cured nano-modified epoxy matrix contained 1 wt% DWCNT-NH_2; good dispersion and wetting of the DWCNT-NH_2 were achieved by using the three-rolled mini-calender shear mixing process.

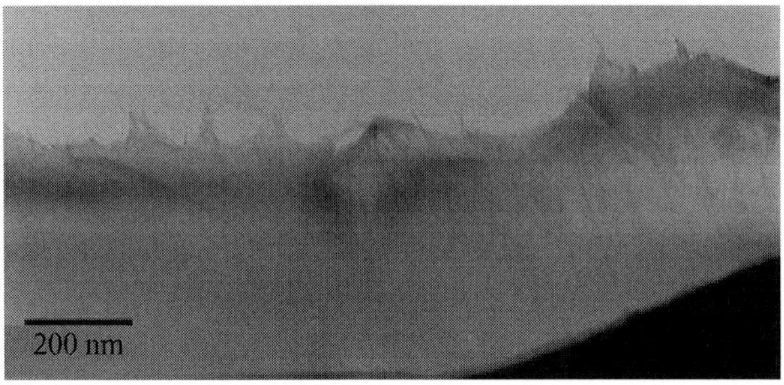

Fig. 15. TEM-picture of the cured DWCNT-NH_2/epoxy specimen processed by the calendering shear-mixing technique; good wetting and exfoliation of DWCNT-NH_2 were observed [59]. (Picture is taken from [59].)

The resin transfer molding (RTM) technique was selected by Gojny *et al.* [59] to manufacture the hybrid carbon nanoparticle and glass fiber reinforced epoxy composites. The carbon nanoparticles/epoxy suspension (matrix) was mixed with the amine hardener and was instantly degassed for 20 minutes. Then the matrix was injected into the mold cavity to impregnate the glass fiber preform with glass fiber volume fraction about 40%. The composite sample was cured at room temperature for 24 hours and followed by a posted curing process with elevated temperature at 60°C for neat epoxy matrix-glass fiber composite and at 80°C for DWCNT-NH_2 modified epoxy matrix-glass fiber composite, respectively. The produced composite samples were in the form of laminate with size of 21 cm (width) by 36 cm (length). Note that the thickness of the composite laminate sample has not been reported by Gojny *et al.* [59] possibly due to the reason that RTM mold filling and the nanoparticle infiltration is primarily in the in-plane directions and the thickness of the part was not considered as a processing issue by them. Nevertheless, the laminate size they produced is comparative to typical

RTM part size for conventional micro-fiber reinforced thermoset polymer composites. They have attempted to increase the weight fraction of carbon nanoparticles (either CB or DWCNT-NH_2) in the epoxy resin. It was found that their RTM setup can produce glass fiber/nano-modified epoxy matrix composite parts with DWCNT-NH_2 weight fraction (with respect to the epoxy) up to 0.3 wt%. However, when the load of DWCNT-NH_2 was raised to 0.5 wt%, the viscosity of the nano-modified epoxy matrix increased dramatically thus failed the RTM process. For the glass fiber composite samples with epoxy containing 0.1 wt% DWCNT-NH_2, no significant filtering effect by the glass fiber bed was observed; however, small agglomerates of DWCNT-NH_2 were found in matrix-rich areas. The DWCNT-NH_2 agglomerates may be due to the re-agglomeration caused by the resin foaming during the degassing before the resin injection (see Figure 16).

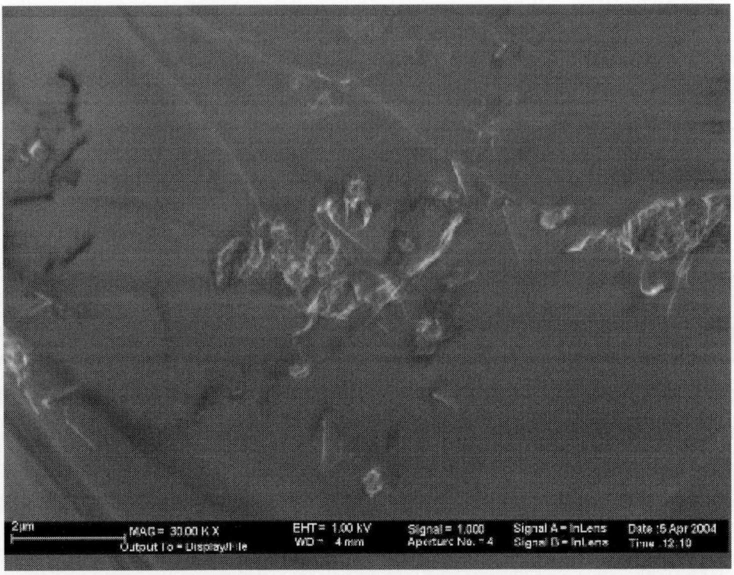

Fig. 16. SEM-picture of a fracture surface of a glass fiber composite sample containing DWCNT-NH_2 modified epoxy matrix. The DWCNT aggregates may be caused by the re-agglomeration during the degassing before the resin injection [59]. (Picture is taken from [59]).

3.1.2. *Vacuum Assisted Resin Transfer Molding (VARTM)*

In addition to typical RTM, other kinds of resin infusion processes can also be used for manufacturing the hybrid nano-/micro- fibers reinforced thermoset polymer composites. Sadeghian *et al.* [60] have shown that vacuum assisted resin transfer molding (VARTM) can be used to produce carbon nanofiber enhanced polyester-glass fiber composite parts with thickness up to 1 cm and without significant CNF filtration. The VARTM process is similar to RTM; however, the fiber preform is placed inside a cavity that is surrounded by a one-side solid mold and a flexible vacuum bag (see Figure 14). Before and during the resin infusion, vacuum is applied to the cavity from the vent(s). Thus the preform and vacuum bag are compacted and pressed against the solid mold surface by the atmospheric pressure. During the resin infusion stage, the resin flows into the mold cavity to impregnate the preform due to the pressure difference between the inlet (injection gate) at the atmospheric pressure and the vent at the vacuum pressure. After resin completely saturates the fiber preform, one keeps the vacuum at the vent and closes the inlet and allows the resin to cure. Since VARTM does not require a high injection pressure and a strong mold like RTM does, it is especially suitable for manufacturing large composite structural parts such as bridge decks, ship hulls, and wind turbine blades with low-to-mid production volume. To accelerate the resin infusion speed especially for manufacturing large composite parts, the flow distribution layer is usually used (see Figure 17). Different from the in-plane only resin filling pattern for most RTM processes, the VARTM mold filling processes usually have a 3-dimensional flow pattern. The resin first flows along the distribution layer and then flows in the thickness direction to saturate the preform under the distribution layer. The resin flow can be modeled by analytical solution [62] or numerical solution [63] based on Darcy's law and the continuity equation. However, for a composite part with complex geometry, the arrangement of distribution layer, injection gates, and vents will be very critical to the success of VARTM mold filling thus will require a reliable process design by either the traditional trial-and-error prototyping approach or the advanced process characterization, prediction, and design techniques [50,61,64].

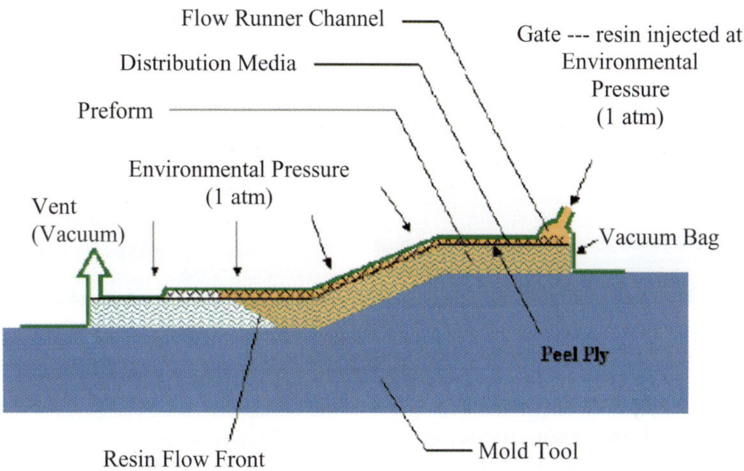

Fig. 17. Schematic of vacuum assisted resin transfer molding (VARTM) [61].

In the experimental study conducted by Sadeghian *et al.* [60], vapor-grown CNFs (Polygraf III, Applied Sciences Incorporated) with diameter of 60–150 nm and length of 30,000–100,000 nm, were dispersed into unsaturated polyester (Eastman Chemical Company) using a three-stage process. First, the CNFs were sonicated for 30 minutes to initially disperse them into acetone. Note that acetone was widely used for the nanotube dispersion process and was reported to have the least negative influence compared with other solvents such as ethanol and dimethylformamide (DMF) [65]. Then, a surfactant, BYK-191 (BYK-Chemie GmbH), at 90 wt% of CNFs was added to coat the CNFs while sonicating for another 30 minutes. A second surfactant, BYK-192, at 90 wt% of CNFs was mechanically mixed with polyester and then the polyester/BYK-192 solution was added to the CNFs/BYK-191/Acetone solution for additional 30 minutes sonication. Finally, the mixture was heated to 60°C at the atmospheric pressure to remove the acetone. The processed CNFs were then mechanically mixed with appropriate amount of polyester to achieve the desired weight fraction of CNFs in the polyester resin. Figure 18 shows the difference between the untreated CNF/resin matrix (Figure 18(a)) and the surfactant-treated CNF/resin

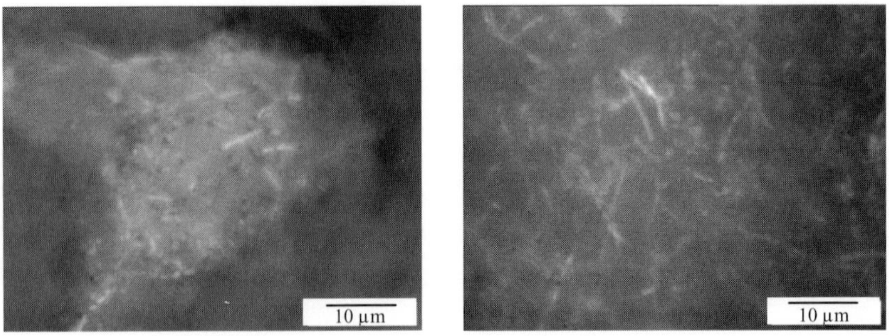

(a) (b)

Fig. 18. 1 wt% CNF dispersion in polyester resin. (a) Untreated CNF aggregate formed in the liquid polyester resin. (b) Surfactant-treated CNFs were uniformly dispersed into the liquid polyester resin. Note that in (a) some resin areas are out of focus due to the significant elevation difference between the CNF aggregate and the liquid resin [60].

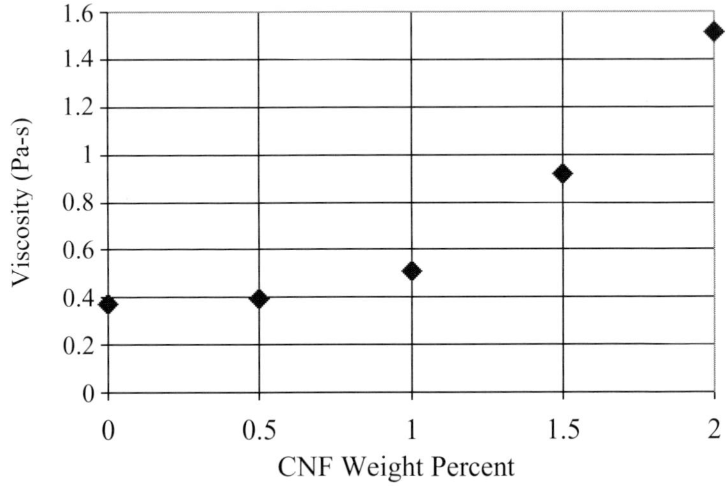

Fig. 19. CNF/polyester suspension viscosity vs. CNF concentration [60].

matrix (Figure 18(b)) observed under an optical microscope with a CVC camera and 100X objective. Large CNF aggregates can be found in the untreated matrix (see Figure 18(a)), while the surfactant-treated matrix shows uniformly dispersed CNFs (see Figure 18(b)). The dynamic viscosity of the surfactant-treated CNF/polyester suspension was measured and plotted against the CNF weight fraction as shown in Figure 19. It was observed that the slope of viscosity against CNF concentration significantly changed when the CNF concentration reached about 1.0 wt%. Such viscosity-CNF concentration slope change indicates that CNFs may begin to contact with each other when the CNF/polyester suspension has the CNF concentration higher than 1 wt%.

After dispersing the CNFs into polyester resin, Sadeghian *et al.* [60] used the VARTM process (see Figure 17) for impregnating random E-glass fiber preform (glass fiber diameter = 20 μm) with the CNF/resin suspension. Due to the large size of CNF aggregates in the untreated CNF/resin matrix (see Figure 18(a)), the VARTM process failed to infuse the untreated CNF/resin into the glass fiber preform. On the other hand, it was quite successful in infusing the surfactant-treated CNF/resin matrix (see Figure 18(b)) into the glass fiber preform and manufactured the CNF toughened polyester/glass fiber composite samples by using VARTM. Figure 20 shows the fracture surfaces of a delaminated glass fiber composite specimen (glass fiber volume fraction was about 43%) with the matrix consisting of 1 wt% CNFs toughened polyester resin. It can be seen that CNFs were well blended with polyester resin and formed a reinforcement network with the glass fibers. The fracture surface pictures were taken from two locations; Figure 20(a) shows the

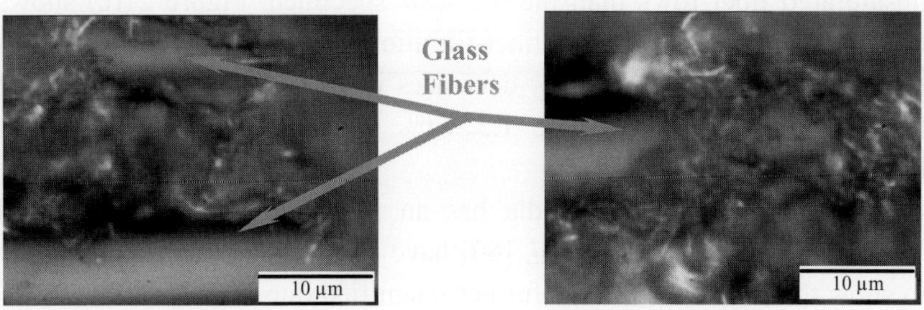

(a) (b)

Fig, 20. Optical microscopy pictures of fracture surfaces of a delaminated 1 wt% CNFs modified polyester/glass fiber composite laminate manufactured by VARTM. (a) fracture surface near the injection gate; (b) fracture surface near the vent. (Note that some out of focus areas have very significant difference in elevation.) [60]

surface near the injection gate and Figure 20(b) shows the surface near the vent. By comparing Figure 20(a) and Figure 20(b), one finds no significant CNF concentration change thus proves that CNFs were not filtered by the glass fiber bed during the VARTM process when 1wt% surfactant-treated CNFs/polyester matrix was used to infuse the glass fiber preform.

Since VARTM is a processing technique capable of producing large and complex composite parts, it will be interesting to understand the manufacturability of using VARTM to fabricate the hybrid CNF and micro-fiber reinforced thermoset polymer composites. Sadeghian *et al.* [60] used different VARTM configurations for the manufacturability study to understand the CNF filtration and micro-void formation. The through-thickness CNF filtration was first studied by conducting a set of VARTM experiments. Pictures of the composite panels of 9.2 mm thickness (16 layers of random glass fiber mats) can be seen in Figure 21 with 0.5 wt%, 1 wt%, and 1.5 wt% CNF concentrations. For manufacturing the thick-section specimens shown in Figure 21, the flow distribution layer was placed on the top of the glass fiber preform so the major component of the resin flow velocity was in the thickness direction according to the analytical VARTM flow model [62]. No CNF filtration can be seen in the 0.5 wt% and 1 wt% specimens (see Figure 21(a) and Figure 21(b) respectively); however, the 1wt% specimen has more unsaturated fiber tows than the 0.5 wt% specimen. Figure 21(c) shows the 1.5 wt% specimen has distinct filtration in the thickness direction as the CNF concentration quickly decreases while traveling away from the flow distribution media. The 1.5 wt% specimen also has the most unsaturated fiber tows.

The flow distribution media has an important role for traditional VARTM [62]. Sadeghian *et al.* [60] have conducted experimental study to show that its importance is further magnified due to the potential CNF filtration in the in-plane directions. Figure 22(a) shows a specially designed VARTM setup used in the in-plane filtration experiment. Note that a VARTM glass fiber preform usually has its in-plane permeability much higher than its through-thickness permeability [62] since the glass fibers are aligned in the in-plane directions. The color difference as a

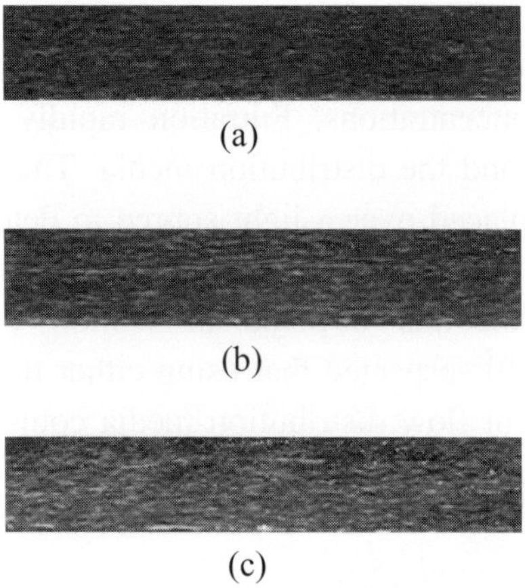

Fig. 21. Filtration in thickness direction: black areas show high CNF concentrations; grey/clear areas show low CNF concentrations due to filtration; white spots are fiber tows with poor saturation; (a) specimen with 0.5 wt% CNFs has no filtration and good saturation of fiber tows; (b) specimen with 1 wt% CNFs has no visible filtration, but has some unsaturated fiber tows; (c) specimen with 1.5 wt% CNFs has significant filtration and many unsaturated fiber tows [60].

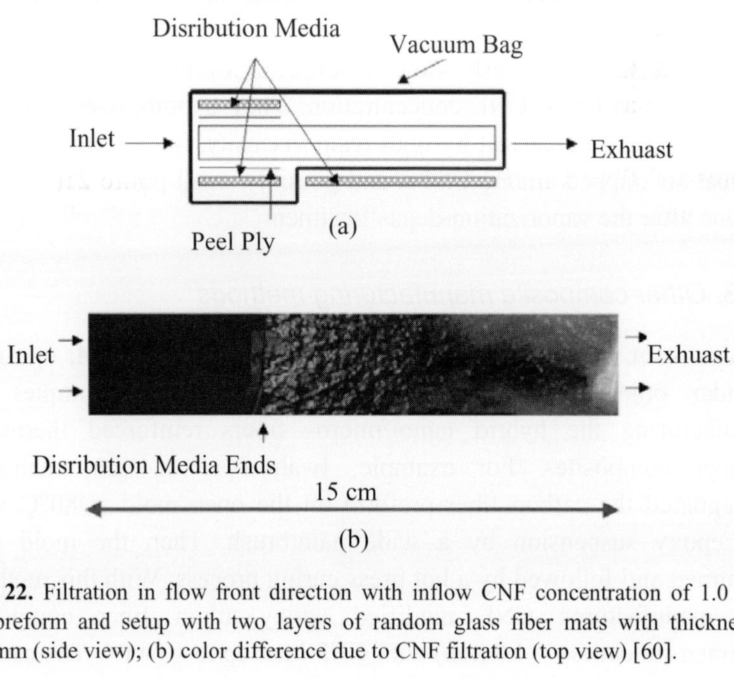

Fig. 22. Filtration in flow front direction with inflow CNF concentration of 1.0 wt%: (a) preform and setup with two layers of random glass fiber mats with thickness of 1.2 mm (side view); (b) color difference due to CNF filtration (top view) [60].

result of CNF filtration can be seen in Figure 22(b); the black area represents high CNF concentration while the clear/gray areas show resin with low CNF concentrations. Filtration rapidly occurs in the flow front direction beyond the distribution media. The thin panel shown in Figure 22(b) was placed over a light source to determine that the CNF only flowed about 2 cm beyond the distribution media without noticeable filtration. Based on this in-plane infiltration experimental results, Sadeghian *et al.* [60] suspected that using either the traditional RTM or the VARTM without flow distribution media could result in composite parts with either very limited dimensions or low CNF weight fraction in the matrix as discussed in the experimental study conducted by Fan *et al.* [44] and Gojny *et al.* [59].

Sadeghian *et al.* [60] also examined the void formation of the composites under an optical microscope and found that more micro-voids formed when CNF concentration increased. Figure 23(a) shows the composite specimen with no CNFs; thus, no visible micro-voids are found in the specimen. Figure 23(b) to Figure 23(d) show the micro-voids of the composite specimens with different CNF concentrations. The micro-voids were concentrated toward the top of the composite specimens. Figure 23 clearly shows that the size of micro-voids increases with the elevation of CNF concentration. They hypothesized that void formation could be caused by high resin viscosity, local CNF filtration, residual air trapped among CNFs, and possibly small portion of residual acetone after the vaporization-degas treatment.

3.1.3. *Other composite manufacturing methods*

Besides resin infusion processes such as RTM and VARTM, one may consider other kinds of composite manufacturing techniques for manufacturing the hybrid nano-/micro- fibers reinforced thermoset polymer composites. For example, Iwahori *et al.* [26] manually impregnated the carbon fiber preform on the open mold at 80°C with CNF/epoxy suspension by a wide paintbrush. Then the mold was vacuumed and followed by a hot press curing process. With this method, they manufactured CNF modified epoxy-carbon fiber composite specimen (thickness = 4.0 mm) with CNF load up to 10 wt%. The carbon

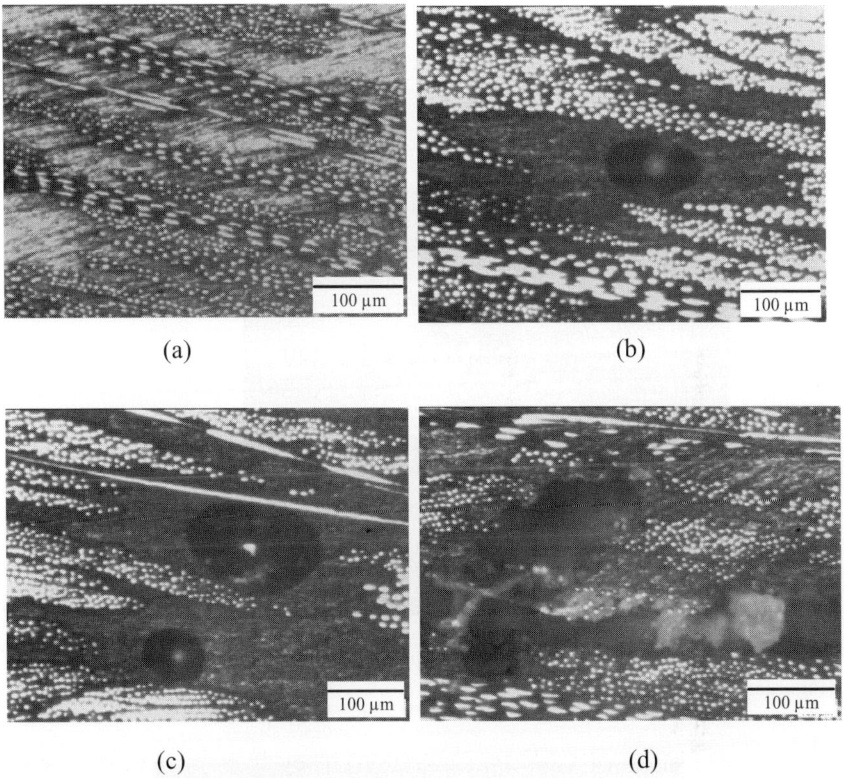

(a) (b) (c) (d)

Fig. 23. Micro-voids (with 5X objective magnification) in the (a) 0 wt% CNFs, (b) 0.5 wt% CNFs, (c) 1.0 wt% CNFs, and (d) 1.5 wt% CNFs toughened polyester-glass fiber composite specimens [60].

fiber volume fraction of the composites was about 56% and the CNFs they used have diameter of 80–100 nm. The CNFs with two different lengths, which were 500–1000 nm (AR-10) and 2500–10000 nm (AR-50), were studied in the hybrid composite manufacturing. Compared with RTM and VARTM, this manual resin impregnation-hot press technique manufactured composite samples with higher carbon nanoparticle weight fraction but with more voids and inhomogeneous CNF dispersion in the resin. Figure 24 shows micro-voids formed due to CNF agglomerates in the highly loaded (10 wt%) CNF/epoxy suspension. However, much larger voids were found in the CNF

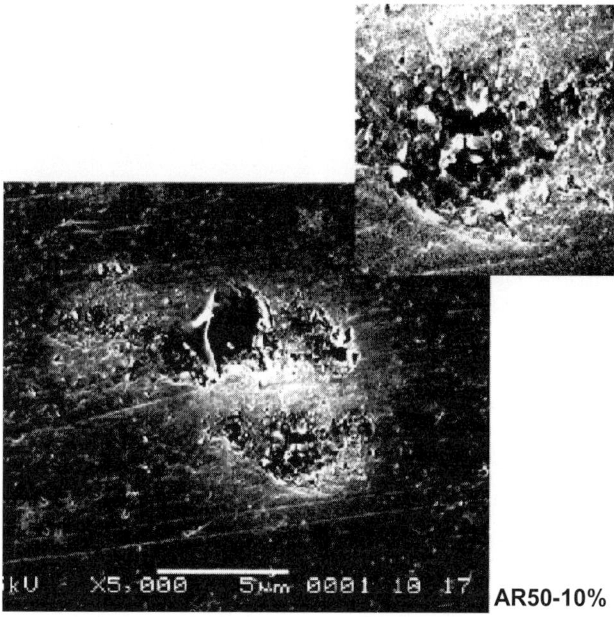

Fig. 24. Micro-voids formed in the CNF/epoxy suspension due to CNF agglomerates. (Figure is taken from [26]).

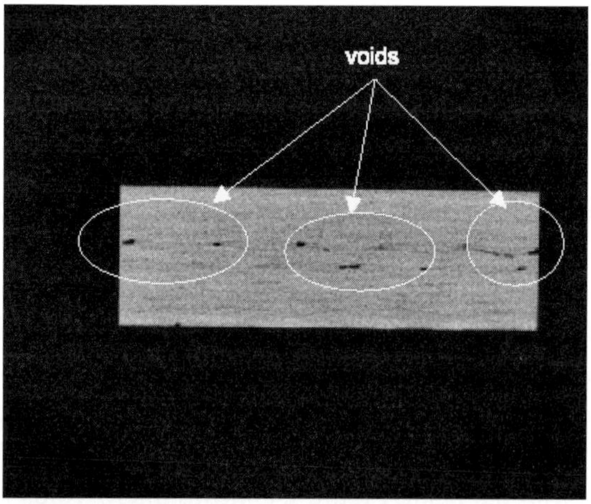

Fig. 25. Large voids in the CNF modified epoxy-carbon fiber composite specimen, which was manufactured by the hand resin impregnation-hot pressed technique, were found under the micro-focus X-ray CT. (Figure is taken from [26]).

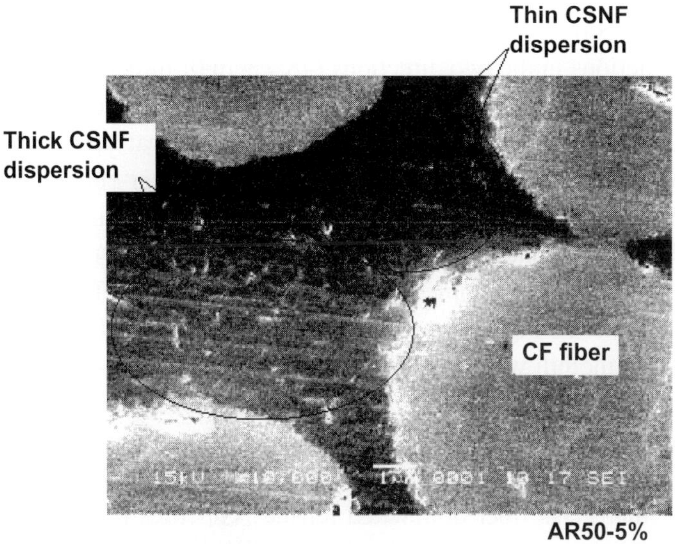

Fig. 26. The CNF dispersion in the cavity between carbon fibers was not uniform when the composite was manufactured using the hand resin impregnation-hot press process. (Figure is taken from [26]).

modified epoxy carbon fiber composites (see Figure 25), which may be introduced during the resin impregnation process they used. In addition to the void formation, it was found non-uniform CNF dispersion in the matrix between carbon fibers (see Figure 26). For future improvement in this kind of hybrid composite manufacturing, Iwahori *et al.* [26] suggested that one should consider the issues regarding the CNF agglomerates, the void introduced during the resin impregnation, and the inhomogeneous CNF dispersion in the matrix.

4. Mechanical Properties Characterization of Hybrid Nano-/ Micro- Fibers Reinforced Thermoset Polymer Composites

4.1. *Tensile properties*

The tensile properties of the hybrid nano-/micro- fibers reinforced thermoset polymer composites can be measured by regular tensile test setup. Gojny *et al.* [59] measured the in-plane tensile strengths and moduli of the 1 wt% DWCNT-NH_2 modified epoxy-glass fiber composite in 0°

and 90° directions. Note that the glass fiber preform has different fiber volume fractions in the two directions (48.7% in 0° direction and 4.8% in 90° direction). In the 0° direction, they recorded average tensile strengths of 573 MPa for the control samples (glass fiber-neat epoxy) and 536 MPa for the nanotube-modified samples (glass fiber-epoxy/DWCNT-NH_2). In the 90° direction, they measured average tensile strengths of 170 MPa for the control samples (glass fiber-neat epoxy) and 170 MPa for the nano-modified samples (glass fiber-epoxy/DWCNT-NH_2). On the other hand, the Young's moduli were also calculated. In the 0° direction, the Young's moduli were 25.5 GPa for the control samples and 26.4 GPa for the DWCNT-NH_2 modified composite samples. In the 90° direction, the Young's moduli were 12.3 GPa for the control samples and 13.2 GPa for the DWCNT-NH_2 modified samples. Based on the data, one can conclude that glass fiber dominated tensile properties (especially in the 0° direction) are not significantly changed by adding DWCNT-NH_2. In the 90° direction, since the glass fiber volume fraction is low (4.8%), the slight improvement of the hybrid composites may be related to the matrix improvement as discussed previously (see Table 2).

Iwahori *et al.* [26] measured the in-plane tensile properties improvement of CNF modified epoxy-carbon fiber (CF) composites. Two different lengths of CNFs were used in their study, i.e. AR10 (CNF length = 500–1000 nm) and AR50 (CNF length = 2500–10000 nm). The in-plane tensile strengths were measured as: control composite samples CF-epoxy (576.1 MPa), CF-epoxy/AR10-5 wt% composites (573.2 MPa), CF-epoxy/AR10-10 wt% composites (601.2 MPa), CF-epoxy/AR50-5 wt% composites (597.5 MPa), CF-epoxy/AR50-10 wt% composites (577.2 MPa). Comparing the data, it was found that the tensile strength was clearly improved for lower CNF loading (5 wt%). Furthermore, for the 5wt% CNF loading, longer CNFs yielded stronger composites. On the other hand, for high CNF loading (10 wt%), the AR50-10 wt% actually weakened the composites since it became more difficult to disperse the long CNFs in the composite system uniformly without forming voids and agglomerates. The in-plane tensile moduli were also characterized as: control composite samples

CF-epoxy composites (57.19 GPa), CF-epoxy/AR10-5 wt% composites (53.20 GPa), CF-epoxy/AR10-10 wt% composites (53.25 GPa), CF-epoxy/AR50-5 wt% composites (55.26 GPa), CF-epoxy/AR50-10 wt% composites (56.24 GPa). Surprisingly, the data shows that addition of CNFs did not improve the in-plane tensile moduli of the composite samples no matter what length and weight fraction of CNFs were mixed in the matrices. The slight decrease of the in-plane tensile moduli may be due to the defects introduced into the composites while CNFs were added. In fact, by mixing different weight fractions and lengths of CNFs with neat epoxy only, Iwahori *et al.* observed about 14%–45% increase in the tensile modulus and 6%–21% increase in the tensile strength in comparison with neat epoxy. Hence, based on the improvements of the epoxy/CNF composites and the hybrid CF-epoxy/CNF composites against their corresponding control samples, the decreasing in-plane tensile moduli of the hybrid CF-epoxy/CNF samples indicates that how to remove or minimize the defects introduced into the hybrid composites through the manufacturing process is a non-trivial issue for producing this kind of hybrid nano-/micro- fibers reinforced thermoset polymer composites.

Table 4 summarizes the in-plane tensile properties we discussed in this section. Based on the experimental results, one can conclude that the improvement in in-plane tensile properties of this kind of hybrid nano-/micro- fibers reinforced thermoset polymer composites is not significant. Regardless of the type, length, weight fraction of carbon nanoparticles added into the hybrid composites, the changes in the in-plane tensile properties are very limited in comparison with the baseline properties of conventional micro-fiber reinforced thermoset polymer composites. If one compares the carbon nanoparticle reinforced polymers (see Table 2) with the commonly used micro-fibers such as E-glass fiber and carbon fiber (see Table 1) in terms of tensile strength and modulus, one can explain why the current hybrid nano-/micro- fibers reinforced thermoset polymer composites do not have much improvement in the in-plane tensile properties from the conventional micro-fiber reinforced thermoset polymer composites.

Table 4. In-plane tensile properties of hybrid nano-/micro- fibers reinforced thermoset polymer composites.

Composite	Tensile Strength (MPa)	Improvement in Tensile Strength	Tensile Modulus (GPa)	Improvement in Tensile Modulus	Source
GF-EP (0 degree-direction)	573.0	–	25.5	–	[59]
GF-EP(90 degree-direction)	170.0	–	12.3	–	[59]
GF-EP/DWCNT-NH2-1 wt% (0 degree-direction)	536.0	–6.5%	26.4	3.5%	[59]
GF-EP/DWCNT-NH2-1 wt% (90 degree-direction)	170.0	0.0%	13.2	7.3%	[59]
CF-EP	576.1	–	57.2	–	[26]
CF-EP/CNF(AR10)-5 wt%	573.2	–0.5%	53.2	–7.0%	[26]
CF-EP/CNF(AR10)-10 wt%	601.2	4.4%	53.3	–6.9%	[26]
CF-EP/CNF(AR50)-5 wt%	597.5	3.7%	55.3	–3.4%	[26]
CF-EP/CNF(AR50)-10 wt%	577.2	0.2%	56.2	–1.7%	[26]

Table 5. In-plane compressive properties of CNF modified epoxy-carbon fiber composites [26].

Composite	Compressive Strength (MPa)	Improvement in Compressive Strength	Compressive Modulus (GPa)	Improvement in Compressive Modulus
CF-EP	446.8	–	47.0	–
CF-EP/CNF(AR10)-5 wt%	452.4	1.3%	50.2	6.7%
CF-EP/CNF(AR10)-10 wt%	510.8	14.3%	50.2	6.7%
CF-EP/CNF(AR50)-5 wt%	513.8	15.0%	49.4	5.0%
CF-EP/CNF(AR50)-10 wt%	500.9	12.1%	48.7	3.7%

4.2. *Compressive properties*

The in-plane compression tests for CNF modified epoxy-carbon fiber composites were conducted by Iwahori *et al.* [26]. Different from the conclusions drawn from the in-plane tensile properties measurement, surprisingly, the hybrid nano-/micro- fibers reinforced polymer composites have considerable improvement from the conventional carbon fiber-epoxy composites. As shown in Table 5, it was recorded that the CNFs can improve the in-plane compressive strength and compressive modulus up to 15% and 6.7%, respectively. From the in-plane compressive modulus comparison, one finds that short CNFs (AR10) yield more improvement. Furthermore, the samples (EP/CNF (AR50)-10 wt%) with higher CNF weight fraction and longer CNFs had dramatic compressive modulus drop. One possible reason for such phenomena is that the defects, including voids and CNF agglomerates, degraded the mechanical performance of the matrix thus affected the total compressive properties of the composite samples. This hypothesis, if is true, again supports that longer CNFs (AR50) will substantially raise the difficulty level in manufacturing quality composite parts.

4.3. *Interlaminar shear strength*

Since the interlaminar properties are usually dominated by matrix, it is reasonable to expect significant improvement in interlaminar properties by adding CNTs or CNFs into the polymer matrix [33,43]. Gojny *et al.* [59] conducted short beam three point bending tests to characterize the interlaminar shear strength (ILSS) for glass fiber reinforced epoxy composites containing carbon black (CB) and amino-functionalized double-walled carbon nanotubes (DWCNT-NH_2). The results listed in Table 6 show that by adding either carbon black or carbon nanotubes into the conventional micro-fiber reinforced polymer composites will positively affect the ILSS. With 0.3 wt% DWCNT-NH_2, the improvement in ILSS can reach 20%. These results suggest that the matrix dominated mechanical properties (such as the ILSS) of the hybrid nano-/micro- fibers reinforced composites should be significantly

Table 6. Interlaminar Shear Strength (ILSS) of glass fiber reinforced epoxy composites containing different carbon nanoparticles [59].

Composite	ILSS (MPa)	ILSS Improvement
GF-EP	31.8	–
GF-EP/CB-0.1 wt%	32.7	2.8%
GF-EP/CB-0.3 wt%	34.8	9.4%
GF-EP/DWCNT-NH_2-0.1 wt%	36.8	15.7%
GF-EP/DWCNT-NH_2-0.3 wt%	38.1	19.8%

improved by adding CNTs or CNFs if such trends have already been observed in nanoparticle/resin composite system and no additional manufacturing difficulty has been experienced while introducing the nanoparticles into the hybrid composite system. In addition to the matrix property enhancement, Gojny *et al.* [59] also suspected that the interface between glass fibers and the carbon nanoparticles (either CB or CNT) could also contribute to the ILSS improvement.

4.4. *Mode-I delamination resistance*

Delamination resistance is another example of matrix dominated mechanical property. Sadeghian *et al.* [0] conducted the double cantilever beam (DCB) delamination tests (see Figure 27) to find the mode-I delamination resistance (G_{IC}) of CNF toughened polyester-glass fiber composites with different CNF weight fractions. The composite specimens were manufactured using a specially designed VARTM setup as shown in Figure 28 to have uniform distribution of CNFs in the specimens and the manufactured notches for delamination to start with. The load "P" applied to the specimen and the crack development length "*a*" were recorded against the displacement "δ" as shown in Figure 29 and Figure 30, respectively.

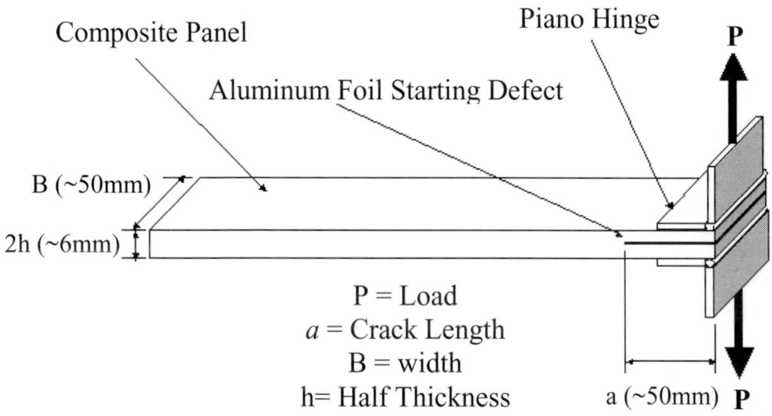

Fig. 27. Mode-I DCB delamination test configuration [60].

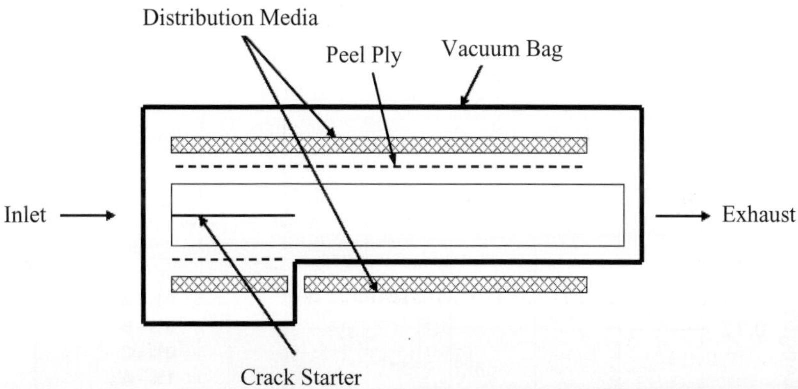

Fig. 28. The VARTM Setup for fabricating mode-I delamination specimens [60].

The DCB test results shown in Figure 29 and Figure 30 were used to evaluate the critical energy release rate, G_{IC}, in which the crack propagates perpendicular to the loads. The critical energy release rate, G_{IC}, can be calculated based on the modified beam analysis [66,67] as:

$$G_{IC} = \frac{3P\delta}{2B(a+D)}, \tag{2}$$

where δ is displacement, P and a are the corresponding load and crack length respectively, B is the specimen width. The correction factor, D, is defined as the x-axis intercept of the $C^{1/3}$ versus a plot as shown in Figure 31. Compliance, C, is calculated by $C := \delta / P$.

By combining, Figure 29, Figure 30, and Figure 31 with Equation (2), the critical energy release rate, G_{IC}, for mode-I delamination (or the resistance to mode-I delamination) can be calculated and plotted against the crack length as shown in Figure 32. In the beginning of the crack development, the G_{IC} value was not stabilized since the manufactured notch could be slightly resin-rich or could have different fiber network than the remaining portion of the composite laminate. After the beginning stage, the G_{IC} value soon stabilized and indicated that the 1 wt% CNFs toughened polyester-glass fiber composites have a higher resistance to mode-I delamination than the conventional glass fiber reinforced polyester composites (0 wt% CNFs). The improvement in G_{IC} was about 100%.

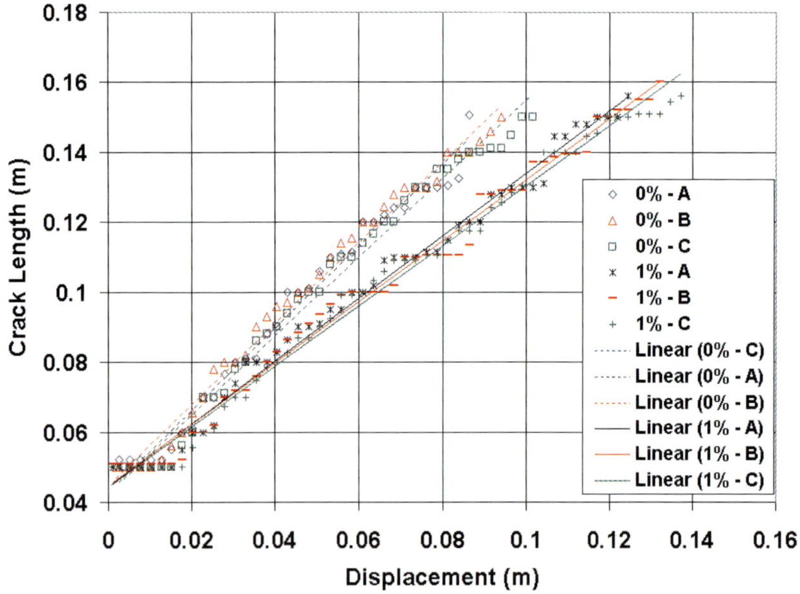

Fig. 29. Crack length-displacement relationship of DCB specimens. (Three specimens (A, B, C) for each of the scenarios, i.e., 0 wt% CNFs or 1 wt% CNFs, were tested for ensuring the measurement consistence) [60].

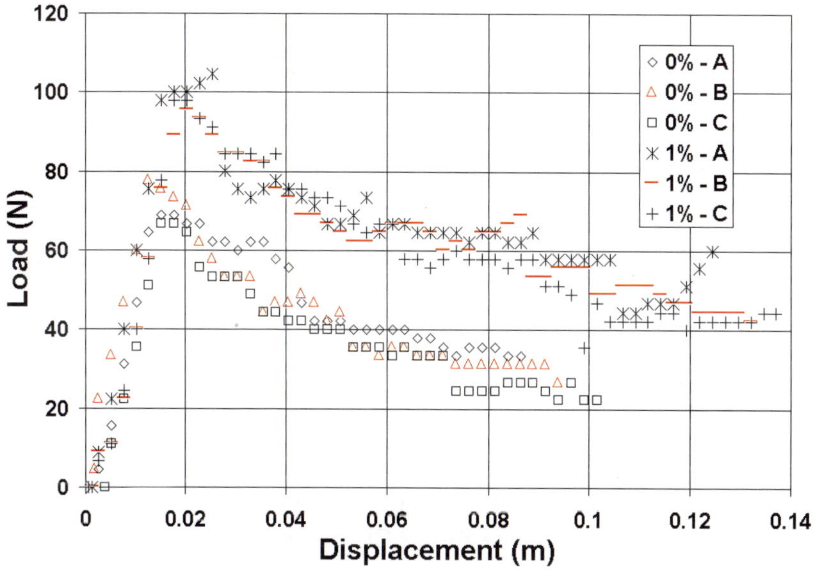

Fig. 30. Load-displacement relationship of DCB specimens. (Three specimens (A, B, C) for each of the scenarios, i.e., 0 wt% CNFs or 1 wt% CNFs, were tested for ensuring the measurement consistence) [60].

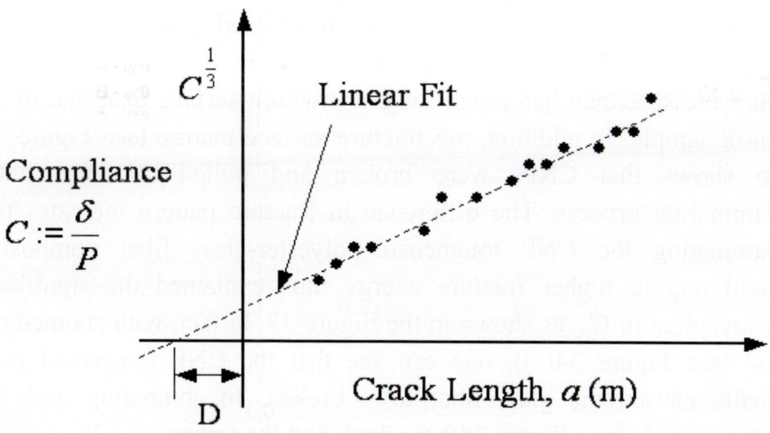

Fig. 31. Correction factor, D.

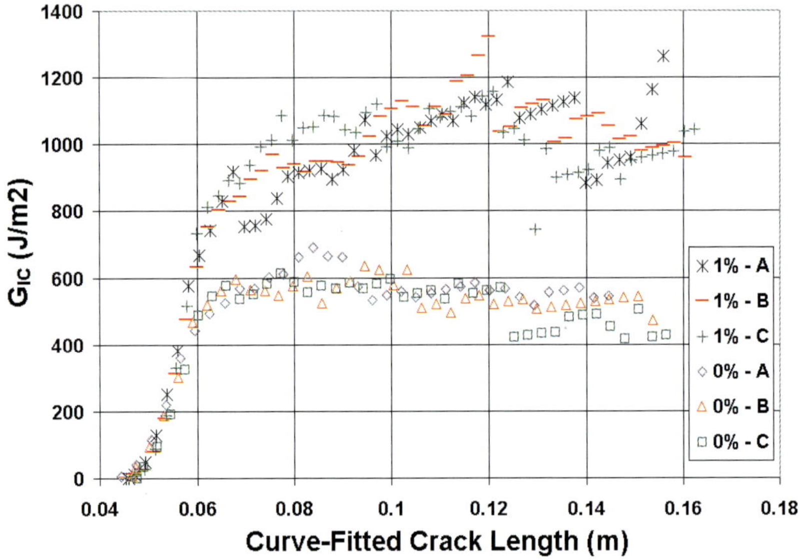

Fig. 32. Mode-I delamination resistance curves based on the curve-fitted crack length. (Three specimens (A, B, C) for each of the scenarios, i.e., 0 wt% CNFs or 1 wt% CNFs, were tested for ensuring the measurement consistence) [60].

Sadeghian *et al.* [60] also examined the fracture surfaces of the control samples and the 1 wt% CNFs toughened polyester-glass fiber composite specimens. As shown in Figure 33, the CNF toughened composite specimen had much rougher fracture surface than that of the control sample. In addition, the fracture surface picture (see Figure 33) also shows that CNFs were broken and pulled out during the delamination process. The difference in fracture pattern indicates that delaminating the CNF toughened polyester-glass fiber composites should require higher fracture energy thus explained the significant improvement in G_{IC} as shown in the Figure 32. In fact, with zoomed-out view (see Figure 34(a)), one can see that the CNF toughened resin actually caused the glass fiber tows broken. In comparing with the control sample (see Figure 34(b)) which had the failure mainly occurred in the polymer matrix, the CNF toughened composite sample should have stronger resistance to mode-I delamination.

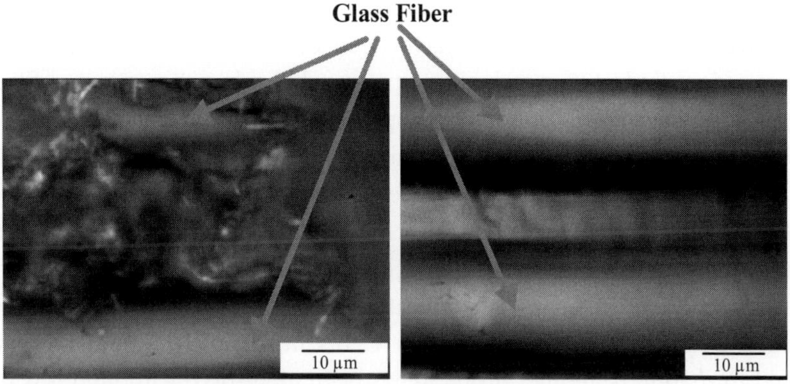

Fig. 33. The fracture surfaces of mode-I delamination of the composite specimens with matrix contains 1 wt% CNFs (left) and the composite specimens with neat resin matrix (right) with 100X objective magnification (Note that some out of focus areas have very significant difference in elevation) [60].

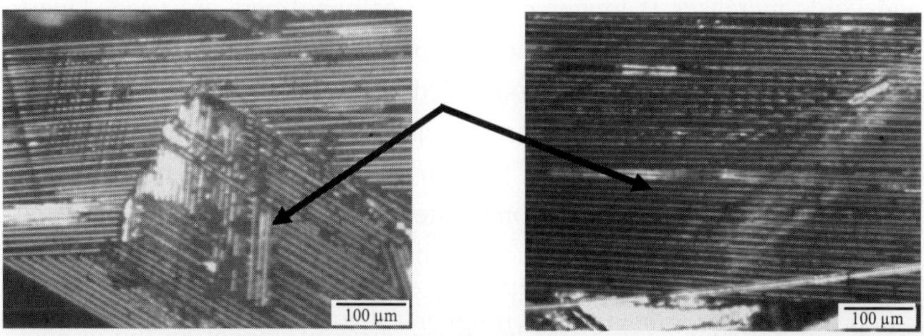

(a) (b)

Fig. 34. (a) Glass fiber tows were broken during delamination test of the 1 wt% CNFs toughened composites. (b) No significant glass fiber tow failure was observed after the delamination test for the pure polymer matrix (0 wt%CNFs) composites [60].

5. Conclusions

In this chapter, we reviewed the use of carbon nanotubes (CNTs) and carbon nanofibers (CNFs) in micro-fiber reinforced thermoset polymer composites. To manufacture such hybrid nano-/micro- fibers reinforced thermoset polymer composites, three different types of manufacturing processes including RTM, VARTM, and manual resin impregnation followed by hot press curing have been discussed. Several challenges were identified for producing quality hybrid nano-/micro- fibers reinforced thermoset polymer composite parts. The agglomeration of CNTs or CNFs in the polymer during the manufacturing process can either cause the nanoparticles to be filtered by the micro-fiber bed or weaken the matrix system and the final composite part. Using appropriate dispersion technique, chemical treatment, and low carbon nanoparticle weight fraction usually reduces the chance to have undesired agglomerates of CNTs or CNFs in the cured composite part. Void formation is another problem has been identified. Although it is not clear what are the exact reasons causing the void formation in the hybrid nano-/micro- fibers reinforced thermoset polymer composites, it is found that the void formation in the composites will be reduced by lowering the nanoparticle weight fraction. Moreover, the chance to have agglomerates and voids increases if longer nanotubes or nanofibers are used. Due to the defects, such as voids and agglomerates, introduced during the manufacturing process, it is concluded that the mechanical properties may not be monotonically improved as one increasing the length and weight fraction of CNTs or CNFs in the hybrid nano-/micro- fibers reinforced thermoset polymer composites.

In terms of in-plane mechanical properties improvement, it was found that the addition of CNTs or CNFs in the micro-fiber reinforced thermoset polymer composites has no significant impact on the in-plane tensile properties. Sometimes, the defects caused by adding CNTs or CNFs could weaken the composites with respect to tensile properties. On the other hand, the in-plane compressive properties will be considerably improved by adding CNTs or CNFs into the matrix system. For instance, by adding 10 wt% CNFs into epoxy-carbon fiber composites, it was measured about 14.3% and 6.7% improvement in in-plane compressive strength and modulus, respectively.

It is expected that significant improvement in the matrix-dominated mechanical properties will be achieved by adding CNTs and CNFs into the micro-fiber reinforced thermoset polymer composites. Experimental studies validated the positive effects of adding CNTs and CNFs into the composite system in terms of interlaminar shear strength (ILSS) and mode-I delamination resistance (G_{IC}). The ILSS was observed to increase about 20% when 0.3 wt% of amine-functionalized double-walled carbon nanotubes (DWCNT-NH_2) were infused together with epoxy into glass fiber preform via RTM. On the other hand, the improvement in mode-I delamination resistance (G_{IC}) can reach about 100% when 1 wt% CNFs toughened polyester was infused into glass fiber preform by VARTM.

The experimental data indicate that CNTs and CNFs can modify the polymer matrix thus change the mechanical properties of the hybrid nano-/micro- fibers reinforced thermoset polymer composites from the baseline properties of the conventional micro-fiber reinforced thermoset polymer composites. The property improvement strongly depends on the role of matrix in the composite system when a mechanical loading condition is applied to the hybrid composite system. It is believed that the mechanical property limits of this kind of hybrid nano-/micro- fibers reinforced thermoset polymer composites have not been reached or confidently assessed since there is no firmly validated process for manufacturing this kind of hybrid composite with high quality and high nanoparticle weight fraction.

6. Outlook

The true advantages of the hybrid nano-/micro- fibers reinforced thermoset polymer composites have not been completely realized since there are many manufacturing issues need to be solved before one can confidently assess the properties of this kind of new material. To improve the manufacturing processes, the discovery of the fundamental phenomena between the nano-scale and the micro-scale is essential. The experiments to observe the rheological, thermal, chemical, and electrical behaviors as well as their interactions are critical to the knowledge development. Supported by the experimental data, theories can be proposed to explain the data and then be used to design better experiments. The accumulated knowledge will be used to achieve the goals in: uniform resin-nanoparticle dispersion, stronger resin-nanoparticle interfacial bonding, better alignment of CNTs or CNFs, higher weight fraction of CNTs or CNFs, and reduced void formation.

To effectively tailor the mechanical properties of the hybrid nano-/micro- fibers reinforced thermoset polymer composites. It is desired to have micro-mechanics models to help us to understand the interactions between the nanoparticles, the resin, the nanoparticle/resin interface, and the micro-fiber system. With improved manufacturing technologies, more material properties will be characterized and then be used to refine the micro-mechanics models. Such micro-mechanics models will be used to predict the mechanical properties and guide one to choose effective combination of nanoparticles, resin, chemical treatment, and micro-fiber system.

With future development in manufacturing process and properties characterization, the advantages of using the hybrid nano-/micro- fibers reinforced thermoset polymer composites will be explored gradually. Based on the extraordinary properties of CNTs and CNFs, the hybrid nano-/micro- fibers reinforced thermoset polymer composites should be able to achieve comparable mechanical properties in all directions (i.e., 1, 2, and 3 directions) if the manufacturing challenges were overcome. All the current matrix-dominated material property limitations such as interlaminar strength, fractural performance, and reliability can be significantly improved. In addition to the mechanical properties, significant improvements in the electrical and thermal properties are also possible. This will provide more degrees of freedom in tailoring the hybrid composites for specific functions. Consequently, the hybrid composite can be used as a multi-functional material that possesses precisely tailored electrical, thermal, and mechanical properties in all directions.

References

1. Iijima, S., Helical microtubules of graphite carbon, Nature, 354, 56–58 (1991).
2. Ebbesen, W.T., Carbon Nanotubes Preparation and Properties, CRC Press, Boca Raton, FL (1997).
3. Mintmire, J.W., White, C.T., "Electronic and structural properties of carbon nanotubes." Carbon, 33, 893–902 (1995).
4. Ajayan, P.M., Nanotubes from carbon, Chem. Rev., 99, 1787–1799 (1999).

5. Yudasaka, M., R. Kikuchi, T. Matsui, Y. Ohki, S. Yoshimura, E. Ota, "Specific conditions for Ni catalyzed carbon nanotube growth by chemical vapor deposition", Applied Physics Letters, 67, 2477–2479 (1995).
6. Hernadi, K., A. Fonseca, J.B. Nagy, D. Bernaertz, J. Riga, A. Lucas, "Catalytic synthesis and purification of carbon nanotubes", Synthetic metals, 31–34 (1996).
7. Chen, P., H.-B. Zhang, G.D. Lin, Q. Hong, K.R. Tsai, "Growth of carbon nanotubes by catalytic decomposition of CH4 or CO on a Ni-MgO catalyst", Carbon, 35, 1495–1501 (1997).
8. Yudasaka, M., T. Komatsu, T. Ichinashi, Y. Achiba, S. Iijima, "Pressure dependence of the structures of carbonaceous deposits formed by laser ablation on targets composed of carbon, nichel and cobalt", Journal of physical chemistry B, 102, 4892–4896 (1998).
9. Kokay, F., K. Takahashi, M. Yudasaka, R. Yamada, T. Ichihashi, S. Iijima, "Growth dynamics of single-wall carbon nanotubes synthetized by CO2 laser vaporization", Journal of physical chemistry B, 103, 4346–4351 (1999).
10. Liu, J., A.G. Rinzler, H. Dai, J.H. Hafner, R.K. Bradley, P.J. Boul, A. Lu, T. Iverson, K. Shelimov, C.B. Huffman, F. Rodriguez-Macias, Y.S. Sheon, T. Randall Lee, D.T. Colbert, R.E. Smalley, "Fullerene pipes", Science, 280, 1253–(1998).
11. Thess, A., R. Lee, P. Nikolaev, H. Dai, P. Petit, J. Robert, C. Xu, Y.H. Lee, S.G. Kim, D.T. Colbert, G. Scuseria, D. Tomanek, J.E. Fischer, R.E. Smalley, "Cristalline ropes of metallic carbon nanotubes", Science, 273, 483–487 (1996).
12. Rinzler, A.G., J. Liu, H. Dai, P. Nikolaev, C.B. Huffman, F.J. Rodriguez-Macias, P.J. Boul, A. Lu, D. Heymann, D.T. Colbert, R.S. Lee, J.E.F ischer, A.M. Rao, P.C. Eklund, R.E. Smalley, "Large scale purification of single-wall carbon nanotubes: process, product and characterization, Applied physics A, 67, 29–37 (1998).
13. Bandow, S., A.M. Rao, K.A. Williams, A. Thess, R.E. Smalley, P.C. Eklund, "Purification of single-wall carbon nanotubes by microfiltration", Journal of physical chemistry B, 101, 8839–8842 (1997).
14. Schlittler, R.R., Seo, J.W., Gimzewski, J.K., Durkan, C, Saifullah, M.S.M., Welland, M.E., "Single crystals of single-walled carbon nanotubes formed by self-assembly," Science, 292, 1136–1139 (2001).
15. Zheng, G.B., Sano, H., Uchiyama Y, "New structure of carbon nanofibers after high-temperature heat-treatment." Carbon, 41, 853–856 (2003).
16. Ma, H., Zeng, J., Realff, M.L., Kumar, S., Schiraldi, D.A., "Processing, structure, and properties of fibers from polyester/carbon nanofiber composites." Composites Science and Technology, 63, 1617–1628 (2003).
17. Tibbetts, G.G., Gorkiewicz, D.W., Hammond, Jr D.C.. Apparatus for forming carbon fibers. U.S. Patent No. 5,024, 818 (1991).
18. Zheng, G.B., Kouda, K., Sano, H., Uchiyama, Y., Shi, Y.F., Quan, H.J., "A model for the structure and growth of carbon nanofibers synthesized by the CVD method using nickel as a catalyst." Carbon, 42, 635–640 (2004).
19. Yoon, Y.J., Baik, H.K., "Catalytic growth mechanism of carbon nanofibers through chemical vapor deposition." Diamond and Related Materials, 10, 1214–1217 (2001).

20. Kamimura, K., Matsumoto, Y., Oo, M.T., Nakao, M., Onuma, Y., "SEM and TEM observation of carbon nano-fibers prepared by hot filament assisted sputtering." Mol Cryst Liq Cryst, 340, 713–717, (2000).
21. Ono, H., Oya, A., "Preparation of highly crystalline carbon nanofibers from pitch/polymer blend." Carbon, 44, 682–686, (2006).
22. Zou, G., Zhang, D., Dong, C., Li, H., Xiong, K., Fei, L., Qian, Y., "Carbon nanofibers: Synthesis, characterization, and electrochemical properties." Carbon, 44, 828–832 (2006).
23. Tu, J.P., Zhu, L.P., Hou, K., Guo, S.Y., "Synthesis and frictional properties of array film of amorphous carbon nanofibers on anodic aluminum oxide." Carbon, 41, 1257–1263, (2003).
24. Ci, L., Li, Y., Wei, B., Liang, L., Xu, C., Wu, D., "Preparation of carbon nanofibers by the floating catalyst method." Carbon, 38, 1933–1937, (2000).
25. Endo, M., Kim, Y.A., Hayashi, T., Yanagisawa, T., Muramatsu, H., Ezaka, M., Terrones, H., Terrones, M., Dresselhaus, M.S., "Microstructural changes induced in ''stacked cup'' carbon nanofibers by heat treatment." Carbon, 41, 1941–1947 (2003).
26. Iwahori, Y., Ishiwata, S., Sumizawa, T., Ishikawa, T., "Mechanical properties improvements in two-phase and three-phase composites using nano-fiber dispersed resin." Composites: Part A, 36, 1430–1439 (2005).
27. Lu, J.P., "Elastic properties of single and multilayered nanotubes." J. Phys Chem Solids, 58, 1649–1652 (1997).
28. Wong, E.W., Sheehan, P.E., Lieber, C.M., "Nanobeam mechanics: elasticity, strength, and toughness of nanorods and nanotubes." Science, 277, 1971–1975 (1997).
29. Estimated CNF Properties: http://www.apsci.com/ngm-pyro1.html.
30. Qian, D., Dickey, E.C., Andrews, R., Rantell, T., "Load transfer and deformation mechanisms in carbon nanotube-polystyrene composites." Appl Phy Lett, 76(20) 2868–2870 (2000).
31. Schadler, L.S., Giannaris, S.C., Ajayan, P.M., "Load transfer in carbon nanotube epoxy composites." Appl Phys Lett, 73, 3842–3844 (1998).
32. Wagner, H.D., Lourie, O., Feldman, Y., Tenne, R., "Stress-induced fragmentation of multiwall carbon nanitubes in a polymer matrix." Appl Phys Lett, 72, 188–190 (1998).
33. Hsiao, K.T., Alms, J., Advani, S.G., "Use of epoxy/multiwalled carbon nanotubes as adhesives to join graphite fibre reinforced polymer composites," Nanotechnology, 14, pp. 791–793 (2003).
34. Cooper, C.A., Ravich, D., Lips, D., Mayer, J., Wagner, H.D., "Distribution and alignment of carbon nanotubes and nano fibrils in a polymer matrix." Composites Science and Technology 62, 1105 –1112 (2002).
35. Gojny, F.H., Wichmann, M.H.G., Kopke, U., Fiedler, B., Schulte, K., "Carbon nanotube-reinforced epoxy-composites:enhanced stiffness and fracture toughness at low nanotube content." Composites Science and Technology, 64, 2363–2371 (2004).

36. Sandler, J., Werner, P., Shaffer, M.S.P., Demchuk, V., Altstädt, V., Windle, A.H., "Carbon-nanofibre-reinforced poly(ether ether ketone) composites." Composites Part A: Applied Science and Manufacturing, 33(8), 1033–1039 (2002).
37. Kuriger, R.J., Alam, M.K., Anderson, D.P., Jacobsen, R.L., "Processing and characterization of aligned vapor grown carbon fiber reinforced polypropylene." Composites Part A: Applied Science and Manufacturing, 33 (1), 53–62 (2002).
38. Halpin, J.C,, Tasi, S.W., "Environmental factors in composite material design." US Air Force Materials Laboratory Report, AFML-TR 67–432 (1969).
39. Yakobson, B.I., Smalley, R.E., "Fullerence nanotubes: C1,000,000 and beyond; Am Sci, 85, 324–337 (1997).
40. Falvo, M.R., Clary, C.J., Taylor, R.M., Chi, V., Brook, F.P., Washburn, S., Superfine, R., "Bending and buckling of carbon nanotubes under large strain." NATURE 389 (6651): 582–584 (1997).
41. Lourie, O., Cox, D.M., Wagner, H.D., "Buckling and collapse of embedded carbon nanotubes." Phys Rev Lett., 81(8): 1638 –1641 (1998).
42. Bower, C., Rosen, R., Jin, L., Han, J., Zhou, O., "Deformation of car bon nanotubes in nanotube-polymer composites." Appl Phys Lett., 74(22): 3317–3319 (1999).
43. Meguid, S.A., Sun, Y., "On the tensile and shear strength of nano-reinforced composite interfaces." Materials and Design 25, 289–296 (2004).
44. Fan, Z., Hsiao, K.T., Advani, S.G., "Experimental investigation of dispersion during flow of multi-walled carbon nanotube/polymer suspension in fibrous porous media." Carbon, 42, 871 –876 (2004).
45. Advani, S., Simacek, P., "Modeling and simulation of flow, heat transfer, and cure (Ch:8). In: T.M. Kruckenberg, R. Paton (eds.), Resin Transfer Moulding for Aerospace Structures. Norwell, MA: Kluwer Academic Publishers, pp. 225–281 (1998).
46. Young, W., "Gate location optimization in liquid composite molding using genetic algorithms", Journal of Composite Materials, Vol. 28, No. 12, pp. 1098–1113 (1994).
47. Mathur, R., Fink, B.K., Advani, S.G., "Use of genetic algorithms to optimize gate and vent locations for the resin transfer molding process" Polymer Composites, Vol. 20, No. 2, pp. 167–178 (1999).
48. Hsiao, K.T., Advani, S.G., "Flow sensing and control strategies to address race-tracking disturbances in resin transfer molding---Part I: design and algorithm development," Composites Part A: Applied Science and Manufacturing, 35(10), 1149–1159 (2004).
49. Devillard, M., Hsiao, K.T., Advani, S.G., "Flow sensing and control strategies to address race-tracking disturbances in resin transfer molding---Part II: automation and validation," Composites Part A: Applied Science and Manufacturing, 36(11), 1581–1589, 2005.
50. Devillard, M., Hsiao, K.T., Gokce, A., Advani, S.G., "On-line characterization of bulk permeability and race-tracking during the filling stage in resin transfer molding process," Journal of Composite Materials, 37(17), pp. 1525–1541 (2003).

51. Li, M., Tucker, III C.L., "Optimal Curing for Thermoset Matrix Composites: Thermochemical and Consolidation Considerations", Polymer Composites, Vol. 23, No. 5, pp. 739–757 (2002).
52. Antonucci, V., Giordano, M., Hsiao, K.T., Advani, S.G., "A methodology to reduce thermal gradients due to the exothermic reactions in composites processing," International Journal of Heat and Mass Transfer, Vol. 45, pp. 1675–1684 (2002).
53. Kranbuehl, D.E., Kingsley, P., Hart, S., Hasko, G., Dexter, B., Loos, A.C., "In situ sensor monitoring and intelligent control of the resin transfer molding process", Polymer Composites, Vol. 15, No. 4, pp. 299–305 (1994).
54. Kessler, M.S., White, S.R., "Cure Kinetics of the Ring-Opening Metathesis Polymerization of Dicyclopentadiene", Journal of Polymer Science: Part A: Polymer Chemistry, Vol. 40, pp. 2373–2383 (2002).
55. Hsiao, K.T., Little, R., Restrepo, O., Minaie, B., "A Study of Direct Cure Kinetics Characterization During Liquid Composite Molding," Composites Part A: Applied Science and Manufacturing, (in press, online available in Science Direct).
56. Gong, X., Liu, J., Baskaran, S., Voise, R.D., Young, J., "Surfactant-assisted processing of carbon nanotube/polymer composites." Chem Mater, 12, 1049–1062 (2000).
57. Shaffer, M.S.P., Fan, X., Widle, A.H., "Dispersion and packing of CNT." Carbon, 36, 1603–1612 (1998).
58. Esumi, K., Ishigami, M., Nakajima, A., Sawada, K., Honda, H., "Chemical treatment of CNT." Carbon, 34, 279–298 (1996).
59. Gojny, F.H., Wichmann, M.H.G., Fiedler, B., Bauhofer, W., Schulte, K., "Influence of nano-modification on the mechanical and electrical properties of conventional fibre-reinforced composites," Composites: Part A, 36, 1525–1535 (2005).
60. Sadeghian, R., Gangireddy, S., Minaie, B., Hsiao, K.T., "Manufacturing carbon nanofibers toughened polyester/glass fiber composites using vacuum assisted resin transfer molding for enhancing the mode-I delamination resistance," Composites Part A: Applied Science and Manufacturing, (in press, online available in Science Direct).
61. Hsiao, K.T., Devillard, M., Advani, S.G., "Simulation Based Flow Distribution Network Optimization for Vacuum Assisted Resin Transfer Molding Process," Modeling and Simulation in Materials Science and Engineering, 12(3), pp. S175–S190 (2004).
62. Hsiao, K.T., Mathur, R., Advani, S.G., Gillespie, Jr. J.W., Fink, B.K., "A Closed Form Solution for Flow During the Vacuum Assisted Resin Transfer Molding Process", ASME J. of Manufacturing Science and Engineering, Vol. 122, No. 3, pp.463–475 (2000).
63. Simacek, P., Advani, S.G., "Desirable features in mold filling simulations for liquid molding processes," Polymer Composites, 25, 355–367 (2004).
64. Hsiao, K.T., Gillespie, Jr. J.W., Advani, S.G., Fink, B.K., "Role of Vacuum Pressure and Port Locations on Flow Front Control for Liquid Composites Molding Processes," Polymer Composites, Vol. 22, No. 5, pp. 660–667 (2001).
65. Lau, K.T., Lu, M., Lam, C.K., Cheung, H.Y., Sheng, F.L., Li, H.L., "Thermal and mechanical properties of single-walled carbon nanotube bundle-reinforced epoxy

nanocomposites: the role of solvent for nanotube dispersion," Composites Science and Technology, 65, 719 (2005).

66. Hashemi, S., Kinloch, A.J., Williams, J.G., "Mechanics and mechanisms of delamination in poly (ether sulphone) fiber composite," Compos Sci & Technol, 37(4), 429 (1990).
67. Kuboki, T., Jar, P.Y.B., Forest, T.W., "Influence of interlaminar fracture toughness on impact resistance of glass fiber reinforced polymers," Compos Sci & Technol, 63(7), 943 (2003).

CHAPTER 5

Shear Rheology of Nanofiber Suspensions and Nanofiber/Polymer Melt Composites

Jianhua Xu[1,*], Yingru Wang[2], Christopher Kagarise[1], Kurt W. Koelling[1] and Stephen E. Bechtel[2]

[1]*Department of Chemical and Biomolecular Engineering*
The Ohio State University, 140W 19th Ave, Columbus, OH 43210, USA
**xu.212@osu.edu*
[2]*Department of Mechanical Engineering*
The Ohio State University, USA

1. Introduction

New composites consisting of nano-sized reinforcing particles in polymer matrices are now commercially available. Compared to composites with traditional (i.e. micron-sized) fillers, polymer nanocomposites exhibit significantly enhanced performance properties, including increased strength, stiffness, barrier properties, and heat resistance, without a significant loss in impact strength [1–19]. However, these improvements come with a cost in processing: small additions of the high surface-area-to-volume-ratio nanoclays and nanotubes have been observed to significantly increase the viscosity and elasticity of polymer melts and solutions, complicating the extrusion and injection molding of polymer nanocomposite products. In addition, processing effects nanoparticle orientation and structure development, and hence performance properties of the product materials.

In this chapter we discuss research that is working toward a comprehensive characterization of the non-linear rheology in shear of nanoparticle suspensions and nanoparticle/polymer melt composites. This characterization will assist in the creation of simulation tools for control and optimization of the processing of nanocomposites.

We focus on suspensions and composites in which the nanoparticles are carbon nanotubes and carbon nanofibers. Carbon nanotubes (CNTs), first studied in detail by Iijima [20], have diameters of approximately 1 nm and lengths on the order of 0.1–10 μm. The modulus of a carbon nanotube is as high as 5.5 TPa, with a 40% elongation before failure [21]. Vapor-grown CNFs, which have typical diameters of 100 to 200 nm, are a lower-cost alternative to carbon nanotubes. Since CNFs have similar structures and physical properties as the nanotubes, the polymer composites filled with CNFs have similar properties as the ones filled with carbon nanotubes [22–29].

Carbon nanotubes and nanofibers have recently been incorporated into several engineering polymers, including polycarbonate and nylon, to produce nanocomposites which maintain high mechanical properties while exhibiting uniform surface conductivity. These polymer/nanotube composites have already started to be incorporated into electronic wafer processing equipment, disk drive components, clean room applications, and automotive applications where static discharge is an important concern. Other promising applications of polymer/nanotube composites are improved electrodes for batteries and conductive fibers with high strength and stiffness. Carbon nanotubes are being used as a conductive filler in plastics. The conductive resin is being used to produce injection molded components which may be painted using electrostatic methods, and computer housing components which must avoid electrostatic buildup.

In contrast to the abundant literature on the performance properties of CNT/polymer and CNF/polymer composites, far less work has been published on their processing properties. There are several reports in the literature describing the rheology of CNFs or CNTs in polymer composites: Lozano *et al.* [30] reported the small amplitude oscillatory shear rheology of CNF/polyethylene composites at CNF concentrations up to 30 wt%. Du *et al.* [31] prepared single-walled CNT/poly(methyl methacrylate) composites and studied their linear viscoelastic response as well as their electrical conductivity. Abdel-Goad *et al.* [32] characterized polycarbonate/ multiwalled CNT composites by also using small amplitude oscillatory shear. Kinloch *et al.* [33] studied the behavior in dynamic and steady shear of CNTs suspended in an aqueous solution.

Hough *et al.* [34] stabilized CNTs in an aqueous solvent using surfactants and used dynamic shear to determine the effect of CNT volume fraction on the onset of rheological percolation. This is the point at which the rheological properties of the suspension change dramatically due to particle interactions.

Although there is a lack of literature for nanoscale particle composites, the area of macroscopic particle composites contains an abundance of research and textbooks, e.g. "The Structure and Rheology of Complex Fluids" by R.G. Larson [35]. In general, the effects of macroscale particles in a composite are much different from those of nanoscale particles.

2. Rheology Background

Rheology is the study of the deformation and flow of matter with both fluid and solid-like behavior. The primary interest of this chapter is how the presence of nanoparticles affects the shear flow of a solvent or polymer matrix. Shear flows are common in the processing of polymers. In this section we discuss shear flows and review six material functions important for the characterization of the shear rheology of nanocomposites. These are the steady and transient viscosities, first and second normal stress differences, and storage and loss moduli.

A shear flow is one in which a fluid has a velocity component in only one direction (without loss of generality the x direction) and a velocity gradient normal to the flow direction (w.l.o.g. the z direction). This flow is accomplished by confining the fluid between two parallel planes, one stationary and the other moving with speed v_0, as illustrated in the figure below. The shear force component in the x direction necessary to create this motion is labeled F_x. That force divided by the area, A, over which it is applied is the shear stress τ_{zx},

$$\tau_{zx} = \frac{F_x}{A}.$$

The flow in shear is characterized by the shear rate, $\dot{\gamma}$, defined as the difference in velocity of the two bounding planes divided by the distance H between the two parallel planes,

$$\dot{\gamma} = \frac{v_0}{H}.$$

Four material functions (transient and steady viscosities, and first and second normal stress differences) are measured in steady shear and start up of steady shear experiments. The ratio of shear stress to shear rate is defined as the shear viscosity, η,

$$\eta = \frac{\tau_{zx}}{\dot{\gamma}}$$

Transient viscosity, η^+, measures the stress response of a fluid as a function of time as indicated in the following equation

$$\eta^+ = \frac{\tau_{zx}(t)}{\dot{\gamma}}.$$

At long times and constant shear rate it is observed that the transient viscosity attains a constant value, called the steady shear viscosity. The time it takes to reach steady state is a function of the material and the shear rate.

A Newtonian fluid is one for which at a given temperature the viscosity η is constant, independent of shear stress, shear rate, and time. Hence for a Newtonian fluid, the shear stress depends only on shear rate, and that dependence is linear. Dependence of viscosity on shear rate and/or time indicates the fluid is non-Newtonian.

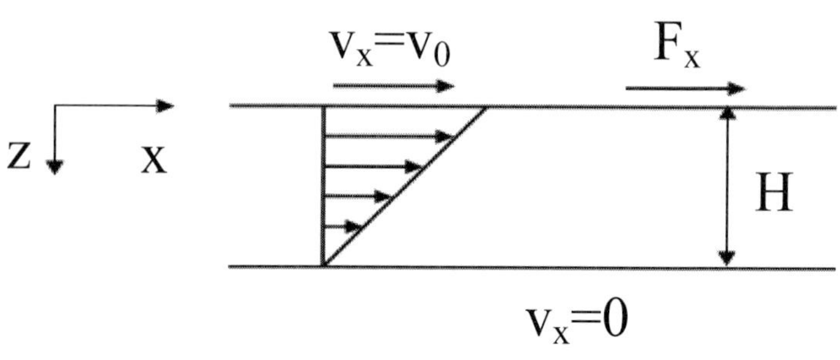

For the choice of coordinate directions in the figure, the first normal stress difference is

$$N_1 = \tau_{xx} - \tau_{yy},$$

and the second normal stress difference is

$$N_2 = \tau_{yy} - \tau_{zz}.$$

For a Newtonian fluid, the first and second normal stress differences are zero, indicating that a shear flow requires only the application of shear stress. An indication of elasticity in a non-Newtonian fluid is the need for non-zero normal stress differences to maintain a shear flow.

Storage and loss moduli are associated with small amplitude oscillatory shear flows, in which the force required to maintain a prescribed harmonic motion of the top confining plane in Figure 1 is separated into its parts in-phase and out-of-phase to the applied motion. The in-phase component is proportional to the storage modulus G′, and the out-of-phase component to the loss modulus G″. For an elastic solid the force is entirely in phase with the motion (shear stress proportional to shear strain) so the G′ is zero. For a Newtonian fluid the force is entirely out of phase with the motion (shear stress proportional to shear rate) so the G″ is zero. G′ and G″ both nonzero is indicative of a non-Newtonian viscoelastic material.

In the remainder of this chapter we compare experimental measurements and model predictions of steady and transient shear viscosities, first and second normal stress differences, and storage and loss moduli for nanocomposites to determine the effect of nanoparticle geometry, concentration, and dispersion on shear rheology.

3. Preparation of Suspensions and Composites

To obtain enhanced performance properties, a limiting step is mixing and uniform dispersion of the composite, without damage to the particles (the aspect ratios of the nanoparticles must be preserved to fully exploit their expected advantages in the composite). Two properties of carbon nanotubes and carbon nanofibers make them difficult to disperse: (i) CNTs and CNFs consist of rolled up graphite sheets which are non-polar by nature and therefore not compatible with polar solvents, and (ii) the small size of the nanoparticles make van der Waals attraction forces between the nanoparticles significant, leading to agglomeration. In order to overcome these issues several methods have been developed to create compatible, highly dispersed nanoparticle suspensions and composites.

One of the most common methods of creating nanoparticle suspensions is to reflux and sonicate the nanoparticles in a strong sulfuric/nitric acid mixture for several hours [36–38]. The acid attacks defect sites in the nanoparticle structure thereby introducing carboxylic acid groups to the surface and often shortening the length of the particle by causing a fracture at the defect site. The added functional groups are covalently bound to the surface and introduce enough polarity to make the particles more compatible with non-polar solvents.

A second method of creating nanoparticle suspensions utilizes the carboxylic acid functional groups introduced in the acid treatment method [39–41]. Once carboxylic acid groups are present on the surface of a nanoparticle they can be further reacted with an oligomer or polymer that is compatible with the desired solvent. The nanoparticle is then covalently bound to a compatible group much larger than the original carboxylic acid group. This method is useful for creating dispersions in both liquid and polymer based suspensions.

A disadvantage of acid treatments is that the particle lengths are shortened, and the particle properties, such as electrical conductivity, can be disrupted by the presence of new covalent bonds. A non-covalent approach to solubilizing and dispersing carbon nanotubes has been developed through the use of surfactants. Several studies have experimented with various combinations of hydrophobic tail lengths and hydrophilic head groups along with different ratios of surfactant to nanoparticle in order to effectively disperse and solvate CNTs in aqueous solutions [34]. This method has been found to be effective to varying degrees depending upon the combination of conditions used.

Acid treatment of carbon surface has been used for traditional carbon fibers to modify the surface chemistry. Herrick [42] and Herrick *et al.* [43] oxidized the surface of carbon fibers using nitric acid and observed increased surface area and surface functionality. Recently, Esumi *et al.* [44] used a mixture of concentrated nitric acid and sulfuric acid to treat carbon nanotubes in order to obtain a better dispersion. Increased dispersion has been attributed to the fact that acidic groups were added to the nanotubes and they readily disperse in polar solvents and form well-dispersed suspensions. Esumi et al. did not report any damage to the nanotubes.

3.1. *Materials*

For the experimental studies reported in this paper that we performed, the nanoparticles were CNFs (Pyrograf® III, type PR-24-PS) made by Applied Sciences, Inc. These nanofibers are produced by decomposing organic vapors at elevated temperature in the presence of metal catalysts. The CNFs have a tree-ring structure with a hollow core. The as-received CNFs are in powder form and contain large agglomerates of mm to cm in size. The CNFs were used directly as received in making CNF/polymer composites. In some of the suspensions the CNFs were chemically treated; the treatment procedures are described in detail in the next section.

For the solvent in our CNF suspensions we used a 90 wt% glycerol in water solution, made by mixing 99.5 wt% pure glycerol with distilled water. This solvent is Newtonian with essentially constant viscosity, zero normal stress differences, and zero storage modulus, so that any non-Newtonian behavior of the suspensions can be attributed to the effect of the CNFs.

We selected polystyrene (PS) as our thermoplastic polymer matrix because its rheology is well characterized, so that any deviation caused by the addition of the CNFs can be readily identified. Another reason is that the molecular structure of polystyrene is simple and no crystalline structure is present in solid phase, so that the structure of the CNFs that has been produced during flow in its molten phase can be preserved after being cooled to its solid phase. The polystyrene we used, produced by Atofina Chemicals (Fina CX5197), has a density of 1000 g/m^3, M_w 200,000 g/mol, and polydispersity index of 2.4.

3.2. *Suspension preparation*

For our experiments we prepared two types of CNF suspensions, one in which the CNFs were chemically treated using strong acids before being placed in suspension, and the other in which the CNFs were left untreated. Detailed procedures of the acid treatment are described elsewhere [45]. The purpose of the acid treatment is to make the surface of the CNFs more hydrophilic so that the interaction with the aqueous solvent is enhanced, leading to improved dispersion of the CNF.

The untreated CNFs were added into the solvent and then the suspension was sonicated using an ultrasound bath for 3 hours. The sonication time was selected such that longer sonication would not provide better dispersion. Untreated suspensions with CNF concentration of 0.5 wt% to 5 wt% were prepared.

The treated suspensions were prepared by adding the treated CNFs into the solvent and then sonicating in the same fashion as in the untreated suspensions. Treated suspensions were prepared at the same CNF concentrations as the untreated suspensions. This is to isolate the effect of acid treatment on the CNF dispersion and the subsequent effect on rheological properties of the suspensions.

3.3. *Polymer composite preparation*

We used two methods to prepare the CNF/polystyrene composites, melt blending and solvent casting, to investigate how different composite preparation methods affect the dispersion and size of the CNFs, and the subsequence effect on the rheology of the melt composites.

For the sake of brevity, the composites made by melt blending method are referred as MB composites, and those made by solvent casting method as SC composites.

In the melt blending process, CNF powder and PS are weighed and then fed into a DACA micro-compounder at elevated temperature. The microcompounder is a small twin-screw extruder in which the compounding time can be controlled. Typically five minutes compounding time at 180°C and 150 rpm achieves the best CNF dispersion with the least polymer degradation, i.e. breaking of polymer chains. Longer compounding time does not generate better CNF dispersion, but can lead to more polymer degradation. The composite is extruded through a 1 mm die and then cut into pellets with the approximate length of 2 mm. Composites containing 2, 5, and 10 wt% CNF were prepared using this method. Pure PS was also compounded using the same procedures, to serve as a control sample (labeled MB0, to indicate a melt-blended polystyrene with 0 wt% CNF).

In the solvent casting process, PS is first dissolved in tetrohydrofuran (THF) to make a 10 wt% solution. Then the CNFs are added into the solution and the suspensions are sonicated for 0.5 hour using a Sonic Dismembrator (Fisher Scientific) at 20 kHz and power level of 760 W/l. After that, the suspensions are heated to boil to drive off most of the solvent. The residual material, still containing about 5 wt% THF solvent, is broken into powder using a blender and then further dried in a vacuum oven at 80°C for 7 days. The dried powder is heated to 200°C and compressed through a 1mm diameter die, and the extruded rods are cut in pellets with length of about 2 mm, the same as the MB composites. This pelletizing process helps to remove the bubbles that would otherwise develop in the compression molding process to make disks, as described below. Composites containing 2, 5, and 10 wt% CNF were prepared using the solvent casting method. Pure PS went through the exact same procedures to serve as a control sample (labeled SC0).

Both the MB and SC composite 2 mm pellets were compression molded into disks (25 mm diameter, 0.6–1 mm thickness) using a hot press at 190°C. These disks served as specimens for the rheometers of section 4 employed to characterize the composite's melt behavior. Although the CNFs are oriented in the flow direction in the pellets, the orientation of the pellets in the compressed disks is random on the scale of the 2 mm pellets. The disks are stored in a vacuum oven at 70°C prior to the measurement.

It is worthwhile to mention that both melt blending process and solvent casting process, as well as the disk making step, cause degradation to the polystyrene. Most likely the degradation is thermo-oxidative, such that the polystyrene chains are shortened and the bulk viscosity decreases. Details of the origin of the degradation are out of the scope of this paper. We have tried our best to achieve the best possible CNF dispersion while minimizing the degradation. The purpose of the two control specimens of pure PS, one undergoing the same melt blending process as the MB composites (MB0) and the other undergoing the solvent casting process (SC0) is to separate the effect of CNFs in the composite from the effect of polymer degradation. We found that the pure polystyrene that went through melt blending and the pure polystyrene that went through solvent casting have about the same viscosity, roughly 40% lower than that of the pristine polystyrene.

4. Morphological Characterization Techniques and Results

The suspension morphology was characterized using scanning electron microscopy (SEM), and optical microscopy. The SEM we used was a Philips XL 30 model equipped with secondary and back-scatter detectors operating at 30 kV. For the as-received nanofibers, conductive glue was used to fix the powder to the SEM stub. For the suspensions, the SEM samples were prepared using the following procedure: One drop of the suspension from the 1 wt% suspensions was diluted with about 2 ml distilled water. Care was taken so that this drop contained fibers that were representative of the whole sample. Several drops of the diluted suspensions were deposited onto aluminum SEM stubs and left in open air for overnight. As the liquid on the stubs contained glycerol, the stubs were put in vacuum oven at 60°C for an hour to completely dry. A thin film of dried nanofibers was formed on each stub surface. The dilution of the suspensions and the drying speed were carefully controlled such that an optimum film thickness of nanofibers was formed (if the suspensions were not sufficiently diluted then the dried film contained too many nanofibers and no individual nanofibers could be identified, and too high a drying rate caused the liquid to contract and eventually form a small dried spot on the stub in which the nanofiber were too tightly compacted to be clearly observed). The stubs with the thin film of dried nanofibers were then sputter coated with platinum at 18 mA for 60 seconds using a Pelco Model 3 sputter coater. Although carbon nanofibers are conductive, the sputtered SEM samples give sharper images. An optical microscope was used to study the CNF morphology at a larger scale. For CNF suspensions, a tiny drop was put on a glass slide and then flattened out by a cover glass. The dispersion of nanofibers in the thin layer was then examined using optical microscopes, with the magnification in the range of 40 to 400 times.

Transmission electron microscopy (TEM) using a Philips CM 12 at 120 kV was employed to observe the CNF dispersion and orientation in the CNF/PS melt composites. TEM samples were prepared by microtoming a 800nm thick using a diamond knife. For the larger scale study of fiber length an optical microscope was used. The CNF/PS composites were dissolved in a THF solvent, and these solutions were viewed optically under a cover glass with a procedure similar to that described above for the CNF/glycerol-water suspensions.

4.1. *Morphological characterization of CNFs in suspensions*

Our morphological characterization of the CNF suspensions summarizes hundreds of SEM images such as those shown in Figure 1, and optical microscopy images such as those in Figure 2. The SEM micrographs are employed in this study to reveal nanofiber length and radius.

They reveal that in the as-received state, individual CNFs entangle with each other and form large aggregations (Figure 1a). In the SEM images of the as-received aggregated CNFs it is difficult to observe both ends of a single CNF, so that an accurate distribution of fiber length is impossible to obtain. After analyzing many SEM micrographs, we can only estimate that more than half of the as-received nanofibers are in the range of 5 to 20 μm in length.

Figure 1b shows the dispersion of untreated CNFs resulting from sonication of the suspension. Large aggregations as seen in the as received state disappear and most CNFs are in a dispersed state. The dispersed CNFs appear to have about the same length as those in the as-received materials. Some of the longer dispersed CNFs tend to entangle to form clusters. Small numbers of clusters smaller than the agglomerates in the as-received state still exist, but none appear in the image in Figure 1b.

The dispersion of chemically treated CNFs placed in suspension and then sonicated are displayed in Figure 1c. There are no CNF aggregations or clusters, and the CNFs are evenly distributed in space. It is important to note, however, that the length of the CNFs in treated sonicated suspensions are almost exclusively less than 5 microns, much shorter than those in untreated sonicated suspensions and as-received materials. From this observation, we conclude that the combination of acid treatment and sonication greatly shorten the length of the CNFs.

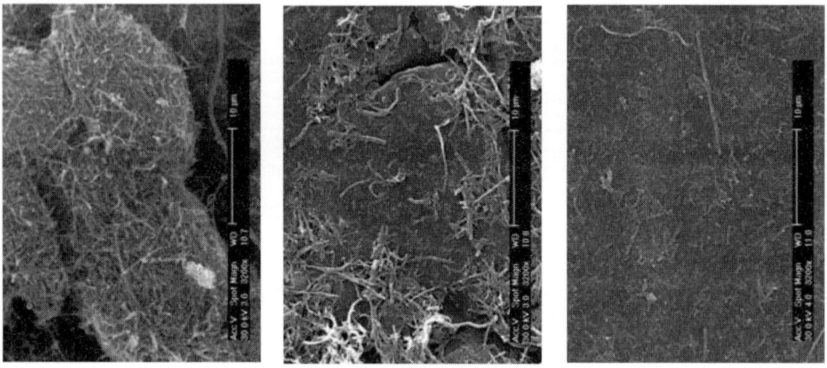

(a) (b) (c)

Fig. 1. SEM micrographs of (a) as-received CNFs, (b) untreated CNFs sonicated in suspension, and (c) treated CNFs sonicated in suspension (scale bar: 10 μm).

Prolonged high-power sonication has been found to cause mechanical damage to carbon nanotubes [46], with researchers observing that the ultrasound causes bending and buckling of the carbon nanotubes. With longer time of sonication, the outer layer of the nanotubes is stripped off to form amorphous graphite. More powerful ultrasound and long sonication time cause more damage to the nanotubes.

The damage to the carbon nanofibers caused by chemical treatment followed by ultrasonication we observed in our study is different from the damage to carbon nanotubes just described. The nanofibers are shortened while the diameters remain unchanged. The nanofiber surface remains smooth with no amorphous graphite observed in the sample. Further analysis of the as-received nanofibers at higher magnification shows that many CNFs contain bends and other curvatures, and some have bamboo-like structures with the distance between joints of about 500 nm. Most of the nanofibers are not straight, leading to structurally weak points. These weak points are attacked by the strong acid and thus become weaker. Subsequent sonication causes the nanofibers to break at the further weakened locations. Liu *et al.* [47] report similar damage when they sonicated single-walled carbon nanotubes in the presence of a mixture of concentrated sulfuric and nitric acid (3:1) and observed that the tubes were reduced from “endless” to several hundred nanometers.

Our morphological characterization indicates that sonication alone does not damage the CNFs. However, treatment with strong acid weakens the nanofibers and the subsequent sonication breaks most of the nanofibers. Hence the acid treatment followed by sonication opens the nanofiber clumps and disperses individual fibers, but also breaks most of them.

It should be noted that in preparing the SEM samples of the suspensions, it is difficult to guarantee that in each micrograph there are the same amount of nanofibers. Even if the same amount of fibers are put on the SEM stubs, it is difficult to ensure the formation of a homogeneous film of uniform thickness.

The dilution, drying, and coating procedures for the preparation of SEM samples may destroy some of the features in the suspension state. Therefore in addition to characterizing the morphology of the suspensions via SEM images, we examined the dispersion of the CNFs in the suspensions directly using optical microscopy. Optical microscopy cannot resolve the individual nanofibers, but does provide images of the undamaged structure developed in the nanocomposites. Figure 2 shows optical micrographs of the 0.5 wt% untreated and treated suspensions. In the treated suspension (Figure 2b), the CNFs are uniformly dispersed with only a few small clumps visible. These observations reflect the

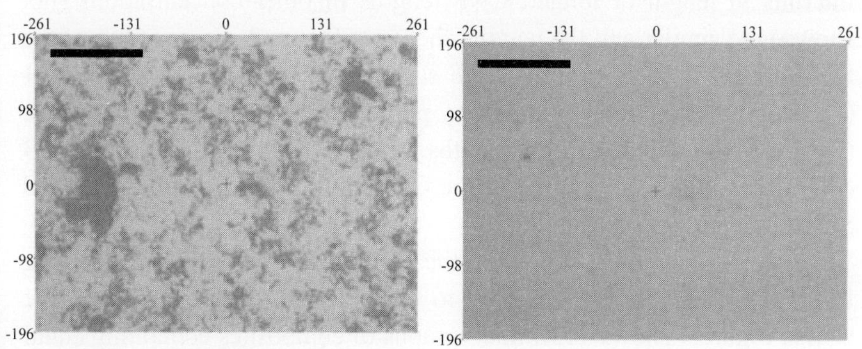

(a) (b)

Fig. 2. Optical microscopy images of (a) 0.5 wt% untreated and (b) 0.5 wt% treated CNF suspensions (scale bar: 100 μm).

same trends as the SEM micrographs. Although the CNFs are too small to be seen individually with optical microscopy at this magnification (100x), in Figure 2a (untreated suspensions) we observe unbroken nanofiber clumps and clouds with partially dispersed CNFs. The CNF dispersion in suspensions with higher concentrations is similar.

4.2. *Morphological characterization of CNFs in polymer melts*

The length and dispersion of CNFs in composites made by the melt blending method and solvent cast method have been characterized using a combination of optical microcopy and TEM. Optical microscopy was used to measure the length of the CNFs after dissolving the composites in a solvent. Due to the limitation of the optical resolution, only CNFs with diameter of 200 nm and larger can be observed. As a result, optical microcopy is useful for measuring size of the large CNFs.

5 wt% MB and SC composites were dissolved in a solvent (THF) with a composite to solvent ratio of about 8:100 by weight. The resulting suspensions were observed under an optical microscope at 400x magnification, and the results are shown in Figure 3. The longest CNFs in melt blended (MB) composites have lengths of about 2–10 micron, while in solvent cast (SC) composites there are many CNFs with 40 microns in length or longer. CNF lengths of other concentrations (not shown) are similar and the lengths shown in Figure 3. Comparison of the SEM micrographs from dried CNF suspensions presented in section 4.1 (Figure 1) to the optical microscopy photos of CNF/polymer composites (Figure 3) shows similar fiber lengths for the treated suspensions and MB composites while the lengths of the CNFs in untreated suspensions and SC composites are also similar.

Since the dispersion of CNF in the polystyrene matrix is lost when the PS is dissolved, TEM was used to characterize the dispersion of the CNFs. Figure 4 shows TEM micrographs of composites containing equal amounts of CNFs, but made using the different methods. It can be seen that the dispersion in the nanometer scale of the CNF is uniform in MB composites, but there are regions containing almost no CNFs in SC composites. The non-uniformity is mostly caused by the CNFs settling

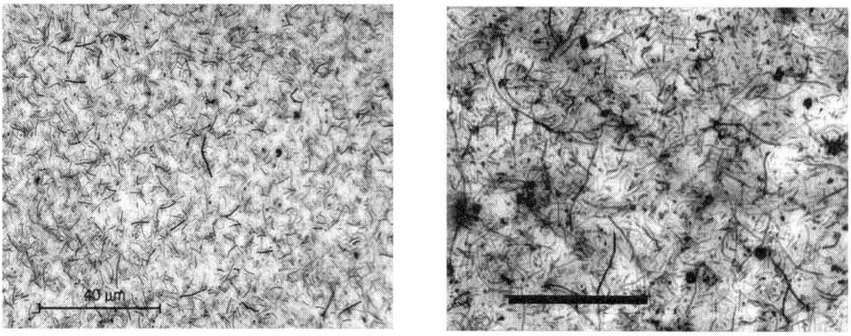

(a) (b)

Fig. 3. Optical microscopy photos of (a) 5 wt% MB composite, and (b) 5 wt% SC composite. Scale bar: 40 microns.

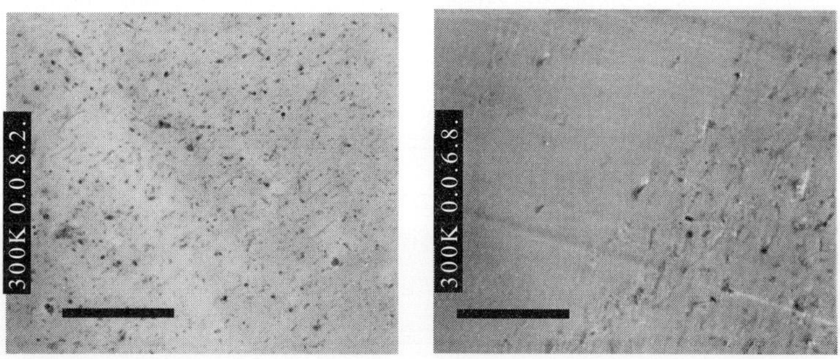

(a) (b)

Fig. 4. TEM photos of (a) 5 wt% MB composite, and (b) 5 wt% SC composite. Scale bar: 5 microns.

during the drying process. However, the SC composites exhibit good uniformity at the millimeter scale. In comparison to the tightly packed agglomerates in the as-received materials (Figure 1a), the CNFs are well dispersed in both types of composites.

In combination of the results from both optical microscopy and TEM, we came to the conclusion that the melt blending method is able to achieve a high degree of uniformity but at the cost of the lengths of the CNFs. The solvent casting method can achieve reasonably good dispersion and the lengths of the CNFs are preserved.

Typically, fiber suspensions are divided into three concentration regimes [48] based on the number of fibers per unit volume n, fiber length L and diameter D. The suspension is in dilute when the value of nL^3 is much less than 1. In this regime, the fibers can rotate freely without touching the neighboring fibers, so that the interactions between fibers can be neglected. When $1 \ll nL^3 \ll L/D$, the suspension is in semi-dilute regime, where the fibers start to collide other fibers during rotating. The suspension is in concentrated regime when $nL^3 \gg L/D$, where the frequent physical touching between fibers may markedly affect the flow properties of the fiber suspension. The characterization of our composite systems are listed in Table 1.

Table 1. Characterization of MB and SC composites.

	MB composites			SC composites		
CNF aspect ratio	5–100			20–500		
CNF density, kg/l	1.8			1.8		
PS melt density, kg/l	1.0			1.0		
	MB2	MB5	MB10	SC2	SC5	SC10
CNF mass fraction	2	5	10	2	5	10
CNF volume fraction	1.12	2.84	5.81	1.12	2.84	5.81
nL^3	1.4	3.6	7.4	35.7	90.4	185.1
regime	Semi-dilute	Semi-dilute	Semi-dilute	Semi-dilute	Concen-trated	Concen-trated

5. Rheological Characterization Techniques and Results

When measuring the shear rheology of Newtonian and non-Newtonian fluids there are several available rheometer types. The most commonly used rheometer geometries include parallel plate, cone and plate, concentric cylinder (Couette), and capillary, shown in Figure 5. If the fluid being studied is highly viscous then the parallel plate, cone and plate, or capillary geometries are most often used; the Couette device does not work as well because of the difficulty of removing high viscosity fluids after the experiment. Parallel plates and cone and plate are limited to low shear rates due to the fact that there is a free surface at their outer radius, and high shear rates can generate enough force to

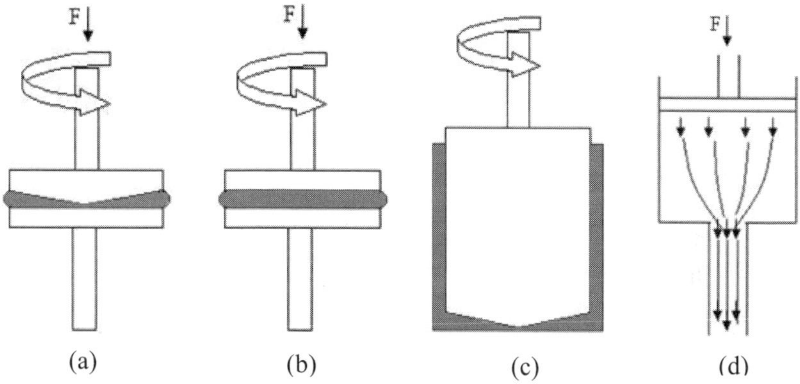

Fig. 5. Four commonly used geometries to measure the shear rheology of fluids: (a) cone and plate, (b) parallel plate, (c) Couette, and (d) capillary.

expel the test fluid radially through this gap, resulting in inaccurate data. Capillary rheometers are useful for determining steady shear viscosities at the high shear rates inaccessible to parallel plat and cone and plate rheometers. Couette devices work well with low viscosity fluids and suspensions over a large range of shear rates. Suspensions tested in a Couette are less prone to settling than in parallel plates or cone and plate. Parallel plate and cone and plate rheometers can measure the normal stresses generated by the flow. It is difficult to measure normal stress differences using capillary and Couette rheometer.

For the shear characterization of our CNF suspensions we used a strain controlled rheometer (Rheometrics RFS II) equipped with an ultra sensitive torque transducer (0.02 g cm to 20 g cm). A Couette geometry (bob diameter 32 mm, bob length 33.3 mm, cup diameter 34 mm) was used to minimize the effect of particle settling, compared to parallel plate and cone and plate geometries. For steady shear viscosity, measurements were made in the shear rate range from 0.02 to 500 s^{-1}. For small amplitude oscillatory shear, a strain sweep was first done to determine the linear viscoelastic regime. Measurements were made in the frequency range 0.1 to 15 Hz. The first normal stress difference was measured using a Rheometrics Mechanical Spectrometer (RMS 800) and a cone and plate (50 mm diameter, 0.04 radian cone angle) geometry. Unless otherwise specified, all suspension measurements were done at 25°C.

Shear rheology of the melt composites was measured using a strain controlled rheometer from TA Instruments (ARES LS2) with both torque transducer (0.02 g cm to 2000 g cm) and normal force transducer (2 g to 2000 g). Parallel plates with 25 mm in diameter were used for all measurements. The gap distance was kept in the range 0.5–0.9 mm for all tests. The molded disks were allowed to rest at the measure temperature for 15 min to relax the residue stress introduced by the compression molding process. Unless otherwise mentioned, all melt measurements were conducted at 200°C.

5.1. *Start-up of steady shear*

In the start-up of steady shear tests performed on our CNF suspensions we observed that the shear stress reached steady state almost immediately after the onset of the shear, showing little time dependent behavior.

For pure polystyrene melt the transient shear viscosity η^+ during start-up of steady shear tests for shear rates 0.01–10 s^{-1} is displayed in Figure 6. It can be seen that the η^+ of PS gradually approaches steady state at small shear rates, and shows an overshoot before the steady state is reached at large shear rates. The η^+ curves at small shear rates (less than 0.01 s^{-1}) fall on to each other and they are not shown for clarity.

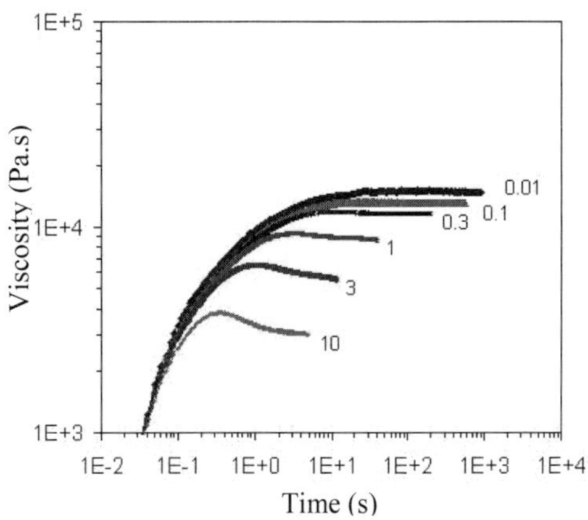

Fig. 6. Transient shear viscosity η^+ in start-up of steady shear at shear rates 0.01–10 s^{-1}.

The transient shear viscosity η^+ of the SC10 composite in the wider range of shear rates shear rates of 0.0001 to 10 s^{-1} is shown in Figure 7. Comparing Figure 6 to Figure 7, the magnitude of η^+ for the nanocomposite SC10 is greater than that of pure PS at the same shear rate, but the shape of the curves are similar for shear rates between 0.1 and 10 s^{-1}. Unlike with the pure PS, the transient viscosities of the composite at small shear rates 0.0001 to 0.01 s^{-1} do not lie on top of each other, but rather the SC10 composite exhibits markedly increased viscosity as shear rate becomes small. Further more, it takes longer time for the viscosity of the composite to reach steady state.

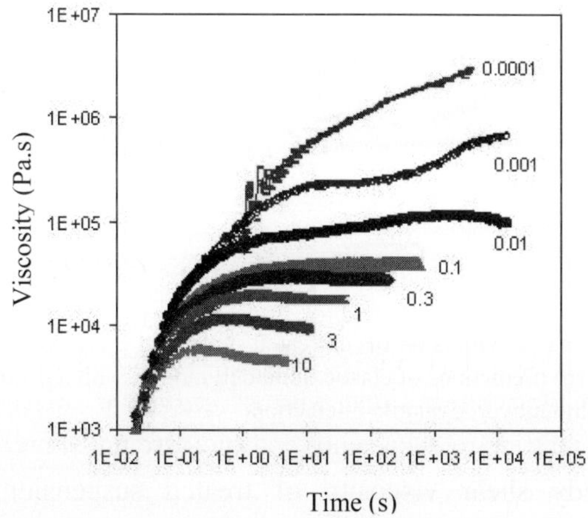

Fig. 7. Transient shear viscosity η^+ in start-up of steady shear at shear rates 0.0001–10 s^{-1}. SC10 composite. Temperature = 200°C.

5.2. *Steady state shear measurements*

5.2.1. *Suspension steady state shear results*

Figure 8a displays the steady state shear viscosity of the untreated CNF suspensions. The viscosity of the suspensions increases monotonically with the CNF concentration by 3 orders of magnitude at low shear rates when concentration increases from 0.5 wt% to 5 wt%. It is clear from the measurements that the suspensions exhibit strong shear thinning behavior. For example, the low shear viscosity of 5 wt% untreated sonicated suspensions increases 3 orders of magnitude and the high shear viscosity increases by a factor of 8. Beyond 5 wt% we found that the suspensions are too viscous to be effectively mixed by sonication.

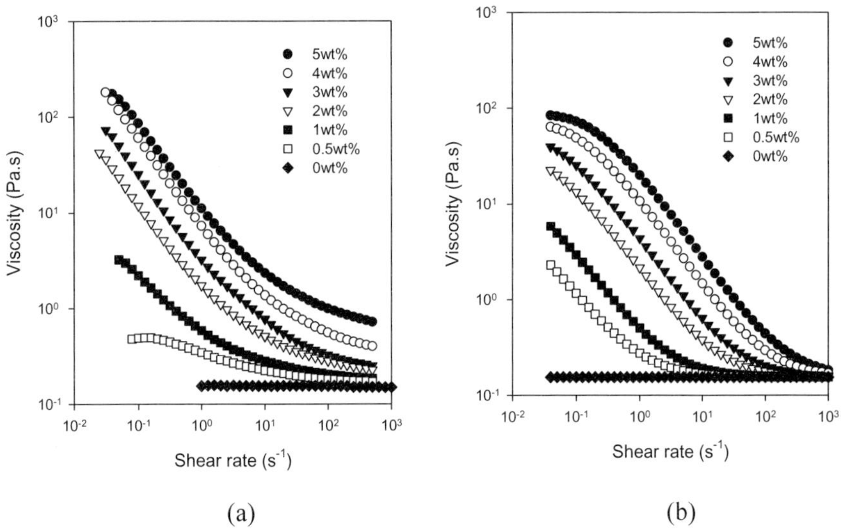

Fig. 8. Steady shear viscosity of untreated sonicated suspensions: (a) experimental measurements, (b) predictions of elastic dumbbell models with anisotropic hydrodynamic drag with or without hydrodynamic interaction.

The steady shear viscosity of treated suspensions is shown in Figure 9a. The viscosity of the suspensions increases with CNF concentration. The viscosity of the 5 wt% treated sonicated suspensions is less than twice as much as that of solvent. However, in contrast to the untreated suspensions, the viscosity of the treated suspensions is almost independent on shear rates.

For the rheology of CNF suspensions, all of the non-Newtonian behavior of we observe in the suspensions is due to the addition of the nanofibers, as the solvent is Newtonian. We conjecture that in the untreated suspensions, the undispersed CNF clumps and partially dispersed CNFs entangle with each other, forming a network structure. The structure is relatively strong at small shear rates and becomes weaker at high shear rates due to the alignment of some fibers induced

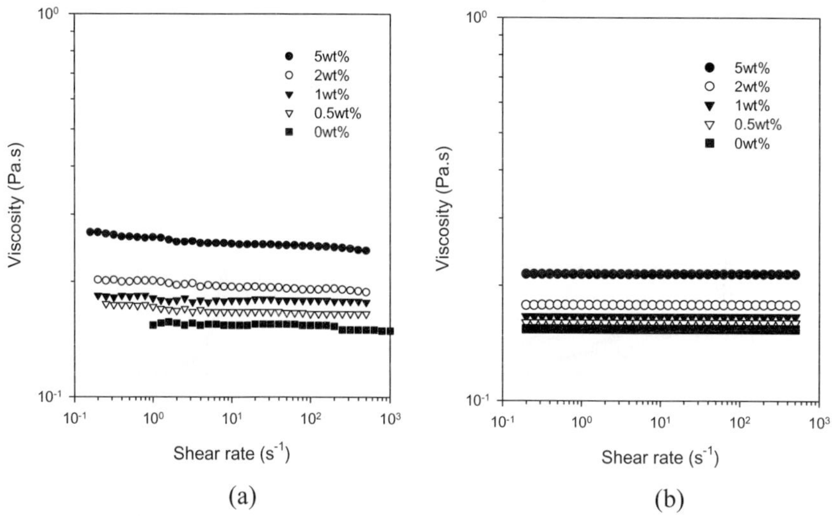

Fig. 9. Steady shear viscosity of treated suspensions: (a) experimental measurements, (b) elastic dumbbell model predictions.

by flow. Hence the untreated sonicated suspension exhibits shear thinning.

Ganani and Powell [49] proposed a mechanism that the shear exerted on some parts of the suspensions is higher than other parts, and this might lead to a drift of low shear Newtonian plateau to lower shear rates. This conjecture of a non-uniform shear mechanism works well for the untreated nanofibers suspensions. The shear stress is transported through both the continuous solvent phase as well as the CNF structure, and both mechanisms may have comparable strengths. However, under stress the CNF structure may break and become aligned to the flow, so that the total stress cannot be transported uniformly, leading to shear thinning in the suspensions.

The untreated CNF suspensions show apparent yield behavior. By plotting the shear viscosity versus stress, Figure 10 shows that the apparent yield stress starts to become apparent when the CNF loading is 1 wt%. At 2, 3, 4, and 5 wt% the suspensions have clearly defined yield stresses of 1, 3, 6, and 9 Pa, respectively. Flocculated suspensions are known to have yield stresses at high particle loadings when interactions between the particles are high enough to form a continuous 3-D network

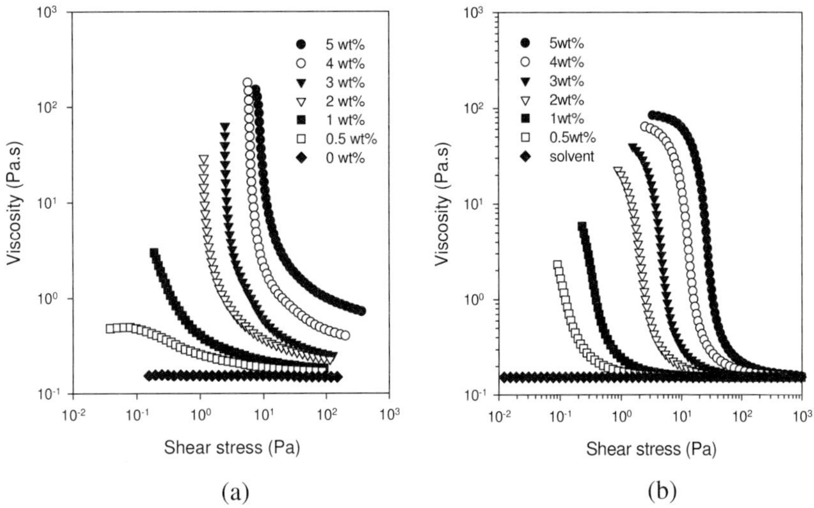

Fig. 10. Apparent yield stress of untreated CNF suspensions: (a) experimental measurements, (b) predictions of elastic dumbbell models with anisotropic hydrodynamic drag with or without hydrodynamic interaction.

structure. The untreated sonicated nanofiber suspensions in this study are of this fluid type. They show no zero-shear viscosity but an apparent yield stress. Therefore it is reasonable to conclude here that the interactions between the nanofibers are large enough in most of the untreated sonicated suspensions such that a continuous 3-D network structure exists and generates yield stresses.

Since the morphology of the treated suspensions is quite different from that of untreated suspensions, large variation in the rheology of the suspensions is expected. As the treated suspensions contain much shorter CNFs, they have more individual particles in a unit volume. One may expect that more individual particles will have more surface area, so that the viscosity would be higher than the untreated suspensions if there were no interactions between particles. However, this is not the case. As analyzed in the morphology section, the major differences between the treated and untreated suspensions are that the latter contain longer CNFs, unbroken nanofiber clumps, and partially dispersed CNFs which form network structure (Figure 2a). Following this reasoning, the structure in the untreated suspension are strong enough to overrule the effects from the increase of the surface area. Further observations supporting the existence CNF interaction is described in the linear viscoelasticity section. The treated suspensions are mainly composed of sub-micron nanofibers and the average aspect ratio is less than 10. As these short fibers are also well dispersed due to the effect of acid treatment, the interactions between them are small even at high concentrations.

We attempted to measure the first normal stress difference N_1 for both treated and untreated suspensions, but the normal force was too small for the rheometer to detect.

5.2.2. *Composite melt steady state shear results*

In the tests of start-up of steady shear for the melt composites a fresh sample was used for each test. From the observation of all the start-up measurements, some of which do not reach an obvious steady state (for example, the polystyrene matrix in Figure 6 and the composite SC10 in Figure 7), viscosities at 20–30 strain units for all shear rates are taken as steady state viscosities for all composites.

Figure 11 displays the steady state viscosities of melt blended and solvent cast composites. Both MB and SC composites show increased viscosities and more pronounced shear thinning behavior with higher CNF concentrations, especially at low shear rates. Of the two, however, greater increase in viscosity is observed in SC composites. Furthermore, the plateau of the viscosity at low shear rates, as seen in the polystyrene, gradually shifts to the lower shear rate region with increasing CNF concentrations. For SC10 composites, there is no plateau at all. The slope of the viscosity curve for SC10 composite at low shear rates is –1, indicating a yield stresses behavior. This is similar to the yield behavior seen in the untreated suspensions.

The effect of CNFs' length on the steady shear viscosity of the CNF/PS composites is illustrated by plotting the relative viscosity η_r of both MB and SC composite in the same figure, as in Figure 12. The relative viscosity is defined as the ratio of the composite viscosity to the matrix polystyrene viscosity. By this definition, any deviation from the $\eta_r = 1$ line can be attributed to the effect of CNFs in the composites. Figure 12 shows that for both MB and SC composites, η_r increases with higher CNF concentration and decreases with higher shear rates. The η_r of SC composites is always higher than that of MB composites, and the difference becomes smaller at higher shear rates. At shear rate of 10 s^{-1}, the η_r of both composites is about the same. This indicates that the longer CNFs in SC composites have more impact on the viscosity of the composites than that of the short CNFs in MB composites, especially at low shear rates.

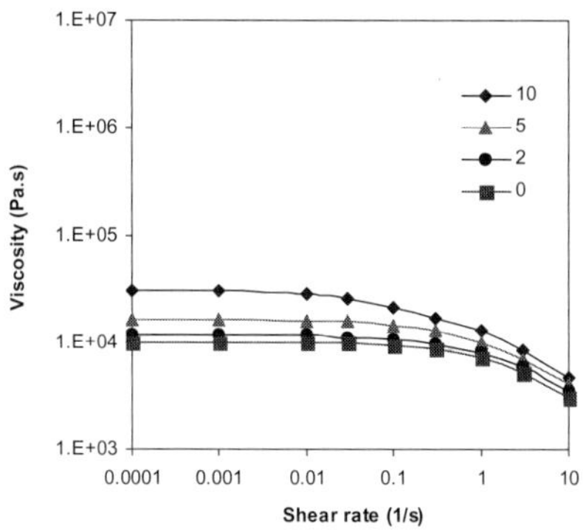

(a)

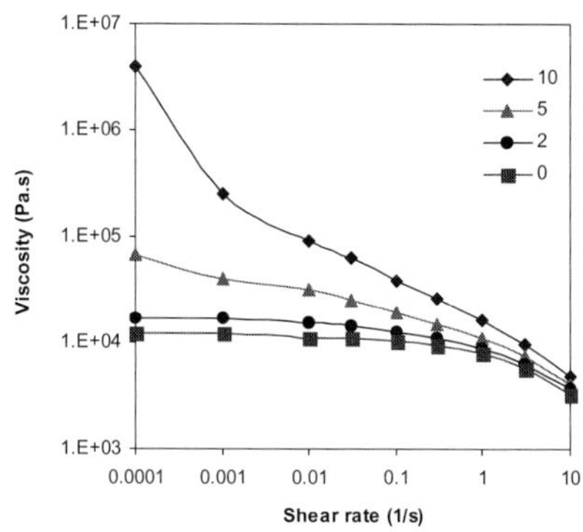

(b)

Fig. 11. Steady shear viscosities of (a) MB composites and (b) SC composites.

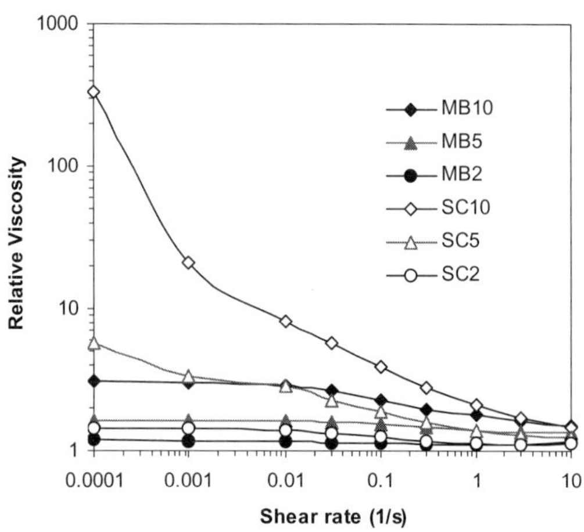

Fig. 12. Effect of CNF length on relative viscosity of MB and SC composites.

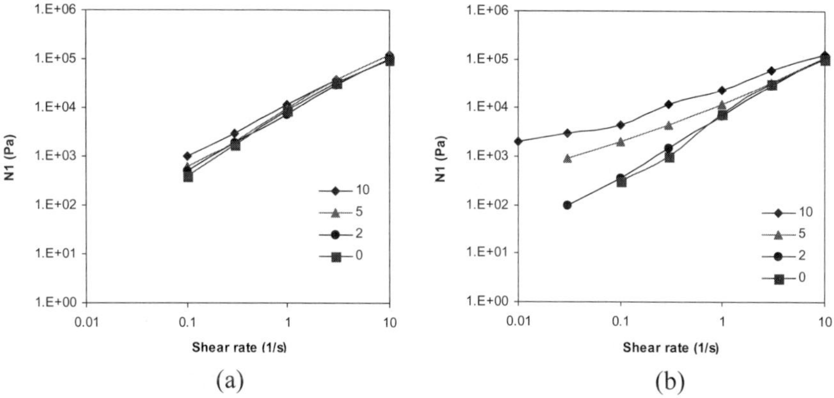

Fig. 13. The first normal stress difference (N_1) under steady shear of (a) MB and (b) SC composites.

First normal stress difference N_1 of both MB and SC composites is displayed in Figure 13. In MB composites, the measured N_1 for all CNF concentrations is about the same as that of pure PS matrix. At the highest CNF concentration we have studied (MB10), the N_1 is only marginally higher than that of PS. In contrast, the N_1 shows much more dependence on CNF concentration in SC composites: higher CNF concentration leads to higher N_1. At shear rate of 0.1 s^{-1}, the N_1 of SC10 is more than one order of magnitude higher than that of pure PS. However, the effect of CNFs on N_1 quickly decreases as the shear rate increases. At 10 s^{-1}, the CNF does not contribute to N_1 at all.

5.2.3. *Comparison of steady shear rheology: CNF suspensions vs. CNF melt composites*

The MB and SC melt composites have a steady shear rheology similar to that of the treated and untreated suspension systems. The major difference between the suspensions and composites are that a relatively low viscosity Newtonian fluid is used as the continuous phase in the suspensions and a polymer melt (polystyrene) is used in the composites. There are many similarities in the morphology and rheology of the systems. The treated suspensions and MB composites contain uniformly dispersed CNFs, with the length of about 1–2 microns, much shorter than that of as received CNFs. Untreated suspensions and SC composites have CNFs of the original length, but the dispersion of the CNFs are not uniform microscopically. As a result, the rheology of shorter CNF systems (treated suspensions and MB composites) is not too far away from the solvent/matrix, due to the small inter-particle interactions. The relative viscosities of MB CNF/polymer melt composites and treated sonicated CNF suspensions are plotted in Figure 14. It is seen that the relative viscosity of both the MB composites and treated suspensions with 2 wt% and 5 wt% CNF concentrations is in the range of 1.2 to 1.8. Further, the relative viscosities of the MB2 and MB5 composites are comparable to that of 2 w% and 5 wt% treated suspensions, respectively. This indicates that the effect of CNFs on the rheology of high viscosity melt systems and low viscosity suspension systems is similar. This effect mainly comes from hydrodynamic interaction, and can be accurately captured by mathematical models presented later in this chapter.

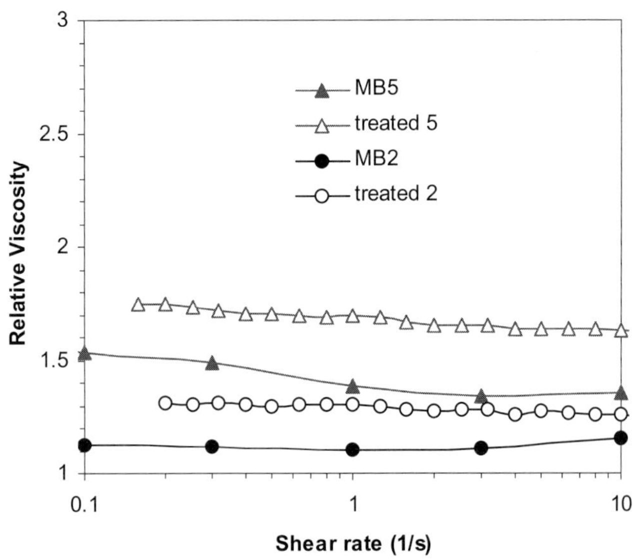

Fig. 14. Relative viscosity of MB CNF/polymer melt composites and treated sonicated CNF suspensions.

For the longer fiber systems, i.e. untreated suspensions and SC composites, the effect of the CNFs on the rheology is much more profound. For SC composites, as the matrix is already non-Newtonian, the addition of long CNFs exaggerates the non-Newtonian behavior. The increased shear thinning behavior in both untreated suspensions and SC composites may very well come from the same origin: physical contact between CNFs. Due to the similar dispersion and CNF length, we can assume that the CNF physical interactions are approximately the same in the untreated suspensions and SC composites at the same CNF concentration. However, effect of the CNF interaction on the rheology of the two systems is very different, as can be determined by the degree of shear thinning in untreated suspensions and SC composites. As the high viscosity of the polymer matrix may dominate the effect of CNF interaction, the degree of shear thinning in SC composites is much less than that in untreated suspensions, where the CNF interactions dominate. The relative viscosities of SC CNF/polymer melt composites and untreated sonicated CNF suspensions clearly show this (Figure 15). The same reasoning also applies to the fact that yield stress appears at 3 wt% for untreated suspensions, while at 10 wt% for SC composites.

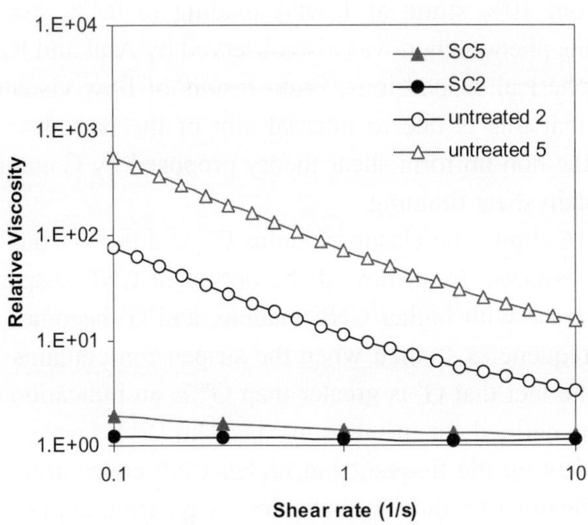

Fig. 15. Relative viscosity of SC composites and untreated suspensions.

5.3. *Small amplitude oscillatory shear measurements*

During a small amplitude oscillatory shear test, a small strain, γ, is applied to a sample in a sinusoidal fashion:

$$\gamma = \gamma_0 \sin(\omega t)$$

The experiments are done at small strain amplitudes in order to remain in the linear viscoelastic region. The stress response is then:

$$\tau = G'\gamma_0 \sin(\omega t) + G''\gamma_0 \cos(\omega t)$$

The storage modulus, G′, is the solid like response of the specimen and corresponds to its storage of elastic energy. The loss modulus, G″, corresponds to the viscous dissipation. Small amplitude oscillatory shear experiments provide indirect measures of structure: in the case of nanoparticle suspensions and nanoparticle/polymer melt composites, the slopes of G′ and G″ when plotted vs. frequency can indicate the presence of a network structure.

For the untreated suspensions, we found that the maximum strain limit for the linear regime decreases with the increase of nanofiber loading, from 10% strain at 1 wt% loading to 0.1% strain at 5 wt% loading. This phenomenon was also observed by Aral and Kalyon [50] in 40 vol% spherical suspensions. From results of flow visualization, they suggested that this is due to internal slip of the particles. This idea is similar to the non-uniform shear theory proposed by Ganani and Powell [49] to explain shear thinning.

Figure 16 shows the elastic modulus G″ and the viscous modulus G″ versus the dynamic frequency of the untreated CNF suspensions. Both moduli increase with higher CNF loading, and G′ becomes greater than G″ at all frequencies studied when the suspension contains 3 wt% CNF or more. The fact that G′ is greater than G″ is an indication of solid-like behavior, as defined for instance in Macosko [51]. Further, G′ becomes less dependent on the frequency at higher CNF concentrations, which is also an indicator of solid-like behavior. A possible explanation for this solid-like behavior is that the nanofibers form a network structure. This structure can be easily broken down by deformation and rebuilt quickly when the deformation is removed. Therefore the suspensions are shear thinning but not thixotropic. At low concentrations, the structure is weak so the solvent dominates the rheology and the suspensions still show liquid-like behavior. Beyond a critical concentration, which we observe is 3 wt% for the system studied here, the network structure becomes stronger and hence elastic solid-like behavior is observed. This type of microstructure is verified from the morphology study, which shows that the dispersed nanofibers are entangled with each other and also with the nanofiber clumps.

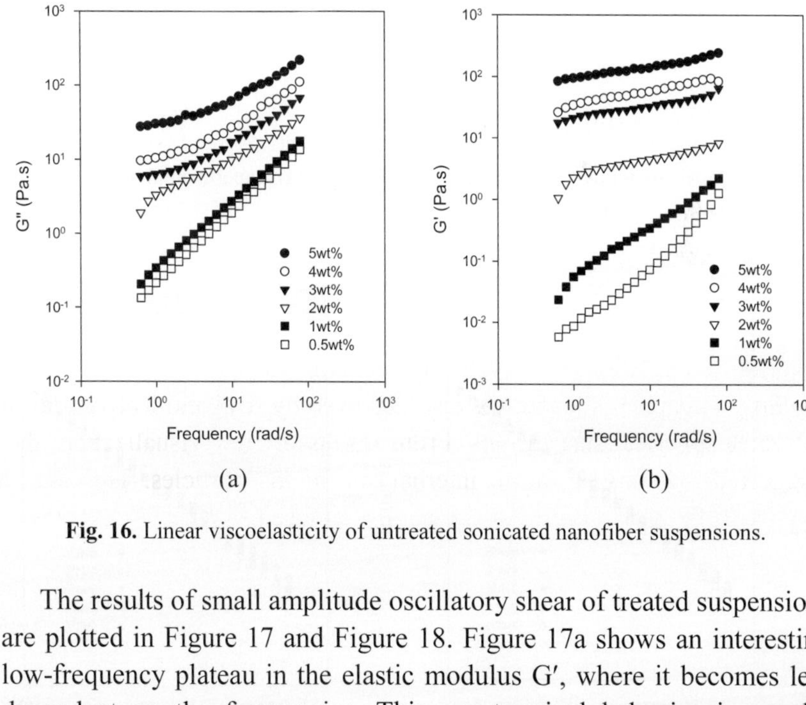

Fig. 16. Linear viscoelasticity of untreated sonicated nanofiber suspensions.

The results of small amplitude oscillatory shear of treated suspensions are plotted in Figure 17 and Figure 18. Figure 17a shows an interesting low-frequency plateau in the elastic modulus G′, where it becomes less dependent on the frequencies. This non-terminal behavior is usually interpreted as an indication of solid-like behavior, as seen in the case of untreated suspensions discussed above. In the range of the frequencies we studied, G′ generally increases with fiber loading. The viscous modulus G″ monotonically increases with the fiber concentration at all frequencies (Figure 18a).

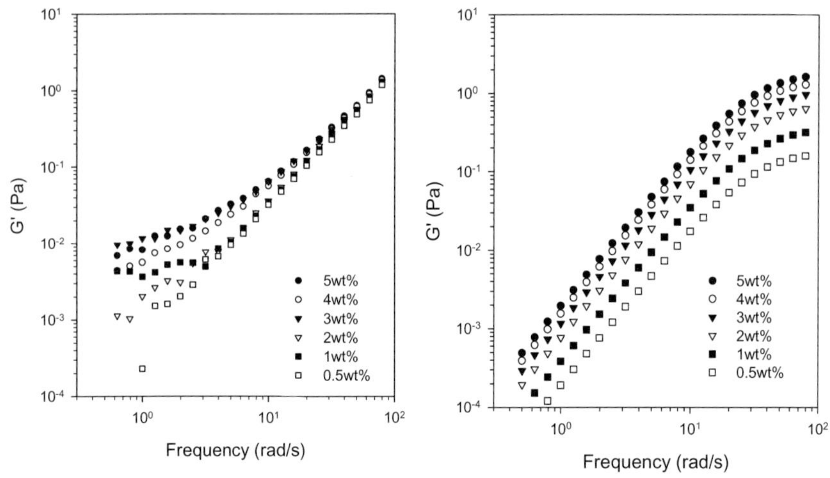

(a) (b)

Fig. 17. Elastic modulus G′ of treated suspensions: (a) experimental measurements, (b) elastic dumbbell model predictions.

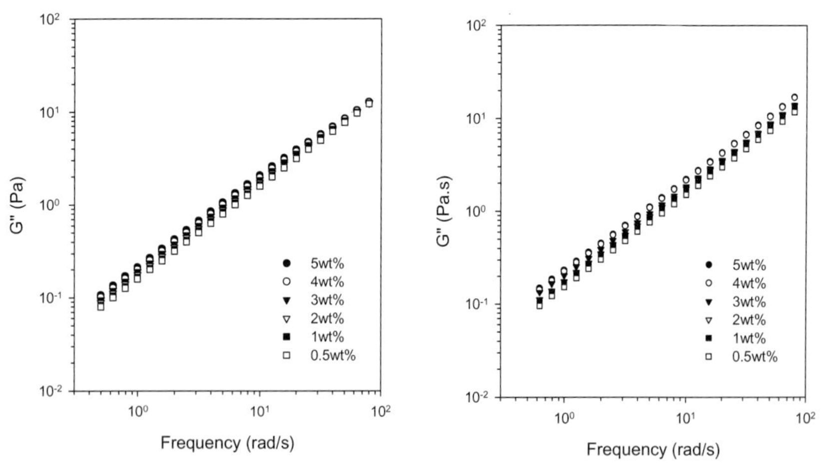

(a) (b)

Fig. 18. Viscous modulus G″ of treated suspensions: (a) experimental measurements, (b) elastic dumbbell model predictions.

5.4. *Measurements of temperature effects*

The measurement of temperature effects was conducted on 2 wt% untreated CNF suspensions. Steady shear viscosity and relative viscosity (ratio of suspension viscosity to solvent viscosity) at various temperatures in the range of 5 to 45°C are plotted in Figure 19 and

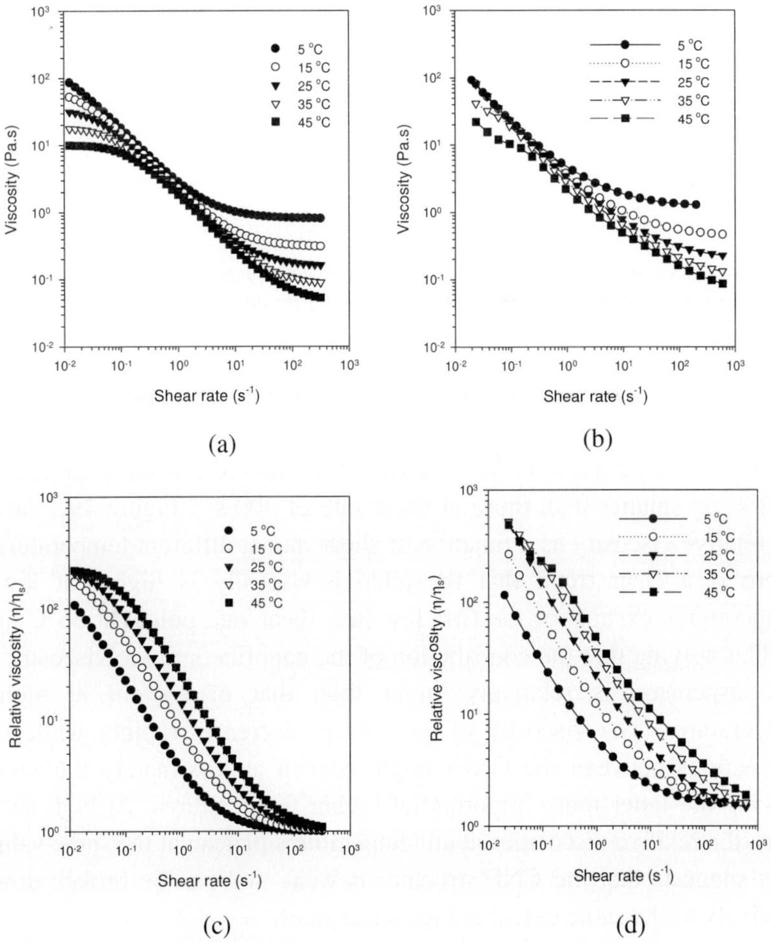

Fig. 19. Effects of temperature on 2 wt% untreated suspensions: (a) experimental measurements of the steady shear viscosity, (b) elastic dumbbell model predictions of the steady shear viscosity, (c) experimental measurements of the relative viscosity, (d) elastic dumbbell model predictions of the relative viscosity.

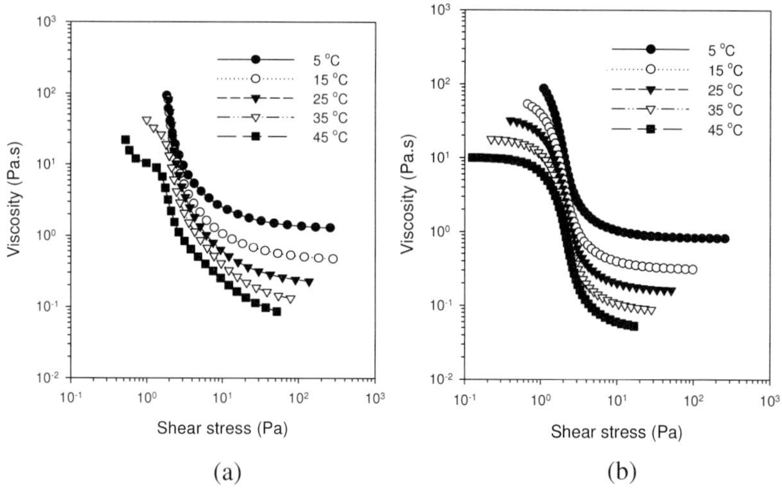

Fig. 20. Effect of temperature on 2 wt% untreated suspensions: (a) experimental measurements showing apparent yield stress, (b) elastic dumbbell model predictions of the apparent yield stress.

Figure 20. As expected, the steady shear viscosity decreases with higher temperature over the entire range of shear rates (Figure 19a). However, the difference between the viscosities at 5°C and 45°C at the shear rate of 0.02 s^{-1} is smaller than those at shear rate of 300 s^{-1}. Figure 19c shows the relative viscosity as a function of shear rate at different temperatures. There is a clear trend that the relative viscosity is higher at lower temperatures except for the first few low shear rate points at 35°C and 45°C, implying that the contribution of the nanofibers to the viscosity of the suspension is relatively larger than that of solvent at higher temperatures. The viscosity of the solvent decreases rapidly while the interactions between the CNFs might remain approximately the same, making the latter more important at higher temperatures. At high shear rates, the relative viscosities at all temperatures plateau at the same value. This suggests that the CNF structure is weak and can be broken down relatively to the same extent at high shear rates.

Whereas Figure 10 displays the dependence of apparent yield stress on concentration, Figure 20 shows the dependence of this apparent yield stress on temperature. The apparent yield stress τ_{ay} decreases with higher temperature (Figure 20a), implying that the strength of network structure originates from both particle-particle interaction and particle-solvent interaction. When temperature increases, while the particle-particle interaction presumably stays the same, the particle-solvent interaction becomes weaker due to the decreased solvent viscosity, resulting in decreased τ_{ay}.

6. Modeling of CNF Suspensions

6.1. *Methodology*

In this section we demonstrate that kinetic theory models, with the parameters in the model specified in terms of primitive properties of the solvent and CNFs, including microscale morphological features measured with SEM, successfully describe the relation between the rheology and morphology of nanofiber suspensions. Kinetic theory-based dumbbell models [20,21,52,53] are customarily applied to polymer solutions, where the dumbbells idealize the polymer molecules and all the parameters (relaxation time λ, particle viscosity η_p, etc.) are obtained by fitting to the rheological measurements of the composite system, rather than the separate constituents.

In our dumbbell models, the carbon nanofiber is idealized as a pair of beads joined by either a massless elastic spring connector (elastic dumbbell) or massless rigid connector (rigid dumbbell). The distributed force from the solvent on the fibers is idealized as being located on the beads. We consider the spring in the elastic dumbbell to be Hookean, i.e. the tension in the spring is proportional to the bead separation. The connector length and bead radius of the idealized particle are calibrated to the measured microstructural dimensions of the physical nanofibers. Macroscale constitutive equations relating flow, stress, and fiber orientation are produced by integrating the coupled equations governing forces, flow, and orientation over a representative volume of fibers and carrier fluid. The bulk (macroscale) rheological properties (moduli, shear viscosity, first normal stress difference, etc.) are deduced from nanostructural measurements of nanofiber microstructure, solvent viscosity and density, and temperature.

Elastic dumbbell models and rigid dumbbell models were employed to predict the macro-scale rheological properties of the nanofiber suspensions [45]. We found that the elastic dumbbell model with anisotropic hydrodynamic drag with or without hydrodynamic interaction (Giesekus model) correctly predicts the observed steady shear behavior and its temperature dependence of the untreated sonicated suspensions if we introduce the concept of an effective dumbbell length to account for the presence of large agglomerates. (Hydrodynamic drag refers to the force acting on the beads from the solvent, and hydrodynamic interaction refers to the effect of velocity disturbance at one bead owing to the movement of the other bead.)

The closed-form constitutive equations [53] of elastic dumbbell models with anisotropic hydrodynamic drag are

$$\tau_{ij} = -p\delta_{ij} + 2\eta_s D_{ij} + \tau_{ij}^{CNF},$$
$$\tau_{ij}^{CNF} + \lambda \frac{D}{Dt}\tau_{ij}^{CNF} + \alpha\frac{\lambda}{\eta_p}\left(\tau_{ik}^{CNF}\tau_{kj}^{CNF}\right) = 2\eta_p D_{ij}. \tag{1}$$

The Hookean elastic dumbbell model with anisotropic hydrodynamic drag, Eq. (1), is also referred to as the Giesekus model. Importantly, we retain the microstructural bases of these models, expressing the relaxation time λ and particle contribution to viscosity η_p of the CNF suspensions are given in terms of the primitive microstructural parameters:

a = length of the spring link (m),
r = radius of the beads (m),
ρ_f = density of the nanofibers (kg/m^3),
ρ_s = density of the solvent (kg/m^3),
c = mass concentration of the suspension (dimensionless),
T = absolute temperature (K),
$\eta_s(T)$ = Newtonian solvent viscosity (Pa·s), (2)
α = mobility factor, only for the elastic models with anisotropic hydrodynamic drag (dimensionless).
D_{ij} = the symmetric part of the velocity gradient $\dfrac{\partial v_i}{\partial x_j}$
τ_{ij} = Cauchy stress
$\dfrac{D}{Dt}$ = the upper convected derivative

In the primitive microstructural parameter list (2), only the mobility factor α is a fitting parameter whose value is obtained by fitting to suspension viscosities. To produce our microstructurally based model for the CNF/glycerol-water suspensions, the length of spring link a and bead radius r are calibrated by morphology measurements of the nanofiber lengths a_f and radii r_f, densities ρ_f and ρ_s mass concentration c, and solvent viscosity η_s are measured directly, and T is controlled in the experiments. Explicitly,

$$\lambda = \text{relaxation time} = \frac{\pi \eta_s r a^2}{2kT\left(1 - h\sqrt{\frac{6}{\pi}}\right)}, \tag{3}$$

η_p = particle contribution to the viscosity

$$= nkT\lambda = \frac{\rho_s c}{\rho_f + c\left(\rho_s - \rho_f\right)} \cdot \frac{\eta_s a}{2r} \cdot \frac{1}{\left(1 - h\sqrt{\frac{6}{\pi}}\right)}, \tag{4}$$

where

k = Boltzmann's constant = 1.3807×10^{-23} J/K,

$$\phi = \text{volume fraction} = \frac{\rho_s c}{\rho_f + c\left(\rho_s - \rho_f\right)},$$

n = number of fibers per unit volume

$$= \frac{\phi}{\pi r^2 a} = \frac{\rho_s c}{\rho_f + c\left(\rho_s - \rho_f\right)} \cdot \frac{1}{\pi r^2 a} \tag{5}$$

h = hydrodynamic interaction parameter, $h = \frac{r}{a}$ if the hydrodynamic interaction is taken into account, and $h = 0$ if the hydrodynamic interaction is neglected. Derivations of the Giesekus model (1) can found in [45,53].

To retain the microstructural basis of the kinetic theory model, the characteristic lengths a and radii r of the model dumbbell are inferred from our morphological SEM measurements, independent of rheological measurements. The relaxation time λ and particle viscosity η_p are then deduced from a and r, and measurements of densities, temperature, and solvent viscosity η_s, also independent of rheological measurements.

6.2. *Modeling predictions for treated sonicated nanofiber suspensions*

Based on the morphological characterization given in experimental section 4.1 that the treated sonicated suspensions contain well-dispersed fibers with radii in the range 100 to 200 nm and lengths mostly in the range of 1 to 5 μm, with few fibers of length 5 to 10 μm and essentially no clumps, we replace the distributions of length and radii with a single length $a_f = 2$ μm and radius $r_f = 100$ nm, which are representative values measured from the SEM images (Figure 1c and hundreds of others analyzed). Other primitive microstructural parameters in list (2) are measured directly: the temperature T = 25°C, solvent viscosity at 25°C η_s = 0.154 Pa·s, fiber density ρ_f =1750 kg/m^3, solvent density ρ_s = 1236 kg/m^3, and fiber concentration c = 0.5, 1, 2, 3, 4, and 5 wt%.

The relation between the length a_f and radius r_f of the physical nanofiber and the connector length a and bead radius r in its dumbbell idealization must be calibrated. We assume that the relation between (a, r) and (a_f, r_f) for the treated sonicated fibers is the same at all concentrations in all flows, and therefore the calibration utilizes only one experiment. This protocol is consistent with dilute suspensions where entanglements do not alter the effective spring constant. We choose to calibrate with the 2 wt% concentration small amplitude oscillatory shear experiment: a small amplitude oscillatory shear experiment so as not to involve the parameter α, and 2 wt% so as to avoid the possible inertia effects at lower concentrations and possible particle interaction at higher concentrations. We perform the calibration as follows [45].

For the elastic models, the linear viscoelastic material functions G′, G″are expressed in terms of the primitive microstructural parameters:

$$G'(\omega)=\omega\eta''(\omega)=\omega\left(\frac{\eta_p\lambda\omega}{1+(\lambda\omega)^2}\right)=\omega\left(\frac{nkT\lambda^2\omega}{1+(\lambda\omega)^2}\right)$$

$$=\omega\left(\frac{\rho_s ckT\ {}_s^2\pi a^3\omega}{\left[\rho_f+c\left(\rho_s-\rho_f\right)\right]\left[4T^2k^2\left(1\text{-}h\sqrt{\frac{6}{\pi}}\right)^2+\pi^2\eta_s{}^2r^2a^4\omega^2\right]}\right),$$

$$G''(\omega)=\omega\eta'(\omega)=\omega\left(\eta_s+\frac{\eta_p}{1+(\lambda\omega)^2}\right)=\omega\left(\eta_s+\frac{nkT\lambda}{1+(\lambda\omega)^2}\right)$$

$$=\omega\left(\eta_s+\frac{2\rho_s ck^2T^2\eta_s a\left(1\text{-}h\sqrt{\frac{6}{\pi}}\right)}{\left[\rho_f+c\left(\rho_s-\rho_f\right)\right]\left[4k^2T^2r\left(1\text{-}h\sqrt{\frac{6}{\pi}}\right)^2+\pi^2\eta_s{}^2r^3a^4\omega^2\right]}\right). \tag{6}$$

Notice that these forms do not contain the primitive parameter α, so that α will not interfere with our calibration relating a and r to the microstructure measurements a_f and r_f. The spring link length a and beads diameter r in the elastic models are obtained by minimizing the error

$$\delta=\sum_{i=1}^{N}\left\{\begin{array}{l}\left[\log_{10}G''_{\exp}(\omega_i)-\log_{10}\left(\left(\eta_s+\frac{\eta_p}{1+(\lambda\omega_i)^2}\right)\omega_i\right)\right]^2\\+\left[\log_{10}G'_{\exp}(\omega_i)-\log_{10}\left(\frac{\eta_p\lambda\omega_i^2}{1+(\lambda\omega_i)^2}\right)\right]^2\end{array}\right\}, \tag{7}$$

where η_p and λ are functions of a and r, $G'_{\exp}(\omega_i)$ and $G''_{\exp}(\omega_i)$ are measured storage modulus and loss modulus at the N discrete frequencies ω_i for a particular concentration c for small amplitude oscillatory shear experiments described in Figure 17 and Figure 18. We use the exhaustive method to search through a and r space. Firstly, we propose a physically reasonable range for each parameter and the search is first carried out in this range using a relatively coarse mesh, after which the parameters are restricted in a smaller range and a refined mesh is employed to search the best fits. The exhaustive method is feasible in our experiments since there are only two parameter searches over a well-defined domain. Finally, we verify that the best fits of parameters are not near the edges of the domains. The searching results using the experimental data at $c = 2$ wt% ($N = 22$) are $a = 0.23$ μm, $r = 10.5$ nm for the elastic models without hydrodynamic interaction ($h = 0$) and $a = 0.22$ μm, $r = 10.7$ nm for the elastic models with hydrodynamic interaction $\left(h = \frac{r}{a}\right)$. Since the hydrodynamic interaction $h = \frac{r}{a} \approx 0.04$ is small, the two relaxation times are nearly equal and the predictions of the elastic models without hydrodynamic interaction are essentially the same as the elastic models with hydrodynamic interaction.

The calibration between the model length a and radius r and the measured fiber length a_f and radius r_f are obtained by comparing a and r to a_f and r_f. Recall from the SEM images of the treated sonicated suspensions, the representative fiber length $a_f = 2$ μm and representative radius $r_f = 100$ nm. Since the ratio $\frac{a}{a_f} \approx \frac{r}{r_f} \approx 0.11$, we conclude the model length a and radius r are proportional to the measured a_f and r_f by essentially the same factor 0.11. This factor, obtained from oscillatory shear data for the 2 wt% concentration, is employed at all concentrations and in both oscillatory and steady shear with no refitting, since the characteristic fiber length and radius are the same at all concentrations, in either type of shear flow [45].

To model the small amplitude oscillatory shear experiments, we apply the fiber length $a = 0.23$ μm, radius $r = 10.5$ nm and different mass concentration c to Eq. (6) to deduce the elastic modulus G′ and viscous modulus G″ of the treated sonicated suspensions shown in Figure 17b and Figure 18b. The comparisons between the measured moduli and predicted moduli show that the elastic models capture the trends exhibited in the measured response: the predicted slopes and orders of magnitude of G′ and G″ are consistent with the experimental measurements; the elastic models predict the that the addition of the particles into the Newtonian solvent creates an elasticity in the composite suspension and the moduli magnitudes increase with the nanofiber concentration. As the elastic models do not take colloidal forces into account, the plateau in the elastic modulus G′ observed in the suspensions at low frequencies could not be captured by the models. At high frequencies, the models overpredict the separation of G′.

To model the steady shear experiments, we employ the Giesekus model (1), which is the elastic model with anisotropic hydrodynamic drag, which predicts shear-rate dependent non-constant viscosity [53,54]:

$$\eta(\dot{\gamma}) = \eta_s + \eta_p \frac{(1-\psi)^2}{1+(1-2\alpha)\psi} = \eta_s + nkT\lambda \frac{(1-\psi)^2}{1+(1-2\alpha)\psi}, \tag{8}$$

where

$$\psi = \frac{1-\chi}{1+(1-2\alpha)\chi},$$

$$\chi = \sqrt{\frac{\left[1+16\alpha(1-2\alpha)\left(\dot{\gamma}\lambda\right)^2\right]^{1/2} - 1}{8\alpha(1-2\alpha)\left(\dot{\gamma}\lambda\right)^2}}$$

$$= \sqrt{\frac{\left[1+16\alpha(1-2\alpha)\left(\dot{\gamma}\frac{\pi\eta_s ra^2}{2kT}\right)^2\right]^{1/2} - 1}{8\alpha(1-2\alpha)\left(\dot{\gamma}\frac{\pi\eta_s ra^2}{2kT}\right)^2}}.$$

In the Giesekus model, the mobility factor α describes the shear thinning property of the suspensions.

The same microstructural parameters a and r inferred from the SEM morphological study that were found to be successful in predicting the small amplitude oscillatory shear behavior of the treated sonicated suspensions at all concentrations also are successful in predicting their steady shear behavior. Inserting model length a = 0.23 μm, radius r = 10.5 nm and mass concentration c = 0, 0.5, 1, 2, 5 wt% into Eq. (8) produces the model predictions of steady shear viscosity of the treated sonicated suspensions (Figure 9b). We deduce $\alpha \approx 10^{-11}$ in the Giesekus model (8), for all mass concentrations by fitting to the experimental measurements in a to minimize the error

$$\delta = \sum_{i=1}^{N} \left[\log_{10} \eta_{\exp}(\dot{\gamma}) - \log_{10} \left(\eta_s + nkT\lambda \frac{(1-\psi)^2}{1+(1-2\alpha)\psi} \right) \right]^2 . \tag{9}$$

where $\eta_{\exp}(\dot{\gamma}_i)$ is the experimental measured steady shear viscosity at N discrete shear rate $\dot{\gamma}_i$ at each particular concentration c. Comparing Figure 9a and Figure 9b, we conclude that the elastic models (with isotropic or anisotropic hydrodynamic drag) are successful in capturing the trends exhibited in the measured response: the elastic models correctly predict that the addition of the particles into the Newtonian solvent increases the steady shear viscosity in the composite suspension. The magnitudes of the steady shear viscosities predicted by the elastic models are also consistent with the experimental measurements.

The comparisons between the experimental measurements and model prediction of the small amplitude oscillatory experiments and steady shear experiments (Figure 9, Figure, and Figure) indicate that the single-mode elastic models based on microstructural measurements successfully model the behavior of treated sonicated suspensions over the range of nanofiber concentrations. We surmise that the success of the elastic models is due to the similarity of the physical system to assumptions of the models, namely a dilute suspension of particles in a Newtonian solvent.

6.3. *Modeling predictions for untreated sonicated nanofiber suspensions*

The morphological study of the untreated sonicated suspensions revealed the presence of many large agglomerates of entangled fibers. In the absence of a microstructurally based model which accounts for both loose fibers and large agglomerates of fibers in a Newtonian solvent, we model the agglomerates in the context of the dumbbell models by keeping the bead radius r unchanged from the models of the treated sonicated suspensions, but propose the concept of an effective connector length. The effective length models both the loose fibers and the large agglomerates [45].

The effective spring length $a_{untreated}$ for the untreated sonicated suspensions is obtained by keeping r = 10.5 nm unchanged and minimizing the error Eq. (9), where $\eta_{\exp}(\dot{\gamma}_i)$ is the experimental measured steady shear viscosity at the N = 42 discrete shear rate $\dot{\gamma}_i$ for a particular concentration c = 5 wt%. By exhaustive searching method, the effective length producing the best fit is $a_{untreated} = 1500a_{treated} = 345$ μm. Since the ratio of particle radius to length is extremely small $\left(h = \dfrac{r}{a} \approx 0\right)$, the influence of the hydrodynamic interactions in the untreated sonicated suspensions is miniscule and the result of elastic models without hydrodynamic interaction are essentially the same as the elastic models with hydrodynamic interaction.

Inserting this effective length $a_{untreated} = 1500a_{treated} = 345$, r =10.5 nm and mass concentration c = 0, 0.5, 1, 2, 5 wt% into Eq. (8) produces the model predictions of steady shear viscosity of untreated sonicated suspensions (Figure 8b). Comparing the experimental measurements (Figure 8a) with Giesekus model predictions (Figure 8b), we conclude that despite its debatable physical basis, the effective length modeling approach in the Giesekus model correctly predict the orders of magnitude of steady shear viscosity and its dependence on fiber concentration. The Giesekus model also successfully captures the significant shear thinning behavior of the untreated sonicated suspensions. The comparison of the experimental measurements and model prediction of shear viscosity as a function of shear stress of the untreated sonicated suspensions (Figure 10a and Figure 10b) indicates the Giesekus model predicts the correct dependence of viscosity on shear stress, except at low shear rates.

In the microstructurally based parameter list (2), the solvent contribution to the viscosity η_s, relaxation time λ, and the particle contribution to the viscosity η_p are explicit functions of temperature T. To study the temperature effects on the viscosity of untreated sonicated suspensions, we insert T = 5°C, 15°C, 25°C, 35°C, and 45°C and c = 2 wt% into Eq. (8), and found that the Giesekus model successfully captures the measured temperature dependence of the untreated sonicated suspensions in shear and accurately predicts the magnitude and shape of the steady shear viscosity (Figure 19a and Figure 19b).

7. Modeling of CNF/PS Melt Composites

7.1. *Model and simulation methods*

In this section, we present models for the CNF/PS melt composites which determine the relationships between the rheological properties of the polymer, the nanostructure of the fibers, and the processing conditions.

The behavior of the nanoparticle/polymer melt composite systems we observe in our experiments is fundamentally different from the behavior of fibers in Newtonian solvent in two significant ways:

a) the polymer melt without nanoparticles exhibits shear thinning, extensional, and viscoelastic tendencies that are not observed in the Newtonian solvent;
b) the interaction between the nanofiber inclusions and the polymer molecules is a significant factor not present in the suspensions.

We model the nanoparticle/polymer melt composites as rigid particles (the carbon nanofibers) in a viscoelastic solvent (the polymer melt), adopting the following coupled constitutive equations (10), (11), (12) [55]:

model for the CNF/polymer composite system:

$$\tau_{ij} = -p\delta_{ij} + 2\eta_s D_{ij} + \tau_{ij}^{CNF} \tag{10}$$

viscoelastic model for polymer matrix:

$$\lambda \frac{D\tau_{ij}^p}{Dt} + \sigma\tau_{ij}^p + \frac{\alpha\lambda}{\eta_p}\left(\tau_{ik}^p\tau_{kj}^p\right) + \frac{m(1-\sigma)}{2}\left(a_{ik}\tau_{kj}^p + \tau_{ik}^p a_{kj}\right) = 2\eta_p D_{ij}, \tag{11}$$

model for CNF behavior:

$$\tau_{ij}^{CNF} = 2\left[\eta_s + \eta\right]\phi\left[AD_{kl}a_{ijkl} + B\left(D_{ik}a_{kj} + a_{ik}D_{kj}\right) + CD_{ij} + 2Fa_{ij}D_r\right]. \tag{12}$$

In Eq. (10), τ_{ij}^c is the total stress in the polymer nanocomposite system, $2\eta_s D_{ij}$ is the stress contribution from a Newtonian solvent, if present, with D_{ij} the symmetric part of the velocity gradient and η_s the solvent viscosity. τ_{ij}^p is the stress contribution from polymer molecules with fiber inclusions, τ_{ij}^{CNF} is the stress contribution from carbon nanofiber, and p is the pressure maintaining incompressibility.

Equation (11), introduced by Azaiez [55], is a modified form of Giesekus model which includes the interaction between polymer matrix and fibers. The fiber/polymer matrix interaction parameter σ characterizes the effect of particle orientation on the hydrodynamic drag acting on the polymer molecules [55]. In the special case $\sigma = 1$, the drag force on the polymer molecules is independent of the fiber orientation, and the usual Giesekus model [53] is obtained. m is the dimension of the space ($m = 2$ for planar flows and $m = 3$ for 3-dimensional flows). Constants η_p, α, λ are the polymer viscosity, mobility factor, and relaxation time, respectively, of the melt phase of the polymer matrix without nanoparticles.

The second order orientation tensor a_{ij} produced by integration over a volume containing a sufficiently large number of nanoparticles describes the orientation state of the entire composite system, just as second order orientation tensor $\langle q_i q_j \rangle$ for the CNF suspensions. Explicitly, the nanoparticle in our model of the composite system is described by a rigid rod with a unit vector **p** along the axis denoting its orientation, as opposed to the dimensional connector vector **q** between the two beads in the dumbbell model employed in the previous CNF suspension systems. $\psi(\mathbf{p})$ is the orientation probability distribution function. The evolution equations of a single nanoparticle orientation vector **p** and the distribution function $\psi(\mathbf{p})$ are [56]:

$$\dot{p}_i = \frac{1}{2}\left(W_{ij}p_j\right) + \chi\left(D_{ij}p_j - D_{kl}p_k p_l p_i\right) - D_r \frac{1}{\psi}\frac{\partial \psi}{\partial p_i}, \tag{13a}$$

$$\frac{\partial \psi}{\partial t} = -\frac{\partial}{\partial p_i}\left(\dot{p}_i \psi\right), \tag{13b}$$

where D_r is the rotary diffusivity due to Brownian motion. Advani and Tucker [56] employs $D_r = 2C_I\left|\mathbf{D}\right|$ as suggested by Folgar and Tucker [57] for concentrated suspensions of large (i.e., non-Newtonian) fibers.

The second order orientation tensor is defined as the dyadic products of **p** averaged over orientation space [56]:

$$a_{ij} = \int p_i p_j \psi(\mathbf{p})\, d\mathbf{p}, \tag{14}$$

with $p_1 = \sin\theta\cos\phi, p_2 = \sin\theta\sin\phi, p_3 = \cos\theta$. Multiplying the kinetic Eq. (1) by $p_i p_j$ and integrating over the entire space gives:

$$\frac{d}{dt}a_{ij} = \int \frac{\partial \psi}{\partial t} p_i p_j d\mathbf{p} = \int \psi\left(\dot{p}_i p_j + p_i \dot{p}_j\right) d\mathbf{p}. \tag{15}$$

Insertion of Eq. (12) into Eq. (15) produces the evolution equation of the second order orientation tensor a_{ij} [56]:

$$\begin{aligned}\frac{da_{ij}}{dt} = &\left(W_{ik}a_{kj} - a_{ik}W_{kj}\right) + \chi\left(D_{ik}a_{kj} - a_{ik}D_{kj} - 2D_{kl}a_{ijkl}\right) \\ &+ 4C_I\left|\mathbf{D}\right|\left(\delta_{ij} - m a_{ij}\right),\end{aligned} \tag{16}$$

where $|\mathbf{D}| = \sqrt{2D_{ij}D_{ji}}$ and χ is a shape parameter related to the particle aspect ratio $h = \frac{L}{D}$,

$$\chi = \frac{h^2 - 1}{h^2 + 1} \tag{17}$$

where L is particle length and D is particle diameter. C_I is the interparticle interaction parameter which measures the intensity of fiber interaction in the suspension and is correlated with the fiber aspect ratio and volume fraction. When modeling injection molding and other similar processes, C_I is commonly in the range 10^{-7} to 10^{-2} [55,58]. For concentrated systems with screening mechanisms, the Bays empirical expression [59] is often used:

$$C_I = 0.0184\exp(-0.7148\phi h), \tag{18}$$

with ϕ the volume fraction of the composite system. Here we adopt $C_I = 10^{-2}$ for MB and SC composite systems since $C_I = 10^{-2}$ gives a highly aligned fiber orientation at steady states as observed in our experiments. In Eq. (16), a_{ijkl} is the fourth order orientation tensor of the nanoparticles defined as [56]:

$$a_{ijkl} = \int p_i p_j p_k p_k \psi(\mathbf{p})\, d\mathbf{p}. \tag{19}$$

Equation (12) is of the form derived by Tucker [60]. In Eq. (12) η is the viscosity contribution from the polymer matrix; in simple shear in the 1-2 plane $\eta = \frac{\tau_{12}^p}{\dot{\gamma}}$ with $\dot{\gamma}$ the shear rate. A, B, C, and F are functions of the particle aspect ratio h of the nanoparticle. In dilute regimes [60]:

$$A = \frac{h^2}{2\left[\ln(2h) - 1.5\right]}, B = \frac{6\ln(2h) - 11}{h^2}, C = 2, F = \frac{3h^2}{\ln(2h) - 0.5}; \tag{20}$$

in semi-dilute regimes [61]:

$$A = \frac{h^2}{3\ln\left(2h_f / D\right)}, \; B = 0, \; C = 0, \; F = 0, \tag{21}$$

with h_f characteristic distance between a fiber and its nearest neighbors:

$$\frac{2h_f}{D} = \sqrt{\frac{\pi}{\phi}} \text{ for aligned fibers,} \tag{22}$$

$$\frac{2h_f}{D} = \frac{\pi}{2\phi h} \text{ for random fibers.} \tag{23}$$

The set of Eqs. (10), (11), (12) and (16) as presented is underdetermined (more unknowns than equations). To restore closure one must adopt an approximation that relates the fourth-order orientation tensor a_{ijkl} in Eqs. (11) and (16) to the second order orientation tensor a_{ij}. The most popular closure approximations are [62] the linear closure approximation,

$$\begin{aligned} \hat{a}_{ijkl} &= -\frac{1}{35}\left(\delta_{ij}\delta_{kl} + \delta_{ik}\delta_{jl} + \delta_{il}\delta_{jk}\right) \\ &+ \frac{1}{7}\left(a_{ij}\delta_{kl} + a_{ik}\delta_{jl} + a_{kl}\delta_{ij} + a_{jl}\delta_{ik} + a_{jk}\delta_{il}\right), \end{aligned} \tag{24}$$

the quadratic closure approximation

$$\tilde{a}_{ijkl} = a_{ij}a_{kl}, \tag{25}$$

and the hybrid closure approximation,

$$a_{ijkl} = \left(1 - f\right)\hat{a}_{ijkl} + f\tilde{a}_{ijkl} \text{ with } f = 1 - 27\left|\mathbf{a}\right|. \tag{26}$$

7.2. *Modeling results and discussion*

To characterize the constitutive model (10)–(12), we need to deduce from our experiments the material constants λ, α, η_p, η_s and σ, and the aspect ratio h. The constants λ, α, η_p, η_s are properties of the polymer melt alone, not the polymer/CNF composite, and determined from experiments on the pure polymer. Zero shear viscosity $\eta_0 = \eta_p + \eta_p$

is the limit of $\frac{\tau_{12}}{\dot{\gamma}}$ as $\dot{\gamma} \to 0$. In our experiments on pure polymer, we choose η_0 to be the measured viscosity at shear rate $\dot{\gamma} = 0.001\ \mathrm{s}^{-1}$. Since there is no solvent in the polymer matrix, the solvent viscosity $\eta_s = 0$ for the polymer melt. Therefore we have $\eta_p = \eta_0$.

For small amplitude oscillatory shear experiments, Eq. (11) with $\alpha = 0$ and $\sigma = 1$ implies that the storage modulus $G' = \eta''\omega$ and loss modulus $G'' = \eta'\omega$ are of the forms

$$G''(\omega) = \left(\eta_s + \frac{\eta_p}{1+(\lambda\omega)^2}\right)\omega,\ G'(\omega) = \frac{\eta_p \lambda \omega^2}{1+(\lambda\omega)^2}. \tag{27}$$

The relaxation time λ of the polymer melts is obtained by minimizing the error

$$\delta_1 = \sum_{i=1}^{N} \left\{ \begin{array}{l} \left[\log_{10} G''_{\exp}(\omega_i) - \log_{10}\left(\left(\eta_s + \frac{\eta_p}{1+(\lambda\omega_i)^2}\right)\omega_i\right)\right]^2 \\ + \left[\log_{10} G'_{\exp}(\omega_i) - \log_{10}\left(\frac{\eta_p \lambda \omega_i^2}{1+(\lambda\omega_i)^2}\right)\right]^2 \end{array} \right\}, \tag{28}$$

where $G'_{\exp}(\omega_i)$ and $G''_{\exp}(\omega_i)$ are measured storage modulus and loss modulus at frequency ω_i. N is the number of measurements, here we use $N = 24$ measurements up to frequency $\omega = 15$ rad/s.

For steady shear at shear rate $\dot{\gamma}$, Eq. (11) with $\sigma = 1$ implies that shear viscosity $\eta(\dot{\gamma})$ and first normal stress coefficient ψ_1 of the pure PS are determined by:

$$\eta(\dot{\gamma}) = \eta_s + \eta_p \frac{(1-\psi)^2}{1+(1-2\alpha)\psi},\ \psi_1(\dot{\gamma}) = 2\lambda\eta_p \frac{\psi(1-\alpha\psi)}{(\dot{\gamma}\lambda)^2 \alpha(1-\psi)}, \tag{29}$$

with

$$\psi = \frac{1-\chi}{1+(1-2\alpha)\chi},\ \chi = \sqrt{\frac{\left[1+16\alpha(1-2\alpha)\left(\dot{\gamma}\lambda\right)^2\right]^{1/2}-1}{8\alpha(1-2\alpha)\left(\dot{\gamma}\lambda\right)^2}}.$$

with λ already determined, the mobility factor α of the polymer melts is obtained by minimizing the error

$$\begin{aligned}\delta_2 &= \sum_{i=1}^{N}\left[\log_{10}\eta_{\exp}(\dot{\gamma}_i) - \log_{10}\left(\eta_p \frac{(1-\psi)^2}{1+(1-2\alpha)\psi}\right)\right]^2 \\ &+ \sum_{i=1}^{N}\left[\log_{10}\Psi_{1_{\exp}}(\dot{\gamma}_i) - \log_{10}\left(2\lambda\eta_p \frac{\psi(1-\alpha\psi)}{(\dot{\gamma}\lambda)^2\alpha(1-\psi)}\right)\right]^2\end{aligned} \tag{30}$$

where $\eta_{\exp}(\dot{\gamma}_i)$ and $\Psi_{1_{\exp}}(\dot{\gamma}_i)$ are the measured viscosity and first normal stress coefficient at shear rate $\dot{\gamma}_i$.

We use an exhaustive method in the searches through λ and α space to minimize errors (28) and (30). Firstly, we propose a physically reasonable range for each fitting parameter, namely $0 < \lambda < 10$ s in the search over λ to minimize δ_1 in Eq. (28), and then $0 < \alpha < 1$ in the search over α to minimize δ_2 in Eq. (30). The searches are first carried out in these ranges using a relatively coarse mesh, after which the parameter is restricted in a smaller range and a refined mesh is employed to search the best fit. The exhaustive method is feasible in our experiments since there are only two decoupled 1-parameter searches over a well-defined domain. Finally, we verify that the best fits of parameters are not near the edges of the domains.

To obtain interaction parameter σ and aspect ratio h, we utilize all four Eqs. (10), (11), (12), and (16). With a_{ijkl} expressed in terms of a_{ij} through a closure approximation such as (24)–(26), the evolution equation (16) for the second order orientation tensor decouples from Eqs. (10)–(12), so that $a_{ij}(t)$ can be solved as a function of aspect ratio h from

Eq. (16) alone. Equations (10), (11), (12) then are a closed set of coupled equations for τ_{ij}^{p}, τ_{ij}^{CNF}, and τ_{ij}^{c}.

The second order orientation tensor a_{ij} is first obtained with initial aspect ratio h based on the experimental observation of $h \approx 10$ for MB composites and $h \approx 50$ for SC composites. Searching through parameter space (h, σ), we solve Eqs. (10), (11) and (12) to minimize the error

$$\delta_3 = \sum_{i=1}^{N} \left[\log_{10} \eta_{\exp}(\dot{\gamma}_i) - \log_{10}\left(\eta_{\text{model}}(\dot{\gamma}_i)\right) \right]^2, \tag{31}$$

and produce new aspect ratio h and σ. Using the new aspect ratio h, we recalculate the orientation tensor a_{ij} with a revised value of χ, and search (h, σ) again until the searching result h is the same as h used to calculate orientation tensor a_{ij}.

Parameter σ measures the interaction between the polymer matrix and carbon nanoparticles. We investigate whether σ is the same for all mass concentrations or a function of mass concentration. Therefore, we deduce the interaction parameter σ by two different methods. In method 1, we assume the interaction parameter σ is independent of the mass concentration. Since σ is the same for all the mass concentrations, we search over σ and aspect ratio h by solving Eqs. (10)–(12) and minimizing the error (31) summed over all 27 experimental measurements of steady shear viscosity η of the CNF/PS melt composites. In method 2, we assume the interaction parameter σ depends on the mass concentration. For each aspect ratio h, which is same for all mass concentrations, we search for a different σ for each of the three mass concentrations by minimizing error (31) summed over the 7 measurements at each mass concentration. We search over aspect ratio h and its corresponding three σ's until the minimum error (31) is obtained.

For the steady shear flow, inserting flow D_{ij}, W_{ij} into Eq. (16), using closure approximation (25), and assuming homogeneous steady behavior (i.e. $\dot{\gamma}$ = constant, $\frac{d}{dt}a_{ij}(t) = 0$) produce the following system of equations for the components of the symmetric tensor a_{ij}:

$$
\begin{aligned}
&0=\dot{\gamma}a_{12}+\chi\dot{\gamma}\left(a_{12}-2a_{12}a_{11}\right)+2C_I\dot{\gamma}\left(1-3a_{11}\right),\\
&0=\frac{1}{2}\dot{\gamma}a_{22}-\frac{1}{2}\dot{\gamma}a_{11}+\chi\dot{\gamma}\left(\frac{1}{2}a_{22}+\frac{1}{2}a_{11}-2a_{12}^2\right)-6C_I\dot{\gamma}a_{12},\\
&0=\frac{1}{2}\dot{\gamma}a_{23}+\chi\dot{\gamma}\left(\frac{1}{2}a_{23}-2a_{12}a_{13}\right)-6C_I\dot{\gamma}a_{13},\\
&0=-\dot{\gamma}a_{12}+\chi\dot{\gamma}\left(a_{12}-2a_{12}a_{22}\right)+2C_I\dot{\gamma}\left(1-3a_{22}\right),\\
&0=-\frac{1}{2}\dot{\gamma}a_{13}+\chi\dot{\gamma}\left(\frac{1}{2}a_{13}-2a_{12}a_{23}\right)-6C_I\dot{\gamma}a_{23},\\
&0=-2\chi\dot{\gamma}a_{12}a_{33}+2C_I\dot{\gamma}\left(1-3a_{33}\right).
\end{aligned}
\tag{32}
$$

Note that with the interparticle interaction parameter C_I independent of shear rate, $\dot{\gamma}$ cancels from the equations, i.e., the second order orientation tensor a_{ij} is independent of the shear rate $\dot{\gamma}$. With C_I specified and χ given, we solve for second order orientation tensor a_{ij} in terms of the aspect ratio h. The coupled Eqs. (10), (11), (12) for τ_{ij}^{P}, τ_{ij}^{CNF}, and τ_{ij}^{c} are then employed to deduce values for σ and h.

We now employ the method described in section 7.1 and 7.2 to explicitly characterize the melt blending and solvent casting CNF/PS composite of this paper.

7.2.1. *Steady state shear flow of melt blended CNF/PS melt composite systems*

For melt blended composite systems, the measured viscosity for the pure polymer at shear rate $\dot{\gamma}=0.001\ s^{-1}$ is 9864 Nm/s^2, so that $\eta_p = \eta_0 = 9864$ Nm/s^2. Minimizing error (28) for the 24 experimental measurements produces $\lambda = 0.329$ s. Minimizing error (30) for the seven experimental measurements produces $\alpha = 0.392$.

According to experimental observation, as an initial value we set the particle aspect ratio $h = 10$. The interparticle interaction parameter is chosen as $C_I = 10^{-2}$, which is commonly used for semi-dilute and concentrated system [55,58]. With these values of h and C_I, we solve Eq. (32) for the second order orientation tensor a_{ij} using quadratic closure approximation (25).

Interaction parameter σ is deduced from the steady shear experimental measurements by the following two methods:

In method 1, which assumes the interaction parameter σ is independent of the mass concentration, there are a total 21 steady shear measurements, with 7 measurements at each mass concentration c = 2 wt%, 5 wt%, and 10 wt%. Since σ is the same for all the mass concentrations, we search σ and aspect ratio h by minimizing error (31) for the total 21 experimental measurements. Searching through parameter space ($0 < h < 200$, $0 < \sigma < 1$), we solve equations (10), (11), and (12) and produce $h = 56.00$ and $\sigma = 0.769$. Using the new aspect ratio $h = 56.00$, we recalculate the orientation tensor a_{ij}, and search the parameter space (h, σ) again. Finally, we obtain aspect ratio $h = 50.56$ and interaction parameter $\sigma = 0.754$ for all mass concentrations.

In method 2, which assume the interaction parameter σ depends on the mass concentration, since σ is different for each mass concentration, we search σ and aspect ratio h by minimizing error (31) for each 7 measurements at mass concentration c = 2 wt%, 5 wt%, and 10 wt%, respectively. Using the similar iteration method, we finally obtain $h = 54.3$ and $\sigma = 0.790$, $\sigma = 0.725$ and $\sigma = 0.540$ corresponding to c = 2%, 5% and 10%, respectively.

In both cases, the values of σ are much less than 1, which indicates the interaction between the nanoparticle and polymer matrix cannot be neglected, i.e. the interaction term $\frac{m(1-\sigma)}{2}\left(a_{ik}\tau_{kj}^{p} + \tau_{ik}^{p}a_{kj}\right)$ in the modified Giesekus model (12) is not trivial. Furthermore, the decreasing of interaction parameter σ in method 2 denotes the fiber/polymer matrix interaction increases with mass concentration.

Figure shows the model prediction of the viscosity and first normal stress difference of the steady shear experiments. The comparison between measured steady shear viscosity of the MB polymer melt composites and predicted viscosity (Figure 21a) shows the rigid rod model of carbon nanofibers combined with the modified Giesekus model of polystyrene melt correctly predicts the magnitude of steady shear viscosity and its dependence on fiber concentration. The model also successfully captures the shear thinning behavior of the MB composites.

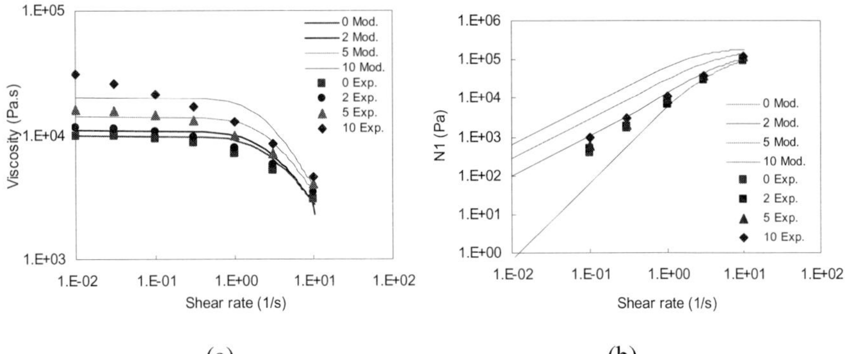

Fig. 21. (a) Model prediction of the viscosity and (b) first normal stress difference (N_1) of the steady shear experiments of melt blended composite systems.

The comparison between the measured composite first normal stress difference (*N1*) and the predicted first normal stress difference (Figure 21b) shows that the predicted slopes of normal stress difference are consistent with the experimental measurements. The model correctly predicts that the addition of the particles into polymer melt increases the first normal stress difference except that the model overpredicts the separations of the first normal stress difference.

7.2.2. *Steady state shear flow of solvent cast CNF/PS melt composite systems*

For the solvent cast composite system, the measured viscosity for the pure polymer at shear rate $\dot{\gamma} = 0.001\ s^{-1}$ is 11160 Nm/s^2, so that $\eta_p = \eta_0 = 11160$ Nm/s^2. Minimizing error (28) for the twenty four experimental measurements produces $\lambda = 0.321$ s. Minimizing error (30) for the seven experimental measurements produces $\alpha = 0.479$.

According to the experimental observation, we initially set the particle aspect ratio h = 50 and again choose $C_I = 0.01$. Solving Eq. (32) using quadratic closure approximation (25) produces the second order orientation tensor a_{ij}.

Interaction parameter σ is also studied by two methods. By method 1, which assumes the interaction parameter σ is independent of the mass concentration, we converge to h = 73.79 and σ= 0.561. By method 2, for each aspect ratio *h,* the interaction parameter σ is different for each mass concentration, we converge to aspect ratio h = 73.79 and interaction parameter σ= 0.688, σ= 0.549 and σ= 0.469 corresponding to c = 2%, 5% and 10%, respectively. Comparing the fitting results of these two methods (Table 2), we conclude there is significant difference between the values of h and σ. Therefore, we conclude h = 76.39 and σ= 0.668, σ = 0.549 and σ = 0.469 corresponding to c = 2%, 5% and 10%, respectively, for the solvent cast composites.

Figure 22 shows the model prediction of viscosity and first normal stress difference (N_1) of the steady shear experiments. The comparison between measured steady shear viscosity of the SC composites and predicted viscosity (Figure 22a) shows the model correctly predict the magnitude of steady shear viscosity and its dependence on fiber concentration. The model successfully captures the shear thinning behavior of the SC composites. At high mass concentration (c = 10 wt%), the model fails to predict the high shear viscosity at low shear rate because the network effect is not included in the model. The comparison between the measured composite first normal stress difference (*N1*) and the predicted first normal stress difference (Figure 22b) shows that the predicted slopes of normal stress difference are consistent with the experimental measurements. The model correctly predicts that the addition of the particles into polymer melt increases the first normal stress difference except that the model overpredicts the separations of normal stress difference at high shear rate.

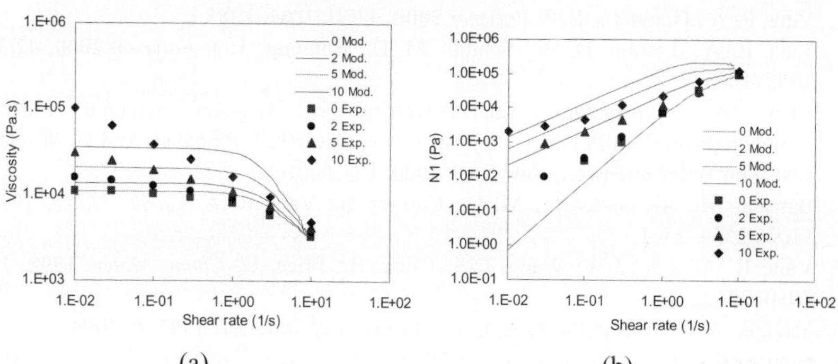

Fig. 22. (a) Model prediction of the viscosity and (b) first normal stress difference (N_1) of the steady shear experiments of solvent cast composite systems.

Table 2. Material constants in MB and SC CNF/PS melt composites.

Parameters		MB		SC	
Pure polymer properties	λ	0.329 s[*]		0.321 s[†]	
	α	0.392[*]		0.479[†]	
	η_p	9864 Pa.s[*]		11160 Pa.s[+]	
	η_s	0 Pa.s[*]		0 Pa.s[†]	
C_1		0.01		0.01	
Fitting parameters of the composites	σ	*if independent on mass concentration*	*if dependent on mass concentration*	*if independent on mass concentration*	*if dependent on mass concentration*
		0.754	0.790 (MB2) 0.725 (MB5) 0.540 (MB10)	0.561	0.688 (SC2) 0.549 (SC5) 0.469 (SC10)
	h	50.56	54.30	73.79	76.39

[*]measured from pure polymer that had undergone the melt blending process.
[†]measured from pure polymer that had undergone the solvent casting process.

Acknowledgements

The authors would like to acknowledge the financial support provided by the National Science Foundation (DMI-0115445).

References

1. Vaia, R. A.; Giannelis, E. P. *MRS Bulletin* **2001,** 26, 394.
2. Vaia, R. A.; Giannelis, E. P. *Polymer* **2000**, 42(3), 1281–1285.
3. Vaia, R. A.; Tomlin, D. W.; Schulte, M. D.; Bunning, T. J. *Polymer* **2000**, 42(3), 1055–1065.
4. Vaia, R. A.“Structural Characterization of Polymer Layered Silicate Nanocomposites” in Polymer-Clay Nanocomposites T. J. Pinnavaia and G. W. Beal eds. John Wiley and Sons, New York **2000**.
5. Banach, M.; Alexander Jr., M. D.; Carraci, S.; Vaia, R. A. *Chem. Mater.* **1999,** 11(9), 2554–2561.
6. Vaia, R. A.; Lee, J-W; Wang, C-S; Click, B.; Price, G. *Chem. Mater.* **1998**, 10, 2030–2032.
7. Vaia, R. A.; Ishii, H.; Giannelis, E. P. *Chemistry of Materials* **1993,** 5, 1694.
8. Giannelis, E. P. *Advanced Materials* **1996**, 8, 29.
9. Vaia, R. A.; Giannelis, E. P. *Macromolecules* **1997,** 30, 8000.
10. Krishnamoorti, R.; Vaia, R. A.; Giannelis, E. P. *Chemistry of Marerials*, **1996,** 8, 1728.

11. Giannelis, E. P.; Krishnamoorti, R.; Manias, E. "Polymer-Silicate Nanocomposites: Model Systems for Confined Polymers and Polymer Brushes", in Advances in Polymer Science, S. Granick, Ed., Springer, 138, **1999**.
12. Shia, D.; Hui, C. Y.; Burnside, S. D.; Giannelis, E. P. *Polymer Composites* **1998**, 19, 608.
13. Nanocomposites 1999: Polymer Technology for the Next Century. Principia Partners, Exton, PA **1999**.
14. Alexandre, M.; Dubois, P. Mater. Sci. Eng. 28, **2000**, 1–63 "Nanocomposites New Low-Cost, High-Strength Materials for Automotive Parts", National Insititutes of Technology, ATP Project, 97-02-0047, 1997.
15. Ruoff, R. S.; Lorents, D. C. *Carbon* **1995**, 33, 925.
16. Mintmire, J. W.; White, C. T. *Carbon* **1995**, 33, 893.
17. Treacy, M. M. J.; Ebbesen, T. W.; Gibson, J. M. *Nature* **1996**, 381, 678.
18. Wong, E. W.; Sheehan, P. E.; Lieber, C. M. *Science* **1997**, 277, 1971.
19. Poncharal, P.; Wang, Z. L.; Ugarte, D.; de Heer, W. A. *Science* **1999**, 283, 1513.
20. Giesekus, H. Journal of Non-Newtonian Fluid Mechanics **1982,** 11, 69–109.
21. Bird, R. B.; Warner, H. R. J.; Evans, D. C. *Advances in Polymer Science* **1971,** 8, 1–90.
22. Glasgow, D. G.; Jacobsen, R. L.; Burton, D. J.; Kwag, C.; Kennel, E.; Lake, M. L.; Brittain, W. J.; Rice, B. P. In *Carbon nanofiber polymer composites*, International SAMPE Symposium and Exhibition, 2003
23. Lake, M. L.; Glasgow, D. G.; Kwag, C.; Burton, D. J. In *Carbon nanofiber polymer composites: electrical and mechanical properties*, International SAMPE Symposium and Exhibition., 2002, 1794–1800.
24. Shi, D.; Lian, J.; He, P.; Wang, L. M.; Xiao, F.; Yang, L.; Schulz, M. J.; Mast, D. B. *Applied Physics Letters* **2003,** 83, (25), 5301–5303.
25. Enomoto, K.; Yasuhara, T.; Kitakata, S.; Murakami, H.; Ohtake, N. *New Diamond and Frontier Carbon Technology* **2004,** 14, (1), 11–20.
26. McKenzie, J. L.; Waid, M. C.; Shi, R.; Webster, T. J. *Biomaterials* **2003,** 25, (7–8), 1309–1317.
27. Lozano, K.; Barrera, E. V. *J Appl Polym Sci* **2001,** 79, 125–133.
28. Lozano, K.; Bonilla-rios, J.; Barrera, E. V. *J Appl Polym Sci* **2001,** 80, 1162–1172.
29. Ma, H.; Zeng, J.; Realff, M. L.; Kumar, S.; Schiraldi, D. A. *Polymeric Materials Science and Engineering* **2002,** 86, 411–412.
30. Lozano, K.; Yang, S.; Zeng, Q. *Journal of Applied Polymer Science* **2004**, 93, 155–162.
31. Du, F.; Scogna, R. C.; Zhou, W.; Brand, S.; Fischer, J. E.; Winey, K. I. *Macromolecules* **2004,** 37, 9048–9055.
32. Abdel-Goad, M.; Potschke, P. *Journal of Non-Newtonian Fluid Mechanics* **2005**, 28, 2–6.
33. Kinloch, I. A.; Roberts, S. A.; Windle, A. H. *Polymer* **2002**, 43,7483–7491.
34. Hough, L. A.; Islam, M. F.; Janmey, P. A.; Yodh, A.G. *Physical Review Letters* **2004**, 93, 168102-1-168-102-4.
35. R. G. Larson, The Structure and Rheology of Complex Fluids, Oxford, **1998**.

36. Niyogi, S.; Hamon, M. A.; Hu, H.; Zhao, B.; Bhowmik, P.; Sen, R.; Itkis, M. E.; Haddon, R. C. *Accounts of Chemical Research* **2002**, 35, 1105–1113.
37. Ago, H.; Petritsch, K.; Shaffer, M. S. P; Windle, A. H.; Friend, R. H. *Advanced Materials* **1999**, 11, 1281–1285.
38. Kymakis, E.; Amaratunga, G. A. J. *Synthetic Metals* **2004**, 142, 161–167.
39. Chen, X.; Yoon, K.; Burger, C.; Sics, I.; Fang, D.; Hsiao, B. S.; Chu, B. *Macromolecules* 2005, 38, 3883–3893.
40. Shaffer, M. S.; Koziol, K. *Chemical Communications* **2002**, 18,2074–2075.
41. Tong, X.; Liu, C.; Cheng, H. M.; Zhao, H.; Yang, F.; Zhang, X. *Journal of Applied Polymer Science* **2004**, 92, 3697 – 3700.
42. Herrick, J. W. In Ann. Tech. Conf., Soc. Plastics Ind., Reinforced Plastic/Composites Div., Section 16A, 1968.
43. Herrick, J. W.; Gruber, J. P. E.; Mansur, F. T. *Surface treatments for fibrous carbon reinforcements*; Air Force Materials Laboratory: 1966.
44. Esumi, K.; Ishigami, M.; Nakajima, A.; Sawada, K.; Honda, H. *Carbon* **1996,** 34, 279–281.
45. Xu, J.; Chatterjee, S.; Koelling, K. W.; Wang, Y.; Bechtel, S. E. *Rheological Acta* **2005**.
46. Lu, K. L.; Lago, R. M.; Chen, Y. K.; Green, M. L. H.; Harris, P. J. F.; Tsang, S. C. *Carbon* **1996,** 34, 814–816.
47. Liu, J.; Rinzler, A. G.; Dai, H.; Hafner, J. H.; Bradley, R. K.; Boul, P. J.; Lu, A.; Iverson, T.; Shelimov, K.; Huffman, C. B.; Rodiguez-Macias, F.; Shon, Y. S.; Lee, T. R.; Colbert, D. T.; Smalley, R. E. *Science* **1998,** 280, 1253–1255.
48. Doi, M.; Edwards, S. F., *The Theory of Polymer Dynamics*. ed.; Clarendon Press: Oxford, 1987.
49. Ganani, E.; Powell, R. L. *J Compos Mater* **1985,** 19, 194–215.
50. Aral, B. K.; Kalyon, D. M. *Journal of Rheology* **1997,** 41, 599–620.
51. Macosko, C. W., Rheology: Principles, Measurements, and Applications. ed.; Wiley-VCH: 1993.
52. Larson, R. G., *The structure and rheology of complex fluids*. ed.; Oxford University Press: Oxford, 1999.
53. Bird, R. B.; Curtiss, C. F.; Armstrong, R. C.; Hassager, O., *Dynamics of polymeric Liquids*. ed.; Wiley&Sons, Inc.: 1987.
54. Macosko, C. W., *Rheology: Principles, Measurements, and Applications*. ed.; Wiley-VCH: New York, 1994.
55. Azaiez, J. Journal of Non-Newtonian fluid mechanics **1996,** 66, 35–54.
56. Advani, S. G.; Tucker, C. L. I. *Journal of Rheology* **1987,** 31, (8), 751–784.
57. Folgar, F.; Tucker, C. L. *Journal of Reinforced Plastics and Composites* **1984,** 3, 98–119.
58. Cintra, J. S.; Tucker, C. L. I. *Journal of Rheology* **1995,** 39, (6), 1095–1122.
59. Chung, D. H.; Kwon, T. H. *Korea-Australia Rheology Journal* **2002,** 14, (4), 175–188.
60. Tucker, C. L., III. *Journal of Non-Newtonian fluid mechanics* **1991,** 39, 239–268.
61. Dinh, S. M.; Armstrong, R. C. *Journal of Rheology* **1984,** 28, 207–227.
62. Advani, S. G.; Tucker, C. L. I. *Journal of Rheology* **1990,** 34, (3), 367–386.

CHAPTER 6

Recent Advances in Polymer/Layered Silicate Nanocomposites: An Overview from Science to Technology

Masami Okamoto

Advanced Polymeric Materials Engineering
Graduate School of Engineering
Toyota Technological Institute
2-12-1 Hisakata, Tempaku, Nagoya 468 8511, Japan
okamoto@toyota-ti.ac.jp

1. Introduction

A decade of research has shown that nanostructured materials have the potential to significantly impact on the growth at every level of the world economy in the 21st century. This new class of materials is now being introduced in structural applications, such as gas barrier film, flame retardant product, and other load-bearing applications (see Table 1 [1]).

Of particular interest is recently developed *nanocomposites* consisting of a polymer and layered silicate (LS) because they often exhibit remarkably improved mechanical and various other properties [2] when compared with pure polymer or conventional composites (both micro- and macro-composites). A primary progress in *polymer/layered silicate nanocomposites* (PLSNCs), a Nylon 6/LS hybrid [3] reported by Toyota Central Research & Development Co. Inc. (TCRD), was successfully prepared by in-situ polymerization of ε-caprolactam in a dispersion of montomorrillonite (MMT). The silicate can be dispersed in liquid monomer or a solution of monomer. It has also been possible to melt-mix polymers with layered silicates, avoiding the use of organic solvents. The latter method permits the use of conventional processing techniques such as injection moulding and extrusion.

Table 1. TPO nanocmposites: Application for automotive parts.

HUMMER H2 SUT: Most Recent Nanocomposite Application	Nanocomposite TPOs Summary of tangible benefits
	Mass savings of 3 to 21% • Specific Gravity of 0.92 vs. 0.96–1.13 g/cm^3 • Lighter weight reduces cost and requires less adhesive for attachment Improved Appearance • Improved Knit Line Appearance • Improved Colorability & Paintability • Sharper Feature Lines & Grain Patterns • Improved Scratch/Mar Performance Large Processing Window • Consistent Physical and Mechanical Properties • Elimination/Reduction of Tiger Striping Reduced Paint Delamination Retains Low Temperature Ductility Improved Recyclability Lower Flammability
Impala: 2nd Nanocomposite Application	M-Van Step Assist: 1st Commercial Launch

*Through the courtesy of M. Verbrugge, General Motors

Continued progress in nanoscale controlling, as well as an improved understanding of the physicochemical phenomena at the nanometer scale, have contributed to the rapid development of novel PLSNCs. This chapter presents current research on PLSNCs with the primary focus of recent advances from basic science to technology.

2. *Historical Point of View*

Earlier attempts of preparing polymer/LS composites are found in almost half-a-century old patent literatures [4,5]. In such cases, incorporation of 40 to 50 wt% clay mineral (bentonite, hectorite, *etc.*) into a polymer was attempted but ended up with unsatisfactory results: The maximal modulus enhancement was only around 200%, although the clay loading was as much as 50 wt%. The failure was obvious because they failed to achieve good dispersion of clay particles in the matrix, in which silicate minerals existed as agglomerated tactoids. Such a poor dispersion of the silicate particles could improve the material rigidity, but certainly sacrificing the strength, the elongation at break and the toughness of the materials [4,5].

A prime reason for this impossibility of improving the tactoids-dispersion into well-dispersed exfoliated monolayers of the silicate is obviously due to the intrinsic incompatibility of hydrophilic layered silicates with hydrophobic engineering plastics. One attempt at circumventing this difficulty was made by Unitika Ltd. [6] about 30 years ago in preparing Nylon 6/LS composites (not nanocomposites) *via in situ* polymerization of ε-caprolactam with montmorillonite (MMT), but the results turned out to be not very good.

The first major breakthrough of the problem was brought about in 1987, when Fukushima and Inagaki of TCRD, *via* their detailed study on polymer/LS composites, persuasively demonstrated that *lipophilization* by replacing inorganic cations in galleries of the native clay with alkylammonium surfactant successfully made them compatible with hydrophobic polymer matrices [7]. The modified clay was thus called *lipophilized* clay, *organo-phillic* clay or simply *organo*-clay (organoclay). Furthermore, they found that the lipophilization enabled to expand silicate galleries and exfoliate the silicate layers into single layers of a nanometer thickness.

Six years later in 1993, Usuki, Fukushima and their colleagues of TCRD successfully prepared, for the first time, *exfoliated* Nylon 6/MMT hybrid *via in situ* polymerization of ε-caprolactam, in which alkylammonium-modified MMT was thoroughly dispersed in advance [3,8].

Table 2. Clay mineral (phyllosilicates) classification.

Type	Group	Groupoid	Species	Tetrahedron	Octahedron	Interlayer cation
2:1 $Si_4O_{10}(OH)_2$	Pyrophyllite Talc (x ~ 0)	di.	Pyrophyllite	Si_4	Al_2	–
		tri.	Talc	Si_4	Mg_3	–
	Sumectite (0.25 < x < 0.6)	di.	Montmorillo-nite	Si_4	$(Al_2, Mg)_2$	Na, Ca, H_2O
		di.	Hectorite	Si_4	$(Mg_2, Li)_2$	Na, Ca, H_2O
		di.	Beidellite	$(Si, Al)_4$	Al_2	Na, Ca, H_2O
		tri.	Saponite	$(Si, Al)_4$	Mg_3	Na, Ca, H_2O
	Vermiculite (0.25 < x < 0.9)	di.	Vermiculite	$(Si, Al)_4$	$(Al, Mg)_2$	K, Al, H_2O
		tri.	Vermiculite	$(Si, Al)_4$	$(Mg, Al)_3$	K, Mg, H_2O
	Mica (x ~ 1)	di.	Muscovite	$Si_3 \cdot Al$	Al_2	K
			Paragonite	$Si_3 \cdot Al$	Al_2	Na
	Brittle mica (x ~ 2)	tri.	Phlogopite	$Si_3 \cdot Al$	$(Mg, Fe^{2+})_3$	K
			Biotite	$Si_3 \cdot Al$	$(Fe^{2+}, Mg)_3$	K
2:1:1 $Si_4O_{10}(OH)_8$	Chlorite (large variation of x)	di.	Donbassite	$(Si, Al)_4$	Al_2	$Al_2(OH)_6$
		di.-tri.	Sudoite	$(Si, Al)_4$	$(Al, Mg)_2$	$(Mg, Al)_3(OH)_6$
		tri.	Clinochlore	$(Si, Al)_4$	$(Mg, Al)_3$	$(Mg, Al)_3(OH)_6$
			Chamosite	$(Si, Al)_4$	$(Fe, Al)_3$	$(Fe, Al)_3(OH)_6$
1:1 $Si_2O_5(OH)_4$	Kaolin-mineral Serpentinite (x ~ 0)	di.	Kaolinite	Si_2	Al_2	—
			Halloysite	Si_2	Al_2	H_2O
		tri.	Chrysotile	Si_2	Mg_3	–
Needle	Sepiolite Palygorskite (x ~ 0)	tri.	Sepiolite Palygorskite	Si_{12} Si_8	Mg_8 Mg_8	$(OH_2)_4 \cdot H_2O$ $(OH_2)_4 \cdot H_2O$
Amorphous ~ low crystalline			Imogolite	SiO_3OH	$Al(OH)_3$	–
			Allophane Hisingerite	$(1–2)\ SiO_2 \cdot (5–6)\ H_2O$ $SiO_2—Fe_2O_3—H_2O$		

x indicates degree of isomorphaus substitution.
di. indicates dioctahedral, tri. indicates trioctahedral.

Apart from this, the intercalation of small molecules into silicate galleries have been found by researchers when studying Mayan archeological sites [9]. Maya blue was used in Mesoamerica and colonial Mexico as late as the 20th century. Maya blue color is resistant to diluted mineral acids, alkalis, moderate heat and even biocorrosion. This blue color contains clay (mainly MMT clay and palygorskite ($Mg_5(Si,Al)_8O_{20}(OH)_28H_2O$) (see Table 2)) and indigo molecules

($C_{16}H_{10}N_2O_2$). Intercalation of indigo molecules into MMT galleries and/or encapsulation in the pores of palygorskite might explain the corrosion resistance in the extreme condition of the rain forest. Up to now Maya blue paint have been understood as an origin of the intercalation , and recognized as an ancient nanostructured materials (see Figure 1).

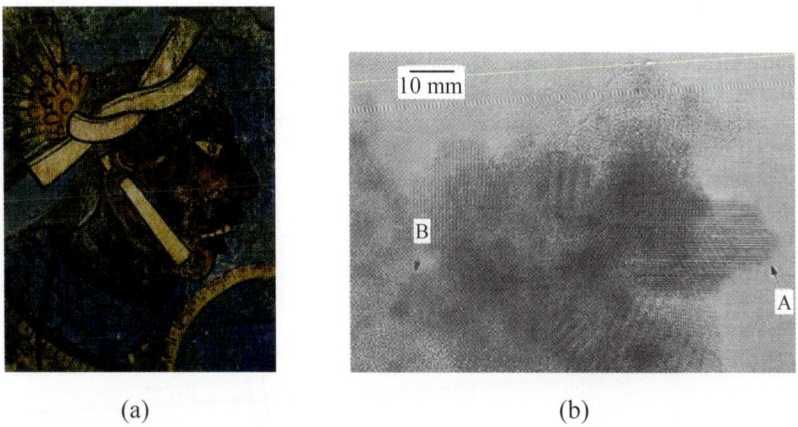

Fig. 1. (a) Maya blue was used in this mural representing a solider from Msoamerica. (b) Needle shape of the palygorskite crystals. Reprinted with permission from [9], Jose-Yacaman M. *et al.*, Science, 273, 223 (1996) © 1996, American Association for the Advancement of Science.

3. Structure of Layered Silicates and Its Modification

The commonly used clays for the preparation of PLSNCs belongs to the family of phyllosilicates. Their crystal structure consists of layers made up of two silica tetrahedral fused to an edge-shared octahedral sheet of either aluminium or magnesium hydroxide. The layer thickness is around 1nm and the lateral dimensions of these layers may vary from 30 nm to several microns and even larger depending on the particular layered silicate. Stacking of the layers leads to a regular van der Waals gap between the layers called the *interlayer* or *gallery*. Isomorphic substitution within the layers (for example, Al^{+3} replaced by Mg^{+2} or by Fe^{+2}, or Mg^{+2} replaced by Li^{+1}) generates negative charges that are counterbalanced by alkali and alkaline earth cations situated inside the galleries, as shown in Figure 2 and Table 2.

The most commonly used layered silicates are montmorillonite (MMT) hectorite and saponite having different chemical formula: $M_x(Al_{2-x}Mg_x)Si_4O_{10}(OH)_2$, $M_x(Mg_{3-x}Li_x)Si_4O_{10}(OH)_2$ and $M_x(Si_{4-x}Al_x)Si_4O_{10}(OH)_2$ (x = 0.25-0.6). The type of clay is characterized by a moderate surface charge (cation exchange capacity) (CEC of 80–120 mequiv/100 g) and layer morphology. These clays are only compatible with hydrophilic polymers, such as poly (ethylene oxide) (PEO), poly (vinyl alcohol) (PVA). To improve compatibility with other

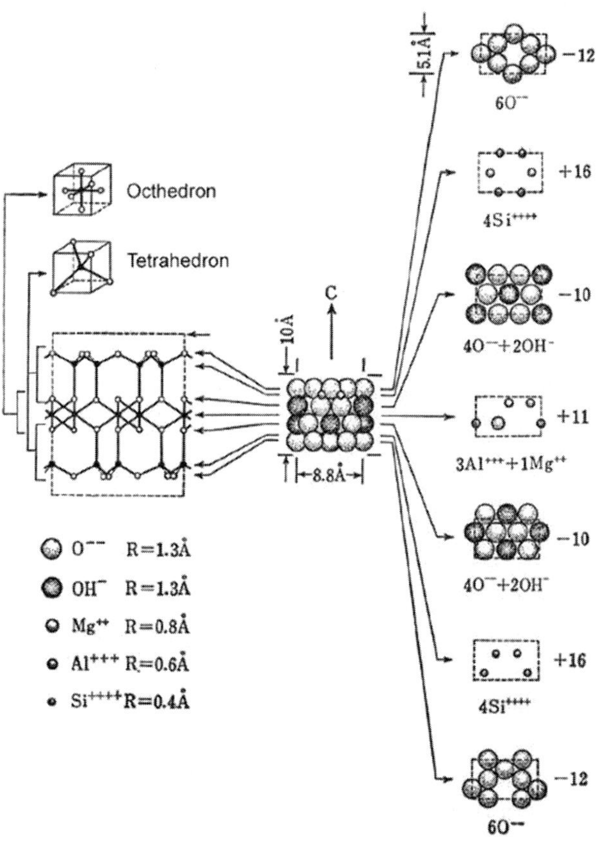

Fig. 2. Structure of 2:1 phyllosilicates (montmorillonite).

polymer matrices, one must convert the normally hydrophilic silicate surface to organophilic, which makes possible intercalation of many engineering polymers. Generally, this can be done by ion-exchange reactions with cationic surfactants including primary, secondary, tertiary, and quaternary alkyl ammonium or alkylphosphonium cations. The role of alkylammonium or alkylphosphonium cations in the organosilicates is to lower the surface energy of the inorganic host and improve the wetting characteristics with the polymer matrix, and results in a larger interlayer spacing. One can evaluate a Na^+ density of $0.7Na^+/nm^2$, i.e., 7000 alkylammonium salt molecules are localized near the individual silicate layers (~100 × 100 nm^2) and active surface area (~700–900 m^2/g as determined by BET). This result indicates the organoclay platelets are *hairy* plates. Furthermore, the surface hydroxyl concentration of clays was determined by titration with triethyl aluminum. Assuming that the hydroxyl groups are randomly distributed on the edge surface, one can calculate a Si-OH density of 5Si-OH/nm^2 [10], i.e., 500 –OH groups are localized near the edge surface of the individual silicate layers (~1 × 100 nm^2). These lipophilic - hydrophobic balance is the key issue of the fine dispersion of the organoclay platelets into the polymeric matrices. Additionally, the alkylammonium or alkylphosphonium cations could provide functional groups that can react with the polymer matrix or in some cases initiate the polymerization of monomers to improve the strength of the interface between the inorganic and the polymer matrix [11,12].

Since the thermal degradation of many organophilic clays beings at temperature more than 200°C, clays with enhanced thermal stability are desired. Recent approach is polymerically modified clays [13–16]. Poly(diallylammonium) salt and oligomeric styrene-based ammonium salt have been prepared and used to produce PLSNCs. More interesting idea of the introducing of a repulsion of the single clay layerds against each other is reported by Fischer (see Figure 3) [17]. The cations located in between the clay platelets are ion-exchanged by one of the functional groups of these organic molecules, e.g., an ammonium group, leaving another functional group, which can be positively or negatively charged, present on the clay layers. 4-amino-1-naphthalenesulphonic acid is one of the candidates.

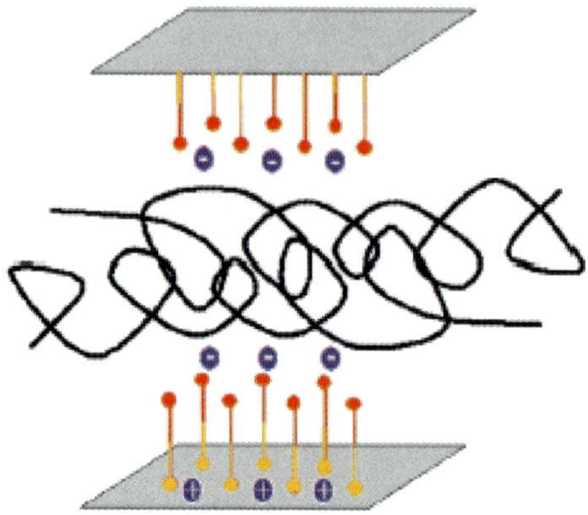

Fig. 3. Principle of a repulsion of sheet-like nanoparticles via surface modification with bifunctional molecules. Reprinted with permission from [17], Fischer H, Mater. Sci. Eng., C23, 763 (2003) © 2003, Elsevier Science.

4. Nobel Compounding Process

Hasegawa and Usuki [18] reported a novel compounding process using the Na^+-MMT slurry and demonstrate preparing a Nylon 6 nanocomposite, where the silicate layers exfoliate and homogeneously disperse at nanometer level. The most merit of this compounding process is that the PLSNC consisted of Nylon 6 and Na^+-MMT is prepared without any surfactants of the clay minerals and additives. But it is difficult to prepare a completely exfoliated nanocomposite by this method (see Figure 4). Originally the perfect exfoliation of the silicate platelets may be impossible because of the strong interaction between hydroxylated edge-edge groups, the clay platelets are sometimes flocculated in any polymer matrices as reported by van Olphen [19] (see Section 5.5).

Recentry, an in-situ polymerization method used supercritical CO_2 (sc-CO_2) as a processing aid to achieve a uniform distribution in a PLSNC at high clay loading (~40 wt%) [20]. Zhao *et al.* [21] presented unambiguous evidence for sc-CO_2-mediated intercalation of poly(ethylene oxide) (PEO) into Na^+-MMT compared with polymer intercalation in solution, in which an entropy-driven process is dominant. This mechanism is probably similar to that in polymer melts. Therefore,

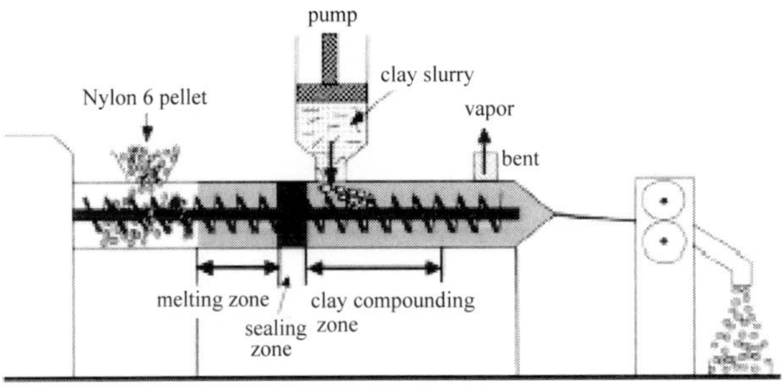

Fig. 4. Schematic representation depicting a compounding process for preparing the nylon 6 nanocomposites using the clay slurry. Nylon 6 was put into the extruder at 2 kg/h and melted in the melting zone. The clay slurry was pumped into the cylinder of the extruder at 2 kg/h and compounded with the melting nylon 6 in the compounding zone at 240–250°C. The screw rotation speed was 200 rpm. The residence time in the cylinder was approximately 10 min. Sealing zone was set by using a sealing ring to prevent water back flow for the hopper. The water of the slurry was removed from the vent by vacuum to obtain the nanocomposites. Reprinted with permission from [18], Hasegawa N *et al.*, Polymer, 44, 2933 (2003) © 2003, Elsevier Science.

the sc-CO_2-mediated intercalation is an enthalpically driven process, deriving from a favorable intercalation between PEO and MMT.

5. Nanostructure-Properties Control

5.1. *Intercalation during crystallization and confinement*

Recentry, Maiti *et al.* [22] reported that in the polypropylene/LS nanocomposites, (PPCN)s at high crystallization temperature (T_c) (≥110°C), where the crystallization rate is low enough to solidify the system, the intercalation should be anticipated in the melt state during crystallization. The driving force of the intercalation originates from strong hydrophilic interaction between the maleic anhydride (MA) group and the polar clay surfaces [23]. With increasing of T_c, the small peak and shoulder shift toward the smaller angle region in the PPCNs, suggesting that the extent of intercalation takes place with crystallization [22].

Figure 5 shows the diffraction peak from the (001) planes ($d_{(001)}$) of the clay gallery quantitatively, as a function of T_c, obtained from their

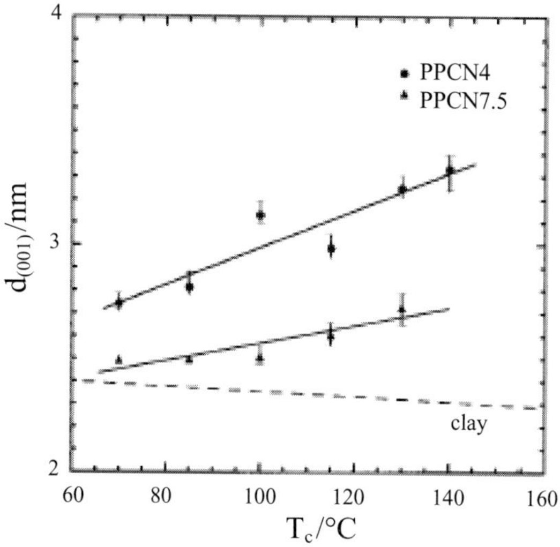

Fig. 5. T_c dependence of the interlayer spacing of PPCN4 and PPCN7.5. The broken line shows the annealing effect on organoclay. Reprinted from [22], Maiti P. *et al.*, Macromolecules 35, 2042 (2002) © American Chemical Society.

respective Bragg reflections. Here, in case of PPCN2 (including 2 wt% of inorganic clay), the peak is not prominent. The dotted line shows the effect of annealing on the $d_{(001)}$ value of organoclay. The $d_{(001)}$ increases with T_c for both PPCN4 and PPCN7.5 systems and PPCN4 exhibits always significantly higher value than that of PPCN7.5. These imply that intercalation proceeds at T_c and increases with decreasing clay content. Further decrease of clay content from 4 to 2 wt% in PPCN2 leads to a partially exfoliated system as discussed in the literature [23]. That is, the PPCN with low clay content crystallized at high T_c (≥110.0°C) exhibits higher amount of intercalation than that with high clay content crystallized at any T_c.

At high T_c (≥110°C) (low crystallization rate), the melt state exists for quite a long time and PP-MA chains have enough time to intercalate before crystallization can occur in the bulk. Then, the enhanced intercalation is produced. The extent of intercalation is strongly dependent on the time of the molten state. In other words, the intercalated PPCNs are not equilibrated. By decreasing the clay content in the nanocomposites, the virtual gallery space in the silicate layers decrease and consequently, the PP-MA molecules would try to accommodate, through interaction, in the minimum space causing higher intercalated species. For sufficiently low clay content, a system, like PPCN2, having less gallery space, is partially exfoliated due to high number density of tethering junction.

There are two possible ways of ordering of polymer chains inside the silicate gallery either by (1) polymer molecules escape from gallery and crystallize outside (*diffuse out*) or (2) molecules may penetrate into the silicate gallery when they are in the molten state (*diffuse in*). When PPCN4 is directly crystallized from the melt at 70.0°C for two different times of 30 min and 17 hrs, the interlayer spacing is the same (2.75 nm). If PPCN4 melt is annealed at 150.0°C, just above T_m (= 145.0°C) for sufficiently long time and then subsequently crystallized at 70.0°C for 30 min, the interlayer spacing increases to 2.96 nm. Furthermore, when PPCN4 is crystallized from the melt at 30°C, where crystallization rate is slow enough, the interlayer spacing becomes 3.08 nm. All these experiments indicate that the extent of intercalation is strongly dependent on the time of the molten state and ordering of polymer chains occurs through a *diffuse-in* mechanism. In other words, a slower crystallization rate makes a more intercalated species as molten polymer molecules have sufficient time to diffuse into the silicate gallery. Based on the WAXD and TEM micrographs, the nature of intercalation has been represented by Maiti and Okamoto [22, 24] in Figure 6. Thus, by suitably crystallizing the PPCNs we can control the fine structure (*confined orientation*) of thePLSCNs.

According to Khare's prediction [25], the confinement of polymer chains increases the viscosity and mechanical properties of the system significantly. One can expect some difference in mechanical properties with the change of the degree of intercalation in the PPCNs vis-à-vis the clay content and T_c (see Table 3). It is clear from the table that for a particular T_c, the dynamic strage modulus (G') increases with increasing clay content. The PP-MA crystallized at 130°C exhibit a 9.9% increase in G' compared to the sample crystallized at 70.0°C. The PPCN7.5 and PPCN 4 show 13.3 and 30.6% increases, respectively, in the same condition. The effect of T_c on the G' is in the order of PP-MA < PPCN7.5 < PPCN4. It may be recalled that the T_c dependence of $d_{(001)}$ showed the order of intercalation PPCN7.5 < PPCN4 in Figure 5. This implies that

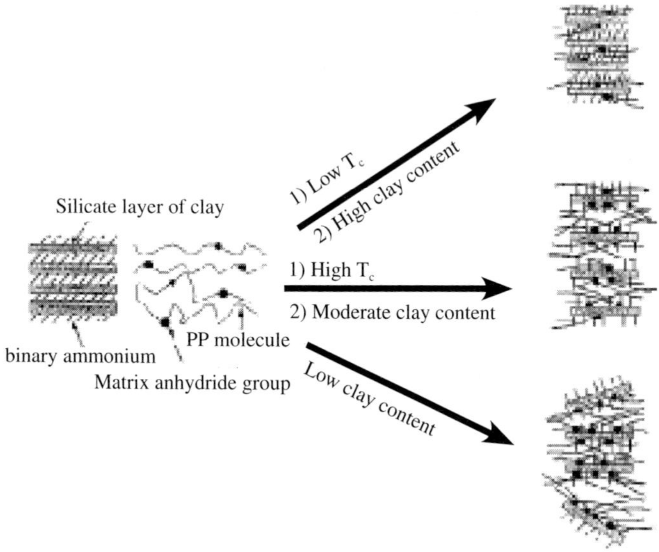

Fig. 6. The illustration for a diffuse-in mechanism by suitably crystallization. Reprinted from [22], Maiti P. *et al.*, Macromolecules 35, 2042 (2002) © 2002 American Chemical Society.

Table 3. Dynamic storage modulus of PP-MA and PPCNs at T = 50°C crystallized at different temperature.

System	T_c/°C	$G' \times 10^{-8}$/Pa	% increase
PP-MA	70	2.92	9.9
	130	3.21	
PPCN2	70	4.79	
	130	4.50	
PPCN4	70	5.16	30.6
	130	6.74	
PPCN7.5	70	7.49	13.3
	130	8.49	

Source: Reprinted from [22], Maiti P. *et al.*, Macromolecules 35, 2042 (2002) © 2002, American Chemical Society.

much higher efficiency of the intercalation for the reinforcement is attained in the PPCN4. For PPCN2, owing to the partial exfoliation, the degree of intercalation decreases and hence the modulus decreases compared to the low T_c condition (= 70°C). Here, it should be mentioned that the crystallinity increases little bit with increasing T_c for both PP-MA and PPCNs and the extent is almost same for all the systems. So, it is believed that not the crystallinity but the degree of intercalation does affect on the storage modulus.

5.2. *Multiscale micromechanical modeling*

Very recently, Sheng and Boyce [26] reported a multiscale modeling strategy to account for the hierarchical morphology of the nanocomposite: at a lengthscale of thousands of microns, the structure is one of high aspect ratio particles within a matrix; at the lengthscale of microns, the clay particle structure is either (a) exfoliated clay sheets of nanometer level thickness or (b) stacks of parallel clay sheets separated from one another by interlayer galleries of nanometer level height, and the matrix, if semi-crystalline, consists of fine lamella, oriented with respect to the polymer/nanoclay interfaces. Models of various representative volume elements of the underlying structure of the clay filled polymer are constructed. Figure 7 represents the influence of internal clay structural parameters (the average number of silicate layers per clay stack: N, $d_{(001)}$) on the macroscopic modulus of the PLSNC. The enhancement of modulus (E_{11}/E_m) is plotted as a fuction of clay content (W_c) and N at fixed $d_{(001)}$. The strong dependence of modulus on N is cleary demonstrared ; at a fixed W_c, modulus increases with decreasing N; the amount of increase gradually expands as $N \rightarrow 1$. On the other hand, the effect $d_{(001)}$ on the modulus for two different values of N ($N = 2$ and 5). Compared with N, the influence of $d_{(001)}$on modulus is rather small, and depends on the specific value of N. This increment is rather negligible when N is small, however, when the nanocomposite is highly intercalated (e.g., $N = 5$), the increase of a few nanometers in $d_{(001)}$ can cause a considerrable increase in modulus.

In the case of Nylon 6, the transcrystallization behavior induced by the nanoclay is taken into account by modeling a layer of matrix surrounding the particle to be highly textured and therefore mechanically anisotropic. Micromechanical models (numerical as well as analytical) based on the "effective clay particle" were employed to calculate the

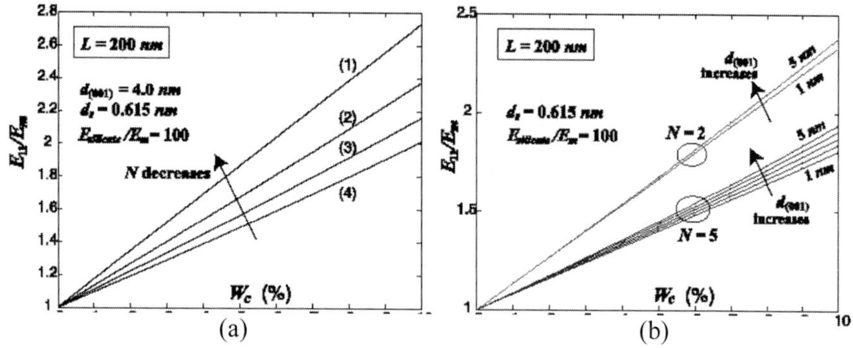

Fig. 7. Effect of clay structural parameters (N, $d_{(001)}$) on the macroscopic modulus predicted by Mori-Tanaka model. (a) Effect of N at fixed $d_{(001)} = 4.0$ nm. (b) Effect of $d_{(001)}$ at two fixed values $N = 2$ and $N = 5$. Reprinted from [26], Sheng N. *et al.*, Polymer, 45, 487 (2004) © 2004, Elsevier Science.

overall elastic modulus of the amorphous and semi-crystalline PLSNCs and to compute their dependence on the matrix and clay properties as well as internal clay structural parameters. The proposed modeling technique captures the strong modulus enhancements observed in elastomer/clay nanocomposites as compared with the moderate enhancements observed in glassy and semi-crystalline PLSNCs. For the case where the matrix is semi-crystalline (like Nylon 6), the proposed approach captures the effect of transcrystallized matrix layers in terms of composite modulus enhancement, however, this effect is found to be surprisingly minor in comparison with the composite-level effects of stiff particles in a matrix. This reason is discussed in Section 5.4.

Note that in order to determine the nanocomposite modulus the modulus of nanoclay of 400 GPa is employed. Most of the nanocomposite researchers believe that the nanoclay has high modulus of 170 GPa [2]. But this value is absolutely acceptable or not even if in the case of mono-layered clay sheet?

5.3. *Flexibility of a single clay layer*

A large degree of flexibility of the mono-layered clay sheet is reported [27]. Two transmission electron microscope (TEM) images are evident

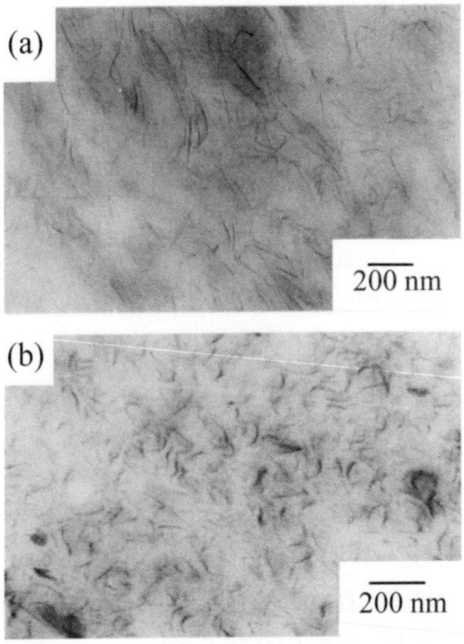

Fig. 8. Bright field TEM images of PPCNs: (a) 2 wt % and (b) 4 wt % MMT. The dark lines are the cross sections of silicate layers and the bright areas are the PP-MA matrix. Reprinted from [23], Nam P. H. *et al.*, Polymer 42, 9633 (2001) © 2001, Elsevier Science.

(see Figure 8). One arises from the clay layers that appear as about 150–200 nm curved sheets. When viewed edge-on as in Figure 8(b), several 5–8 nm stacked sheets are apparent. The curved nature of the sheet is observed, for it is well known that smectite clay sheet have a large degree of flexibility [27]. Sato and Kawamura [28] reported the study of the flexibility of smectite clay minerals by using molecular dynamics (MD) simulations. They took into account the quantitative understanding of the mechanical behavior of a single clay layer in a completely exfoliated state. The repeating unit of a layer is taken to be a_0 = 0.52 nm and b_0 = 0.902 nm with formula of $2Na_{1/3}\ Al_2[Si_{11/3}\ Al_{1/3}]O_{10}(OH)_2$ which corresponds to that of beidellite (see Table 2). When the size of the basic cell (A = 9.3 nm, B = 2.6 nm, and C = 5 nm) (A-type cell) is reduced by 3–40% in the A-direction, the stationary structure of a clay layer is obtained as a curved sheet with a 2:1 smectite-type layer. In such curved state, the layer experiences a external stress of 0.5–0.7 GPa. The layer structure of a clay fractures when the size of the same basic cell is reduced by more than 40%. This value is much lower than that of moscovite (~2 GPa) which is also reported by same authors [29]. The simulation has also done by reducing the size of the basic cell (A = 3.1 nm, B = 10.7 nm, and C = 5 nm) (B-type) in the B-direction. The clay layer is found to be more flexible along the A-direction than the B-direction. When the microscopic structure of a curved clay layer is examined, it is concluded that the main origin of the flexibility lies in the change of Si-O-Si angle in the silicate tetrahedral sheets rather than in the change of bond lengths. These simulation results agree with the atomic force microscopy (AFM) observations [30].

5.4. *Higher-order structure development and crystallization controlled by silicate surfaces*

The formation of γ-form in presence of clay in Nylon 6/LS nanocomposite is well known [31]. The essential difference between γ-form and α-form is the molecular packing; in the α-form hydrogen bondings are formed between antiparallel chains while the molecular chains have to twist away from the zigzag planes to form the hydrogen bonds among the parallel chains in γ-form giving rise to lesser inter-chain interaction as compared to α-form.

The lamellar morphology and distribution of clay particles in the Nylon 6 nanocomposite (N6CN3.7) (MMT = 3.7 wt%), crystallized at 170 and 210.0°C, have been reported by Okamoto and Maiti [32] in Figure 9. The white strips (Figure. 9a) represent the discrete lamellar pattern, and after a close look, a black clay particle inside the lamella is clearly observed. In other words, lamellar growth occurs on both sides of the clay particles *i.e.* the clay particle is sandwiched by the formed lamella. This is a unique observation of lamellar orientation on the clay layers. In the semi-crystalline polymer generally the stacked lamellar orientation takes place. The lamellar pattern at high T_c (Figure. 9b) is somehow similar but along with the sandwiched structure, branched lamellae are formed which are originated from the parent sandwiched lamella. There are no clay particle found inside the branch lamella and the γ-phase having irregular chain packing with distortion (γ^*-phase) is

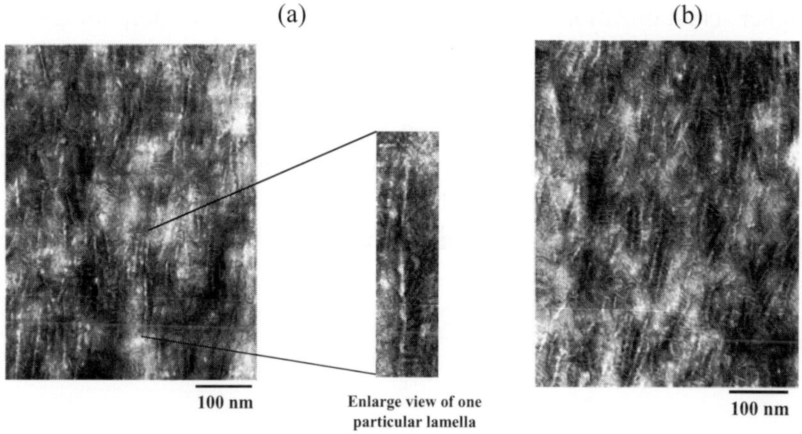

Fig. 9. TEM micrographs of N6CN3.7 crystallized at (a) 170°C and (b) 210°C. The black strip inside the white part is clay. Figure b shows the typical shish-kebab type of structure. Reprinted from [32], Maiti P. *et al.*, Macromole. Mater. Eng., 288, 440 (2003) © 2003, Wiley-VCH.

formed as revealed by wide-angle X-ray diffraction (WAXD) which one can observe only in case of high T_c crystallized N6CN3.7. This epitaxial growth (γ^*-phase) on the parent lamella forms the *shish-kebab* type of structure, which virtually enhance the mechanical properties of the nanocomposites like a bone material which consists of collagen fibrils reinforced with tiny mineral particles, a few nanometers in thickness [33].

From this sandwiched structure the accurate determination of long spacing and lamellar thickness of N6NC3.7 from small angle X-ray scattering (SAXS) is questionable [34,35]! It has to be remembered that Nylon 6 has the highest capability of forming hydrogen bonding to form hydrogen-bonded sheet. The pseudohexagonal packing is favored with the hydrogen-bonding between the silicate layers and Nylon 6 as a result the induction time of N6CN3.7 becomes very short, as compared to neat Nylon 6. Once one molecular layer is nucleated on the clay surface, other molecules may form the hydrogen-bonding on the already formed hydrogen-bonded molecule to the silicate surface giving rise to the discrete lamellar structure on both side of the clay. The nucleation and growth process has been demonstrated in Figure 10, following direct observation by TEM [32]. This unique mechanism can well explain the higher crystallization rate of PLSNCs along with morphology and developed internal structure. This sandwiched structure (each silicate layer is strongly covered by polymer crystals) makes the system very rigid as a result the heat distortion temperature (HDT) increases up to 80°C but the surrounding excess amorphous part (lower crystallinity of N6CN3.7 as compared to neat Nylon 6) can easily retain the polymeric properties like impact strength, ultimately makes a improved/perfect system in PLSNCs.

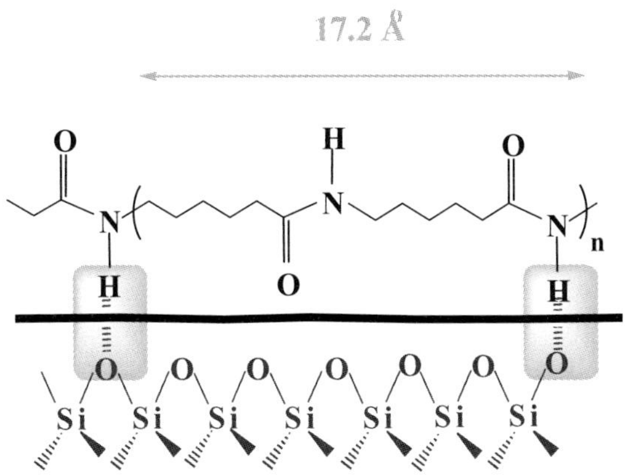

Fig. 10. Schematic view of the nucleation and growth mechanism in N6CN3.7. Reprinted from [32], Maiti P. *et al.*, Macromole. Mater. Eng., 288, 440 (2003) © 2003, Wiley-VCH.

Kim and Kressler [36] also reported that the fine lamellae of Nylon 12 crystals are oriented perpendicularly to the Nylon 12/LS interface, i.e., on planes lying normal to the injection molding direction. This interfacial ordering may be a result of the crystallization process and its similar to the well-known transcrystallization [37]. The nanocomposites consist of a nanostructured network finely dispersed and uniformly oriented silicate layers are aligned perpendicular to lamellae, and the two materials are strongly bonded to each other (see Figure 11). In these Nylon nanocomposites, the nanoclay particles with the sandwiched-and-network (shish-kebab) structure have a high function as micro-void initiation sites, which are necessary for high toughness during deformation. Very recently, Fratzl P. *et al.* [38] reported that the mechanical behavior of the collagen mineral nanocomposite in bone depends crucially on both components, mineral and wet collagen, and on their interaction (see Figure 12).

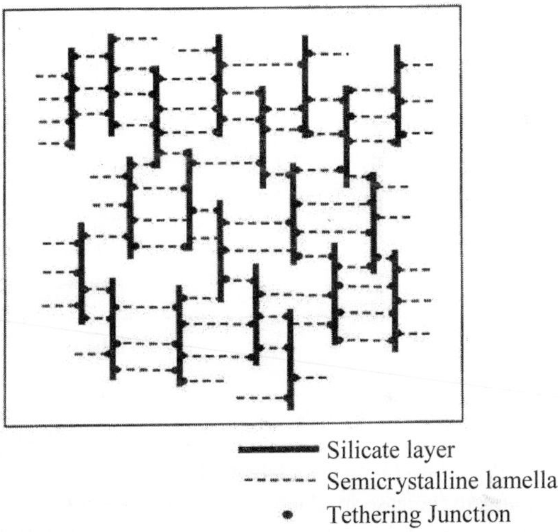

Fig. 11. Nanostructured network in Nylon 12/LS nanocomposite. Reprinted from [36], Kim G. M. *et al.*, Polymer 42, 1095 (2001) © 2001, Elsevier Science.

Fig. 12. Schematic arrangement of mineral and organic phase in a mineralized collagen fibril subjected to tensile load. Mineral particles are shown in dark gray and should be imagined as platelets viewed edge on. The horizontal white lines in the light gray matrix (left part in figure) are not indicating any physical reality, they are just drawn to visualize the shear deformation in the matrix between the particles as a consequence of tensile deformation of the tissue (right part of figure). Reprinted from [38], Fratzl P. *et al.*, J. Mater. Chem., 14, 2115 (2004) © 2004, The Royal Society of Chemistry.

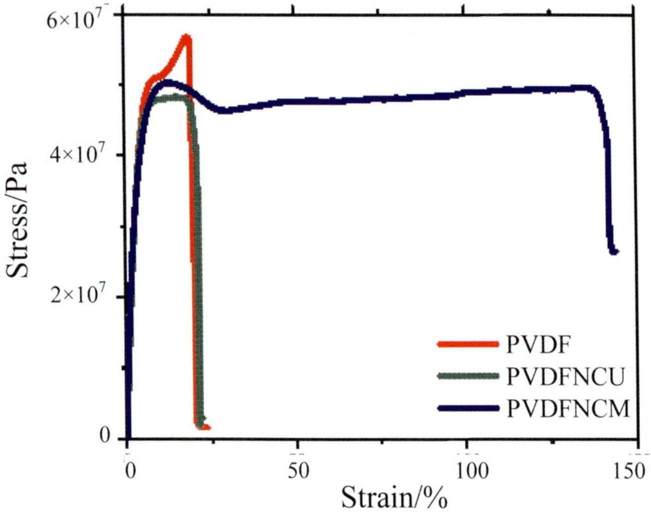

Fig. 13. Stress-strain curves for neat PVDF (α-phase crystallite), PVDFNCU (micro composite by using MMT) and PVDFNCM (nanocomposite by using organoclay) showing the dramatic increase in elongation at break for the nanocomposite. Reprinted from [40], Shah D. *et al.*, Adv. Mater., 16, 1173 (2004) © 2004, Wiley-VCH.

In case of polyvinylidene (PVDF)/LS nanocomposites, the formation of β-form have been observed [39,40]. Shah *et al.* [40] reported a remarkable order of magnitude enhancement in toughnes of the nanocomposites (see Figure 13). They postulated that nucleation of the fiber-like β-phase (more ductile than the α-phase) on the surface of individual silicate layers leads to a structure conductive to plastic flow under applied stress. Energy dissipation could be further enhanced due to the presence of more mobile β crystallites which have a potential of acting-like rigid fillers surrounded by the crystalline phase of PVDF. Thus the crystallization control by silicate surfaces may provide not only a new approach for toughening of polymers but also the way towards a novel approach for the design of new materials.

In case of polyester systems, Yamada *et al.* [41] examined the HDT of various polylactide (PLA)/LS nanocomposites (PLACNs) with different load conditions. In case of PLACN (MMT = 5 wt%), there is marked increase of HDT with intermediate load of 0.98 MPa, from 76°C for the neat PLA up to 111°C for PLACN (see Figure 14). In case of high load (1.81 MPa), however, it is very difficult to achieve high HDT enhancement without strong interaction between polymer matrix and organoclay like Nylon systems [32]. So the improvement of HDT with intermediate load (0.98 MPa) originates from the better mechanical stability of the PLACNs due to mechanical reinforcement by the dispersed clay particles with higher value of the degree of crystallinity χ_c and intercalation. This is qualitatively different from the behavior of Nylon systems, where the MMT layers stabilize in a different crystalline phase (γ-phase) [32] with the strong hydrogen bondings between the silicate layers and Nylon 6 as a result the discrete lamellar structure on both side of the clay (see Figures 9 and 10). Nylon /LS nanocomposites are successfully prepared without strategy for designing of materials with desired properties of the PLSNCs.

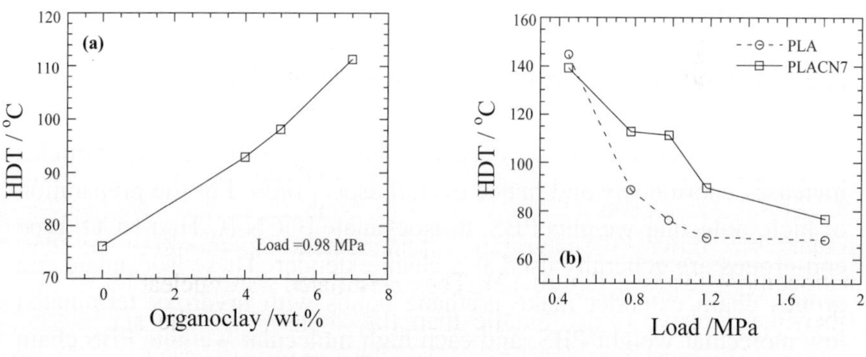

Fig. 14. (a) Organoclay (wt%) dependence of HDT of neat PLA and various PLACNs. (b) Load dependence of HDT of neat PLA and PLACN7. Reprinted from [41], Sinha Ray S. *et al.*, Polymer 44, 857 (2003) © 2003, Elsevier Science.

5.5. *Flocculation control and modulus enhancement*

Most of the nanocomposite researchers obdurately believe that the preparation of completely exfoliated structure is the ultimate target for better overall properties. However, these significant improvements are not observed in every nanocomposite systems, including systems where the silicate layers are near to exfoliated [42]. While from the barrier property standpoint, the development of exfoliated nanocomposites is preferred always. On the other hand, Nylon /LS nanocomposite systems are completely different from other nanocomposite systems as discussed before.

In Figure 15, Okamoto summarized the clay content dependence of dynamic storage modulus (G′) of various types of nanocomposites obtained under well below T_g of the matrices. Einstein coefficient (k_E) derived by using Halpin and Tai's theoretical expression modified by Nielsen is shown in the Figure, and represents the aspect ratio (L_{clay}/d_{clay}) of dispersed clay particles without intercalation. From this Figure, it is clearly observed that poly(butylene succinate)(PBS)/LS nanocomposites (PBSCNs) show very high increment in G′ compared to other nanocomposites having same content of clay in the matrix. PPCNs are well known for intercalated systems, N6CNs are well-established exfoliated nanocomposites, PLACNs are going to establish intercalated-and-flocculated nanocomposites, while PBSCNs are intercalated-and-extended flocculated nanocomposites systems [43,44]. Due to the strong interaction between hydroxylated edge–edge groups, as mentioned in Section 3, the clay particles are some time flocculated in the polymer matrix. As a result of this flocculation the length of the clay particles increases enormously and hence overall aspect ratio. For the preparation of high molecular weights PBS, di-isocyanate [OCN-(C_6H_{12})-NCO] type end-groups are generally used as a chain extender. These isocyanate end groups chain extender make urethane bonds with hrydroxy terminated low molecular weight PBS, and each high molecular weights PBS chain contain two such kind of bonds (see schematic illustration in Figure 16). These urethane type bonds lead to the strong interaction with silicate surface by forming hydrogen bonds and hence strong flocculation (see Figure 17). For this reason, the aspect ratio of dispersed clay particles is much higher in case of PBSCNs compared to all nanocomposites, and hence high enhancement of modulus.

We can explain this behavior with the help of classical rheological theory of suspension of conventional filler reinforced systems. According to this theory [45], the rotation of filler is possible when volume fraction of clay $\phi_{filler} < \phi_{critical} \cong (\text{aspect ratio})^{-1}$. All PBSCNs studied here follow this relation except PBSCN4 (MMT = 3.6 wt%), in which $\phi_{filler} >> (\text{aspect ratio})^{-1}$. For this reason in PBSCN4 rotation of

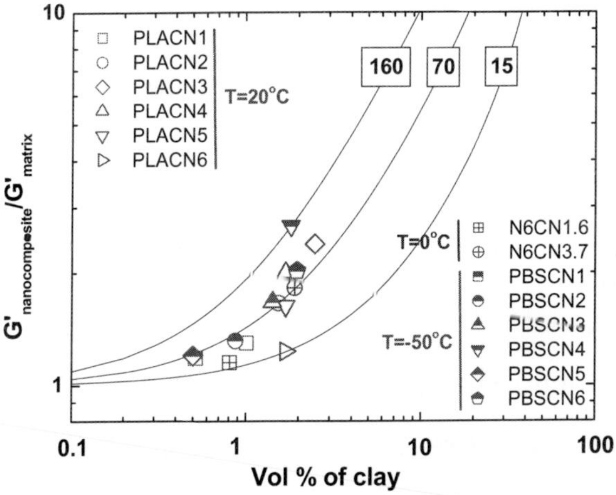

Fig. 15. Plots of $G'_{nanocomposite} / G'_{matrix}$ vs. volume% of clay for various nanocomposites. The Einstein coefficient k_E is shown with the number in the box. The lines show the calculated results from Halpin and Tai's theory with various k_E.

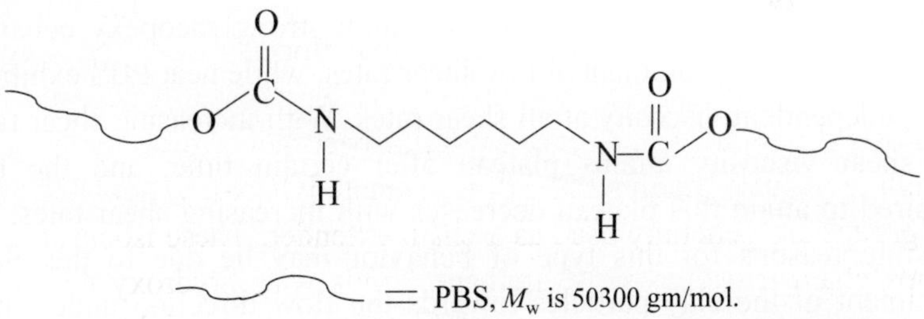

Fig. 16. Formation of urethane bondings in high molecular weight PBS.

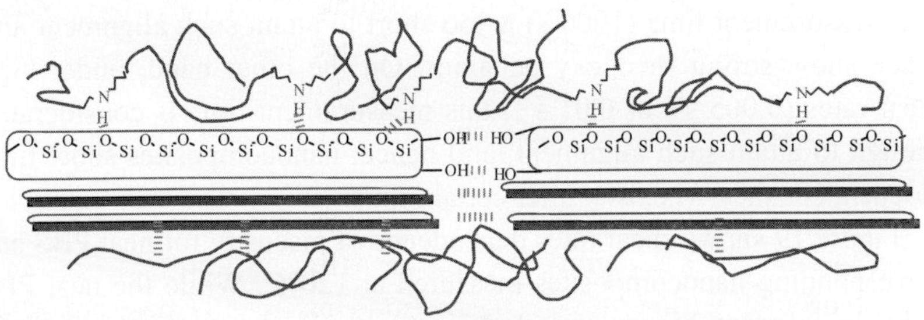

Fig. 17. Formation of hydrogen bonds between PBS and clay, which leads to the flocculation of the dispersed silicate layers. Reprinted from [43], Sinha Ray S. *et al.*, Macromolecules, 36, 2355 (2003) © 2003 American Chemical Society.

dispersed intercalated with flocculated stacked silicate layers is completely hindered and only translational motion is available, and hence show very high modulus. This behavior is clearly observed in dynamic storage modulus measurements under molten state [43]. In case of N6CN3.7 (MMT = 3.7 wt%) we can see same high increment in G′ as well as PBSCNs. The development of the flocculated structure is occurred even if N6CNs are well-established exfoliated nanocomposite systems.

6. Flow-Induced Structure Development

6.1. *Steady shear flow*

The steady shear rheological behavior of neat PBS and various PBSCNs are shown in Figure 18. The steady viscosity of PBSCNs is enhanced considerably at all shear rates with time, and at a fixed shear rate increases monotonically with increasing silicate loading [44]. On the other hand, all intercalated PBSCNs exhibit strong rheopexy behavior, and this becomes prominent at low shear rates, while neat PBS exhibits a time independent viscosity at all shear rates. With increasing shear rates, the shear viscosity attains plateau after certain time, and the time required to attain this plateau decreases with increasing shear rates. The possible reasons for this type of behavior may be due to the planer alignment of the clay particles towards the flow direction under shear. When the shear rate is very slow (0.001 s^{-1}), clay particles take longer time to attain complete planer alignment along the flow direction, and this measurement time (1000 s) is too short to attain such alignment and hence show strong rheopexy behavior. On the other hand, under high shear rates (0.005 s^{-1} or 0.01 s^{-1}) this measurement time is considerable enough to attain such alignment, and hence, nanocomposites show time independent shear viscosity after certain time.

Figure 19 shows shear rates dependence of viscosity for neat PBS and corresponding nanocomposites measured at 120°C. While the neat PBS exhibits almost Newtonian behavior at all shear rates, whereas nanocomposites exhibited non-Newtonian behavior. At very low shear rates, shear viscosity of nanocomposites initially exhibits some shear-

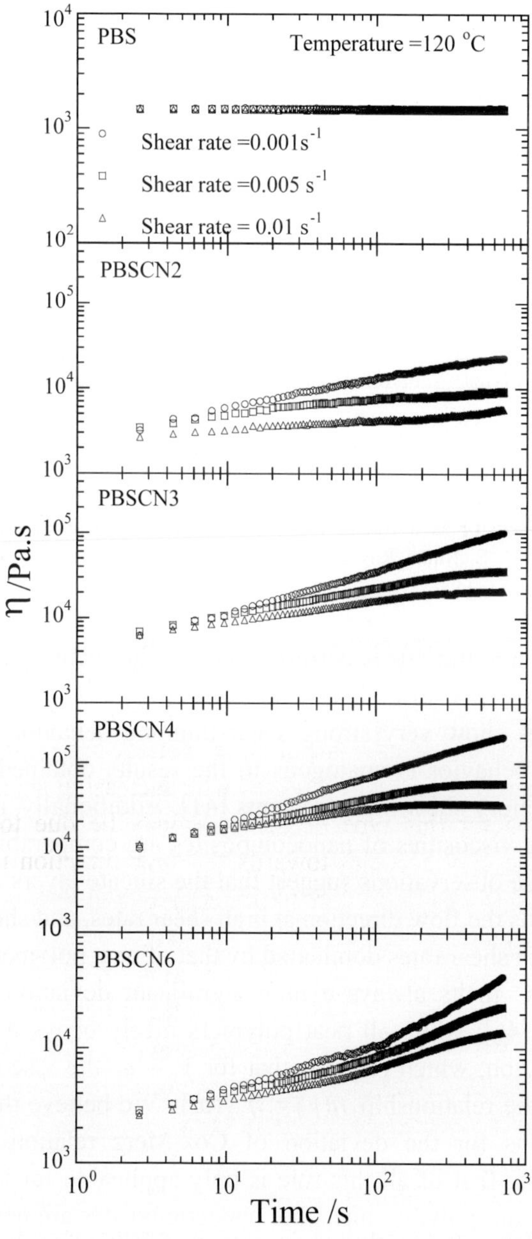

Fig 18. Time variation of shear viscosity for PBSCN. Reprinted from [43], S. Sinha Ray *et al.*, Macromolecules, 36, 2355 (2003) © 2003 American Chemical Society.

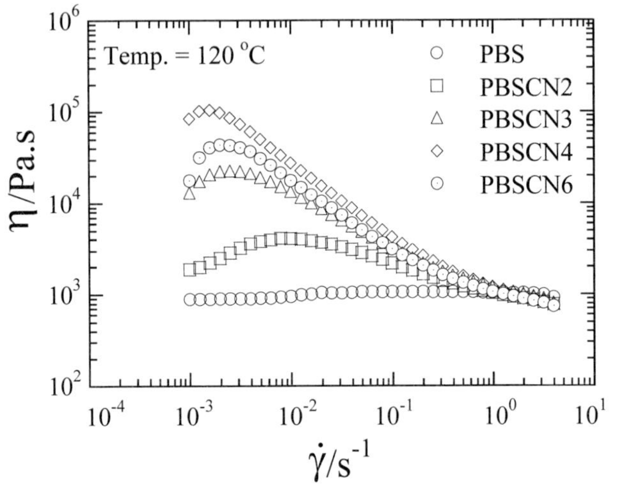

Fig. 19. Shear viscosity as a function of shear rates for the shear rate sweep test. Reprinted from [43], S. Sinha Ray *et al.*, Macromolecules, 36, 2355 (2003) © 2003 American Chemical Society.

tickening behavior and this is corresponding to the rheopexy behavior as we observed at very low shear rates (see Figure 18). After that all nanocomposites show very strong shear thinning behavior at all shear rates and this behavior is analogous to the results obtained in case of dynamic oscillatory shear measurements [41]. Additionally, at very high shear rates, the viscosities of nanocomposites are comparable to that of neat PBS. These observations suggest that the silicate layers are strongly oriented towards the flow direction at high shear rates, and shear thinning behavior at high shear rates dominated by that of neat polymer.

The PLSNC melts always exhibit significant deviation from Cox-Merz relation [46], while all neat polymers nicely obeys the empirical Cox-Merz relation, which requires that for $\dot{\gamma} = \omega$, the viscoelastic data should obeys the relationship $\eta(\dot{\gamma}) = |\eta^*|(\omega)$. We believe there are two possible reasons for the deviation of Cox-Merz relation in case of nanocomposites: first of all this rule is only applicable for homogenous systems like homo-polymer melts but nanocomposites are heterogeneous systems. For this reasons this relation is nicely obey in case of neat polymer [43]. Second, the structure formation is different when nanocomposites are subjected to dynamic oscillatory shear and steady shear measurements.

Okamoto *et al.* [47] constructed unique rheo-optical device, angle light scattering apparatus (Rheo-SALS), which enables us to perform time-resolved measurements of light intensity scattered from the internal structure developed under shear flow. Figure 20 shows a schematic illustration of the apparatus: Plane polarized light normal to the Ox-axis (flow direction) was applied vertically to the parallel-plate type shear cell along the velocity gradient (Oy-axis). Scattering profiles were observed either under Vv mode (depolarized geometry in which the optical axis of the analyzer was set parallel to that of the polarizer) or Hv (the cross-polarized geometry with the two axes being set perpendicular to two

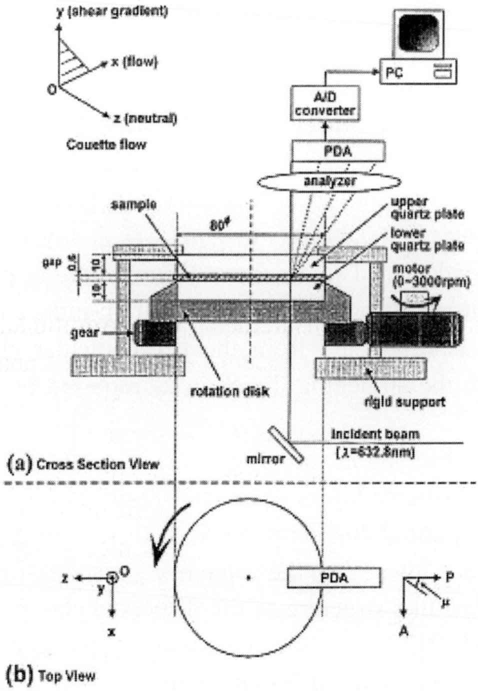

Fig. 20. A schematic illustration of Rheo-SALS apparatus: (a) cross-sectional and (b) top views. Reprinted from [47], M. Okamoto *et al.*, Nippon Rheology Gakkaishi, 28, 201 (2000) © 2000 The Society of Rheology, Japan.

axes) optical alignment at azimuthal angle μ of 0°. They reported the time variation of the mean-square density fluctuation $\langle\eta^2\rangle$, the mean-square anisotropy $\langle\delta^2\rangle$ and the relevant value of correlation distance (ξ_η and ξ_δ) upon imposition/cessation of steady shear flow at both low shear rate ($\dot{\gamma}$) (0.5 s^{-1}) and high $\dot{\gamma}$ (60 s^{-1}).

6.2. *Elongational flow and strain-induced hardening*

Okamoto *et al.* [48] first conducted elongation test of PP/LS nanocomposites (PPCN4) under molten state at constant Hencky strain rate, $\dot{\varepsilon}_0$ using an elongation flow optorheometry [49] and also they attempted to control the alignment of the dispersed silicate layers with nanometer dimensions of an intercalated PPCNs under uniaxial elongational flow.

Figure 21 shows double logarithmic plots of transient elongational viscosity $\eta_E(\dot{\varepsilon}_0;t)$ against time t observed for Nylon 6/LS system (N6CN3.7) and PPCN4 (MMT = 4 wt%) with different Hencky strain rates $\dot{\varepsilon}_0$ ranging from 0.001 s^{-1} to 1.0 s^{-1}. The solid curve represents time development of three-fold shear viscosity, $3\eta_0(\dot{\gamma};t)$, at 225°C with a constant shear rate $\dot{\gamma}$ = 0.001 s^{-1}. In $\eta_E(\dot{\varepsilon}_0;t)$ at any $\dot{\varepsilon}_0$, N6CN3.7 melt shows a weak tendency of *strain-induced hardening* as compared to that of PPCN4 melt. A strong behavior of strain-induced hardening for PPCN4 melt was originated from the perpendicular alignment of the silicate layers to the stretching direction as reported by Okamoto *et al.* [48].

From TEM observation (see Figure 9), the N6CN3.7 forms a fine dispersion of the silicate layers of about 100 nm in L_{clay}, 3 nm thickness in d_{clay} and ξ_{clay} of about 20–30 nm between them. The ξ_{clay} value is a one order of magnitude lower than the value of L_{clay}, suggesting the formation of spatially-linked like structure of the dispersed clay particles in Nylon 6 matrix. For N6CN3.7 melt, the silicate layers are so densely dispersed into the matrix and hence difficult to align under elongational flow. Under flow fields, the silicate layers might translationally move, but not rotationally in such a way that the loss energy becomes minimum. This tendency was also observed in PPCN7.5 melt having higher content of MMT (= 7.5 wt%) [50].

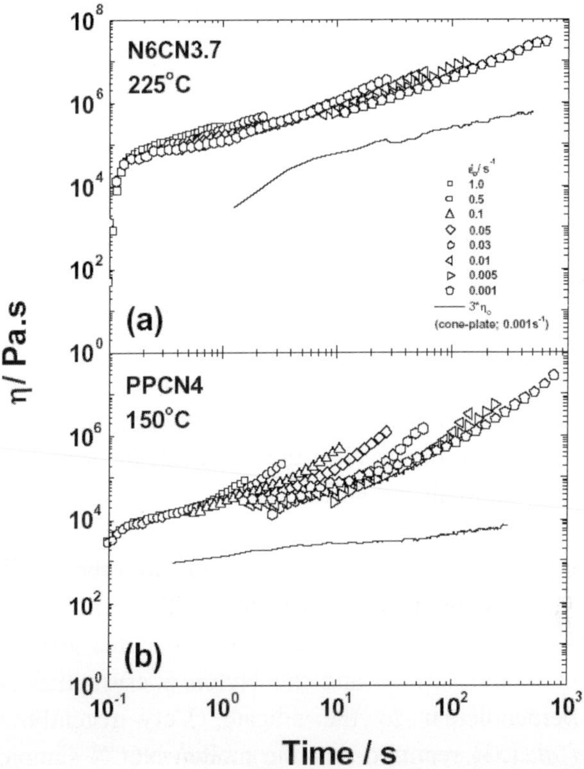

Fig. 21. Time variation of elongational viscosity $\eta_E(\dot{\varepsilon}_0;t)$ for (a) N6CN3.7 melt at 225°C and for (b) PPCN4 at 150°C. The solid line shows three times the shear viscosity, $3\eta_E(\dot{\gamma};t)$, taken at a low shear rate $\dot{\gamma}$ = 0.001 s^{-1} on a cone-plate rheometer. Reprinted from [2], M. Okamoto, Rapra Review Report No 163, Rapra Technology Ltd., London (2003) © 2003 Rapra Technology Ltd.

On the other hand, one can observe two features for the shear viscosity curve. First, the extended Trouton rule, $3\eta_0(\dot{\gamma};t) \cong \eta_E(\dot{\varepsilon}_0;t)$, does not hold for both N6CN3.7 and PPCN4 melts, as opposed to the melt of ordinary homo-polymers. The latter, $\eta_E(\dot{\varepsilon}_0;t)$, is more than 10 times larger than the former, $3\eta_0(\dot{\gamma};t)$. Second, again unlike ordinary polymer melts, $3\eta_0(\dot{\gamma};t)$ of N6CN3.7 melt increases continuously with *t*, never showing a tendency of reaching a steady state within the time span (600s or longer) examined here. This *time-dependent thickening* behavior may be called *anti-thixotropy* or *rheopexy*. *Via* slow shear flow

($\dot{\gamma}$ = 0.001 s^{-1}), $3\eta_0(\dot{\gamma};t)$ of N6CN3.7 exhibits a much stronger rheopexy behavior with almost two order of magnitude higher than that of PPCN4. This reflects a fact that the shear-induced structural change involved a process with an extremely long relaxation time as well as for other PLSNCs having rheopexy behavior [43], especially under weak shear field.

6.3. *Alignment of silicate layers*

The orientation of silicate layers and Nylon 6 crystallites in injection molded N6CN using WAXD and TEM is examined [51,52]. Kojima and his colleagues have found three regions of different orientations in the sample as a function of depth. Near the middle of the sample, where the shear forces are minimal, the silicate layers are oriented randomly and the Nylon 6 crystallites are perpendicular to the silicate layers. In the surface region, shear stresses are very high, so both the clay layers and the Nylon 6 crystallites are parallel to the surface. In the intermediate region, the clay layers, presumably due to their higher aspect ratio, still orient parallel to the surface and the Nylon 6 crystallites assume an orientation perpendicular to the silicate. Very recently, Medellin-Rodriguez *et al.* [53] reported that the molten N6CN samples showed planar orientation of silicate layers along the flow direction, which is strongly dependent on shear time as well as clay loading, reaching a maximally orienting level after being sheared for 15 min with $\dot{\gamma} = 60\ s^{-1}$.

Okamoto and his colleagues conducted the TEM observation for the sheared N6CN3.7 with $\dot{\gamma} = 0.0006\ s^{-1}$ for 1000 s [2]. The edges of the silicate layers laying along the z-axis (marked with the arrows (A)) or parallel alignment of the silicate edges to the shear direction (x-axis) (marked with the arrows (B)) rather than assumingly random orientation in the Nylon 6 matrix is observed, but in fact, one cannot see these faces in this plane (Figure 22). Here, it should be emphasized that the planar orientation of the silicate faces along the x-z plane does not take place prominently. For the case of rapid shear flow, the commonly applicable conjecture of the planar orientation of the silicate faces along the shear direction first demonstrated to be true by Kojima and his colleagues [51].

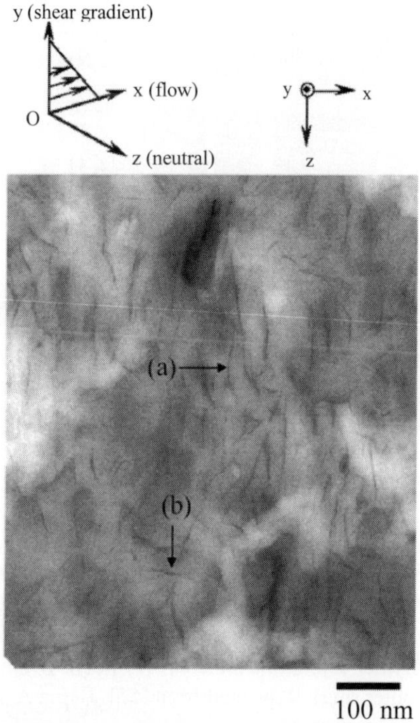

Fig. 22. TEM micrograph in the x-z plane showing N6CN3.7 sheared at 225°C with $\dot{\gamma} = 0.0006\ s^{-1}$ for 1000 s. The x-, y- and z-axes correspond respectively to flow, shear gradient and neutral direction. Reprinted from [2], M. Okamoto, Rapra Review Report No 163, Rapra Technology Ltd., London (2003) © 2003 Rapra Technology Ltd.

In uniaxial elongational flow (converging low) for a PPCN4, the formation of *a house-of-cards* structure is found by TEM analysis [48]. The perpendicular (but *not* parallel) alignment of disk-like clay particles with large anisotropy toward the flow direction might sound unlikely but this could be the case especially under an elongational flow field, in which the extentional flow rate is the square of the converging flow rate along the thickness direction, if the assumption of *affine* deformation without volume change is valid. Obviously under such conditions, energy dissipation rate due to viscous resistance between the disk surface and the matrix polymer is minimal, when the disks are aligned perpendicular to the flow direction.

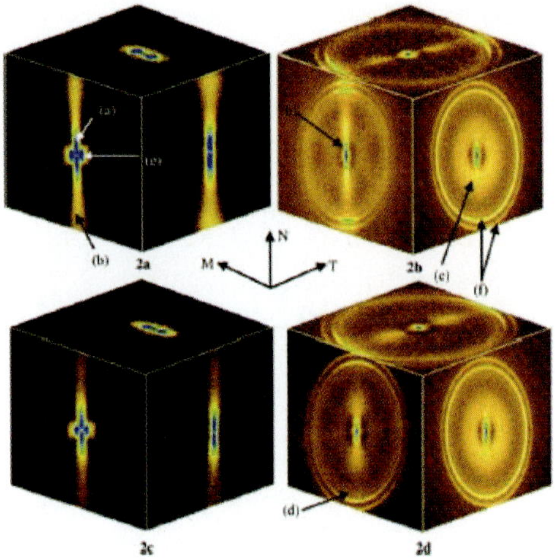

Fig. 23. 2D SAXS (a and c) and WAXS (b and d) patterns for orientation MN (left face), NT (right face) and MT (top face) of films HD603 (a and b) and HD612 (c and d). The numbers in the parenthesis represent the reflections from the following: (a) clay tactoids, (b) modified/intercalated clay (002) plane, (c) unmodified clay (002) plane, (d) clay (110) and (020) plane, (e) polymer crystalline lamellar, (f) polymer unit cell (110) plane (inner ring) and (200) plane (outer ring). Reprinted from [55], Bafna A. *et al.*, Polymer 44, 1103 (2003) © 2003, Elsevier Science.

Moreover, Lele and his colleagues [54] recently reported the in situ Rheo-X-ray investigation of flow-induced orientation in syndiotactic PP/layered silicate nanocomposite melt.

Very recently, Bafna *et al.* [55] developed a technique to determine the three-dimensional (3D) orientation of various hierarchical organic and inorganic structures in a PLSNCs. They studied the effect of compatibilizer concentration on the orientation of various structures in PLSNCs using 2D small angle X-ray scattering (SAXS) and 2D WAXD in three sample/camera orientations. Reflections and orientation of six different structural features were easily identified: (a) clay clusters/tactoids (0.12 μm), (b) modified/intercalated clay stacking period (002) (2.4–3.1 nm), (c) stacking period of unmodified clay platelets (002) (1.3 nm), (d) clay (110) and (020) planes, normal (b) and (c), (e) polymer crystalline lamellae (001)(19–26 nm), long period ((001) is an average crystallographic direction), and (f) polymer unit cell (110) and (200) planes. The corresponding identified reflections are presented in Figure 23. A 3D study of the relative orientation of the above-mentioned structures was carried out by measuring three projections of each sample. Quantitative data on the orientation of these structural units in the nanocomposite film was determined through calculations of the major axis direction cosines and through a ternary, direction-cosine plot called a "Wilchinsky triangle" [56], previously proposed in lamellar orientation studies [57]. It allows a direct comparison of average preferred orientations for different structural features. In this way it is conceptually more useful than stereographic projections involving orientation density maps for a single WAXD reflection.

Some 20 years ago van Olphen [19] pointed out that the electrostatic attraction between the layers of natural clay in aqueous suspension arises from higher polar force in the medium. The intriguing features such as yield stress thixotropy and/or rheopexy exhibited in aqueous suspensions of natural clay minerals may be taken as a reference to the present PLSNCs.

6.4. *Electrorheology*

Electrorheological fluids (EFR)s, sometimes referred to as "smart fluids", are suspensions consisting of polarizable particles dispersed in insulating media. A mismatch in conductivity or dielectric constant between the dispersed particle and the continuous medium phase induces polarization upon application of an electric filed. The induced particle dipoles under the action of an electric field tend to attract neighboring particles and cause the particles to form fibril-like structures, which are aligned to the electric field direction.

Among various materials [58–61], semiconducting polymers are one of the novel intrinsic ER systems since it has the advantage of a wide range of working temperature, reduced abrasion of device, low cost, and relatively low current density. As a result, development of a high-performance ER fluide followed by conducting polymer optimization and tuning has been the subject of considerable interest for practical applications as a new electomechanic interface. Nevertheless, the yield stress and modulus of ER fluids is lower than those of magnetorheological fluids. Thus the performance of conducting polymer-based ER fluids is still insufficient for the successful development of specific application devices.

On this basis of this information's, Kim *et al.* [58] first introduced nanocomposite as EFRs using polyaniline (PANI)/LS nanocomposites with intercalated structure. Though PANI/clay intercalated nanocomposites are a new material for application of ER materials, yield stresses of the system showed less than 100 Pa at 1.2 kV/mm (20 wt% suspensions). This value is a little lower than the yield stress of pure PANI particle system [59]. In other words, no synergistic effect of clay on yield stress was shown.

Recently, Park and his colleagues [60] have observed remarkable enhancement of yield stress for electrorheological fluids in PANI-based nanocomposites of clay. In their further study [61], they fabricated three kinds of EFRs containing different content of PANI/clay nanocomposite and pure PANI particles in order to investigate the effect of nanocomposite particles on the enhancement of yield stress more systematically. They observed that there is an optimum content ratio between nanocomposite and pure PANI particles to produce minimum yield stress. Details regarding data collection and explanations are presented in reference [61].

7. Processing Operations

The flow-induced internal structural change occurs in both shear and elongational flow, but almost differs from each other, as judged from the above results on $\eta_E(\dot{\varepsilon}_0;t)$ and $3\eta_0(\dot{\gamma};t)$ (see Figure 21). Thus, with these rheological features of the PCNs and the characteristics of each processing operation, a tactics of which process type should be selected for a particular nanocomposite for the enhancement of its mechanical properties?

For example, the strong strain-induced hardening in $\eta_E(\dot{\varepsilon}_0;t)$ is requisite for withstanding the stretching force during the processing, while the rheopexy in $3\eta_0(\dot{\gamma};t)$ suggests that for such PLSNC a promising technology is the processing in confined space such the injection molding where shear force is crucial.

7.1. *Foam processing using sc-CO_2*

Very recently, the first successful nanocomposite foam, processed by using supercritical CO_2 as a physical foaming agent, appeared through a pioneering effort by Okamoto and his colleagues [62,63].

Figure 24 shows the typical results of SEM images of the fracture surfaces of the intercalated polycarbonate (PC)/layered silicate nanocomposites (PCCNs) and PC/SMA blend (Matrix) without clay foamed at 160°C under different isobaric saturation condition of supercritical CO_2 (10, 14 and 18 MPa) [64]. PC/SMA foams exhibit the polygon closed-cell structures having pentagonal and hexagonal faces, which express the most energetically stable state of polygon cells. Such foam structure was obtained probably because these foams belong to the polymeric foams having high gas phase volume (>0.6) [65]. Obviously, under low saturation CO_2 pressure (~10 MPa) both PCCN foams exhibit large cell size, indicating the dispersed clay particles hinder CO_2 diffusion by creating a maze or a more tortuous path as discussed in the literatures [2]. However, High CO_2 pressure (~18 MPa) provide a large supply of CO_2 molecules, which can subsequently form a large population of cell nuclei upon depressurization. The PC/SMA/MAE1 ($2C_12C_{18}$-fluorohectrite) (including 1 wt% of organoclay) foam shows smaller cell size, i.e., larger cell density compared to PC/SMA foam, suggesting that the dispersed clay particles act as nucleating sites for cell formation and lowering of *d* with clay. The incorporation of nano-clay hinder CO_2 diffusion and simultaneously induce heterogeneous nucleation because of a lower activation energy barrier compared to homogeneous nucleation [66]. They conducted the characterization of the interfacial tension between bubble and matrix by using modified classical nucleation theory [66].

In Table 4, the interfacial tension of the systems are summarized [67]. We can see that PC/SMA system has slightly large value (17.3 mJ/m^2) compared to that of PC/SMA/MAE1 (14.6 mJ/m^2) and PC/SMA/MTE1 (C_13C_8-fluorohectrite) (16.0 mJ/m^2) in case of low CO_2 pressure (10MPa). These estimated values of γ are in good agreement with that of other poly(methyl methacrylate) (PMMA)-CO_2 system (10–20 mJ/m^2) [68]. The value for PC/SMA system decreases with increasing CO_2

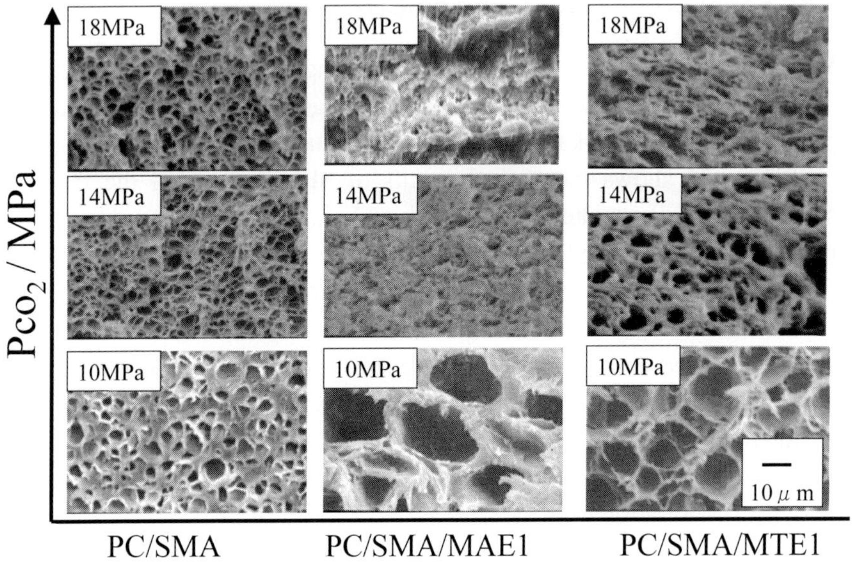

Fig. 24. Typical SEM images of the fracture surfaces of the PCCNs and PC/SMA blend without clay foamed under at 160°C under different isobaric condition (10, 14 and 18 MPa).

Table 4. Interfacial tension of the systems calculated using Suh and Colton's theory.

System	P_{CO2}/MPa	γ/mJ/m^2
PC/SMA-CO_2	10	17.3
PC/SMA/MAE-CO_2*	10	14.6
PC/SMA/MTE-CO_2*	10	16.0

*) MAE; dimethyl dioctadecyl ammonium cation-modified synthetic fluorohectorite
MTE; methyl trioctil ammonium cation-modified synthetic fluorohectorite.

pressure, as expected, while PC/SMA/MAE1 show a constant value with increasing pressure. This trend reflects the relative importance of heterogeneous nucleation, which dominates over homogeneous one in the event that the amount of CO_2 available for bubble nucleation is limited.

Figure 25 shows the stress-strain curves and the strain recovery behavior of the PP/LS nanocomposite (PPCN) foams [62] in the

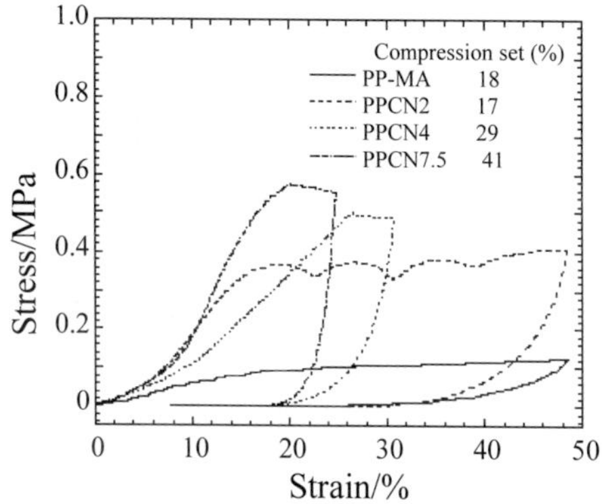

Fig. 25. Stress-strain curves and strain recovery behavior of the PP-based nanocomposites (PPCNs).

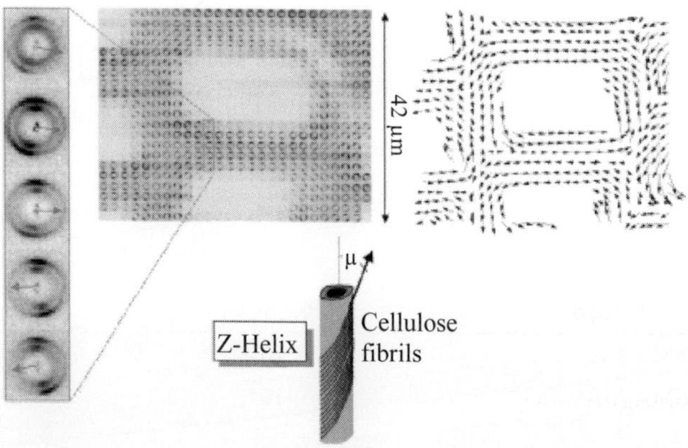

Fig. 26. X-Ray microdiffraction experiment with a 2-μm-thick section of spruce wood embedded in resin. Note the asymmetry of the patterns in the enlargement (far left) which can be used to determine the local orientation of cellulose fibrils in the cell wall (arrows). The arrows are plotted in the right image with the convention that they represent the projection of a vector parallel to the fibrils onto the plane of the cross-section. The picture clearly shows that all cells are right-handed helices. Reprinted from [63], P. Fratzl, Current Opinion in Colloid Interface Sci., 8, 32 (2003). © 2003, Elsevier Science.

compression mode at a constant strain rate of 5% min^{-1}. The nanocomposite foams exhibit high modulus compared to neat PP-MA foam. The residual strain is 17% for PPCN2 (including 2 wt% of organoclay) as well as neat PP foam, providing the excellent strain recovery and the energy dissipation mechanism, probably with the house-of-cards structure formation in the cell wall [62], which enhance the mechanical properties of the nanocomposites like a spruce wood which is close to right-handed helix (see Figure 26) [63].

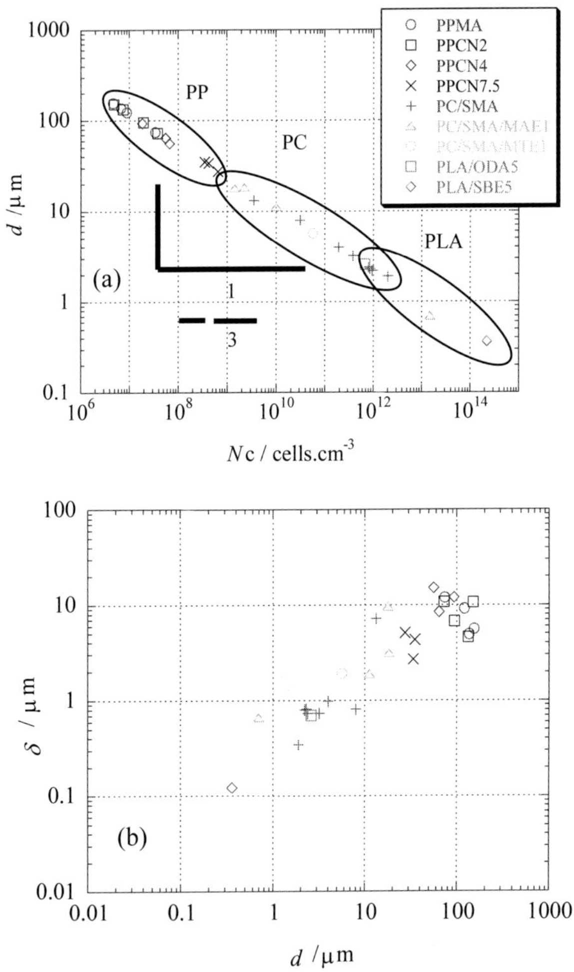

Fig. 27. (a) Cell size versus cell density and (b) cell wall thickness (δ) versus cell size for PCCN systems. For the comparison, here we show other nanocomposite foams obtained by a series of our recent studies [64,70]; PP: polypropylene-based , PLA: polylactide-based nanocomposite foams.

Figure 27 shows the cell size (d), the cell density (N_c) and the mean cell wall thickness (δ) relations of the nanocomposite foams obtained by a series of their recent studies [64,70]. In case of nanocomposite foams, the cell wall thickness becomes 2–6 times compared to that of neat polymer foams due to the spherical cell shape [63] caused by the high modulus of the materials during processing. The controlled structure of the nanocomposite foams from microcellular ($d \cong 20$ μm and $N_c \cong 1.0 \times 10^9$ cell·cm^{-3}) to nanocellular ($d \cong 200$ nm and $N_c \cong 1.0 \times 10^{14}$ cell·cm^{-3}).

Recently, some literatures are also available [71–74] related to the reactive extrusion foaming of various nanocomposites.

7.2. *Electrospinning processing*

Fibers and nanofibers of N6CN (diameter of 100–500 nm) were electrospun from HFIP solution and collected as non-woven fabrics or as aligned yarns [75]. The electrospinning process resulted in highly aligned MMT particles and Nylon 6 crystallites. The cylindrical shaped fibers and nanofibers, ribbon shaped fibers were also found in the products (Figure 28). The electrospinning can be expected to align other nanofillers such as carbon nanotubes.

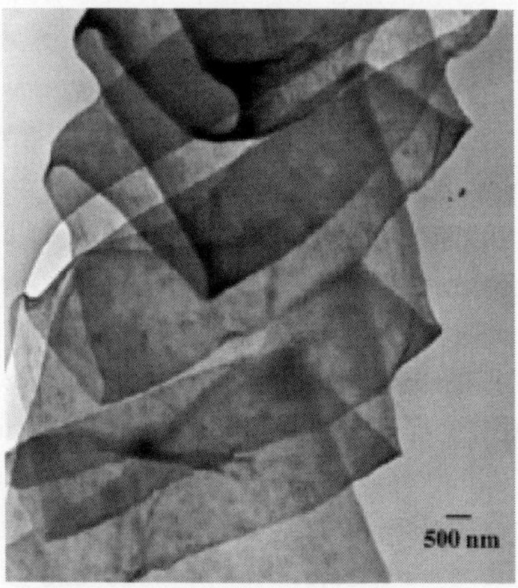

Fig. 28. TEM micrograph of a ribbon shaped nanofiber. Reprinted from [75], P. H. Fong *et al.*, Polymer 43, 775 (2002) © 2002, Elsevier Science.

7.3. *Porous ceramic materials via PLSNCs*

Very recently, a new route for the preparation of porous ceramic material from thermosetting epoxy/clay nanocomposite was first demonstrated by Brown *et al.* [76]. This route offers attractive potential for diversification and application of the PCNs. Sinha Ray and coworkers have reported the results on the novel porous ceramic material via burning of the PLA/LS system (PLACN) [77]. In the PLACN containing 3.0 wt% inorganic clay. The SEM image of fracture surface of porous ceramic material prepared from simple burning of the PLACN in a furnace up to 950°C is shown in Figure 29. After complete burning, as seen in the figure, the PLACN becomes white mass with porous structure. The bright lines in the SEM image correspond to the edge of the stacked silicate layers. In the porous ceramic material, the silicate layers form a cards-house structure, which consist of the large plates having length of ~1000nm and thickness of ~30–60 nm. This implies that the further stacked platelet structure is formed during burning. The material exhibits the open-cell type structure having 100–1000 nm diameter void, BET surface area of 31 m^2 g^{-1} and low density of porous material of 0.187 g ml^{-1} estimated by the buoyancy method. The BET surface area value of MMT is 780 m^2/g and that of the porous ceramic material is 31 m^2/g, suggests about 25 MMT plates stacked together. When MMT is heated above 700°C (but below 960°C) first all OH groups are eliminated from the structure and thus MMT is decomposed into that of a non-hydrated aluminosilicate. This transformation radically disturbs the crystalline network of the MMT, and resulting diffraction pattern is indeed often typical of an amorphous (or non-crystalline) phase. The estimated rough value of compression modulus (*K*) is in the order of ~1.2 MPa, which is five orders of magnitude lower than the bulk modulus of MMT (~10^2 GPa) [2]. In the stress-strain curve, the linear deformation behavior is nicely described in the early stage of the deformation, i.e., the deformation of the material closely resembles that of ordinary polymeric foams [78]. This open-cell type porous ceramic material consisting of the cards-house structure is expected to provide the strain recovery and excellent energy dissipation mechanism after unloading in the elastic region up to 8% strain, probably each plate bend like leaf spring. This

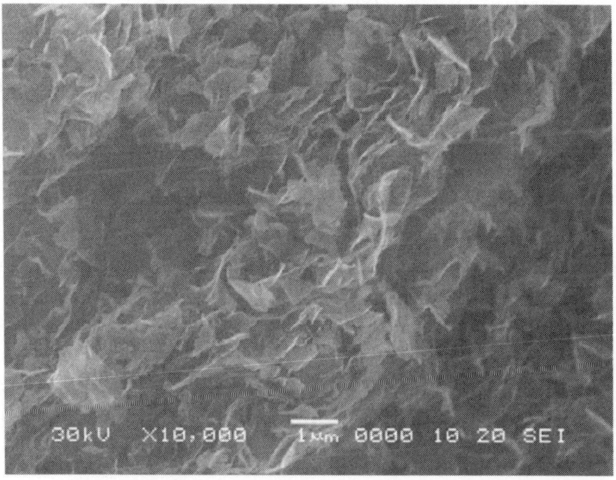

Fig. 29. SEM image of porous ceramic material after coated with platinum layer (~10 nm thickness). Reprinted from [77], S. Sinha Ray *et al.*, Nano Letts. 2, 423 (2002) © 2002 American Chemical Society.

porous ceramic material is a new material possessing feature of elastic and very lightweight. This new route for the preparation of porous ceramic material via burning of nanocomposites can be expected to pave the way for much broader range of applications of the PLSNCs. This porous ceramic material closely relates an excellent insulator property for flame reterdant of PLSNCs [2]. The flame behavior must derive from the morphological control of the shielding properties of the graphitic/clay created during polymer ablation.

8. Physicochemical Phenomena

8.1. *Biodegradability*

Another most interesting and exciting aspect of nanocomposite technology is the significant improvements of biodegradability of biodegradable polymers after nanocomposites preparation with organoclay. Aliphatic polyesters are among the most promising materials for the production of environmentally friendly biodegradable plastics. Biodegradation of aliphatic polyester is well known, in that some bacteria degrade them by producing enzymes, which attack the polymer. Tetto and his colleagues [79] first reported some results about the biodegradability of nanocomposites based on PCL, where authors found that the poly(ε-caprolactone) (PCL)/LS nanocomposites showed improved biodegradability compared to pure PCL. According to them, the improved biodegradability of PCL after nanocomposites formation may be due to the catalytic role of the organoclay in the biodegradation mechanism. But still it is unclear how the clay increases the biodegradation rate of PCL.

Recently, Lee *et al.* [80] reported the biodegradation of aliphatic polyester based nanocomposites under compost. Parts a and b of Figure 30 respectively represents the clay content dependence of biodegradation of APES-based nanocomposites prepared with two different types of clays. They assumed that the retardation of biodegradation due to the improvement of the barrier properties of the aliphatic APSE after nanocomposites preparation with clay. However, there is no data about permeability.

Very recently, Yamada and Okamoto *et al.* [81–84] first reported the biodegradability of neat PLA and corresponding nanocomposites prepared with trimethyl octadecylammonium modified MMT ($3C_1C_{18}$-MMT) with details mechanism. The used compost was prepared from food waste and tests were carried out at temperature of 58 ± 2°C. Figure 31a shows the real picture of the recovered samples of neat PLA and PLACN4 ($3C_1C_{18}$-MMT = 4 wt%) from compost with time. The decreased molecular weight M_w and residual weight percentage R_w of the initial test samples with time also reported in Figure 31b. The biodegradability of neat PLA is significantly enhanced after PLSNC preparation. Within one month, both extent of M_w and extent of weight loss are almost same level for both PLA and PLACN4. However, after one month, a sharp change occurs in weight loss of PLACN4, and within two months, it is completely degrade in compost. The degradation of PLA in compost is a complex process involving four main phenomena, namely: water absorption, ester cleavage and formation of oligomer fragments, solubilization of oligomer fragments, and finally diffusion of soluble oligomers by bacteria [85]. Therefore, the factor, which increases

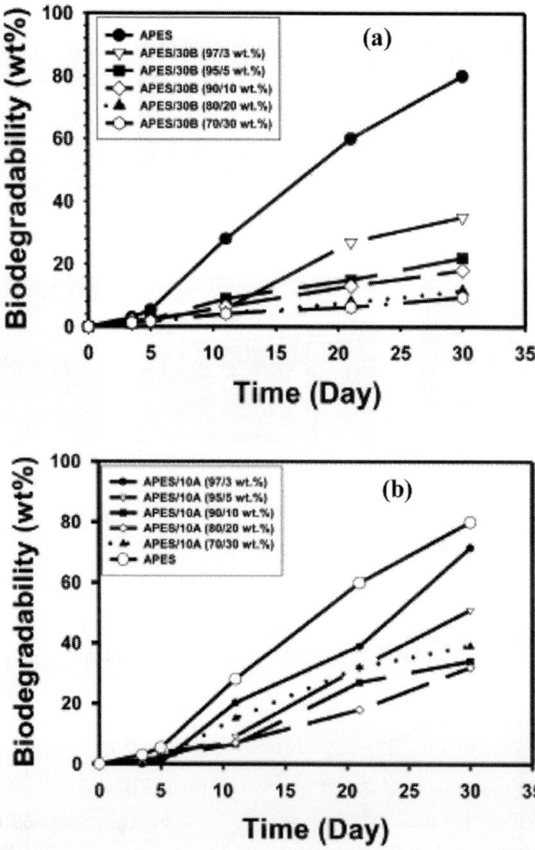

Fig. 30. Biodegradability of APES nanocomposites with: (a) Closite 30B and (b) Closite 10A. Reprinted from [80], S. R. Lee *et al.*, Polymer 43, 2495 (2002) © 2002, Elsevier Science.

the hydrolysis tendency of PLA, ultimately controls the degradation of PLA. They expect the presence of terminal hydroxylated edge groups of the silicate layers may be one of the responsible factors for this behavior. In case of PLACN4, the stacked (~4 layers) and intercalated silicate layers are homogeneously dispersed in the PLA matrix (from TEM image [83]) and these hydroxy groups start heterogeneous hydrolysis of the PLA matrix after absorbing water from compost. This process takes

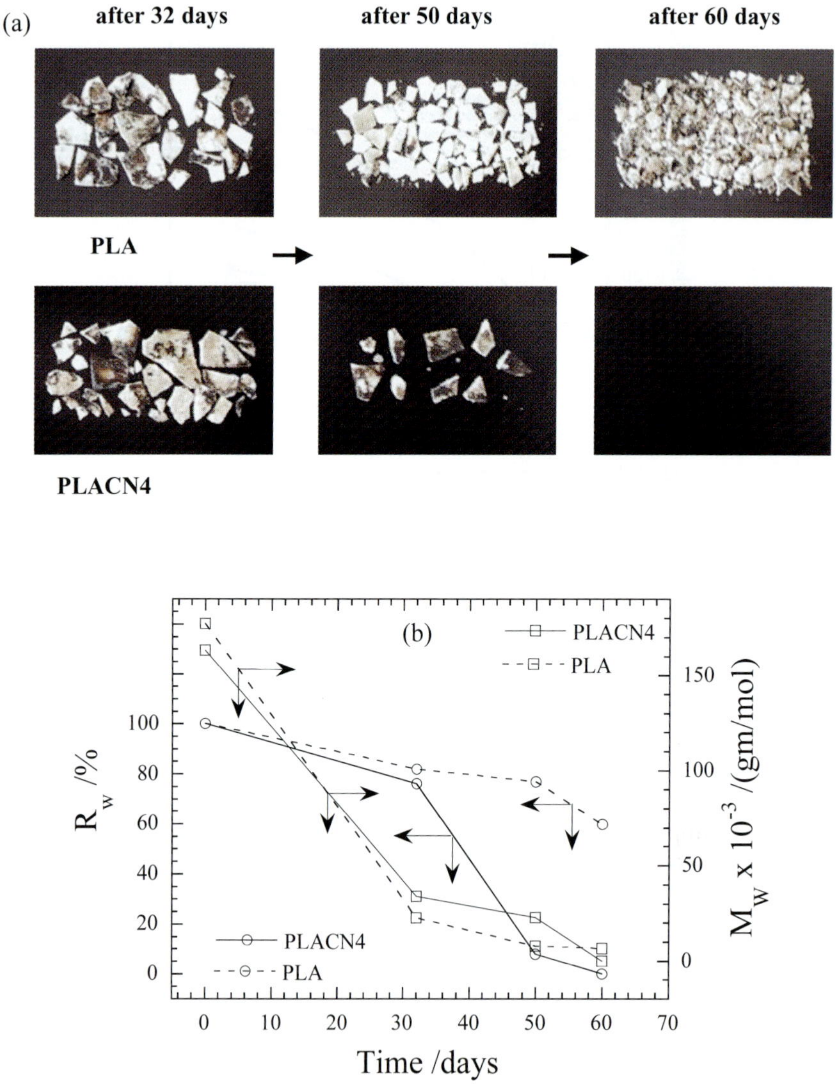

Fig. 31. (a) Real picture of biodegradability of neat PLA and PLACN4 recovered from compost with time. Initial shape of the crystallized samples was 3 × 10 × 0.1 cm^3. (b) Time dependence of residual weight, R_w and of matrix, M_w of PLA and PLACN4 under compost at 58 ± 2°C. Reprinted from [81], S. Sinha Ray *et al.*, Nano. Lett., 2, 1093 (2002) © 2002 American Chemical Society.

some time to start. For this reason, the weight loss and degree of hydrolysis of PLA and PLACN4 is almost same up to one month (see Figure 31b). However, after one month there is a sharp weight loss in case of PLACN4 compared to that of PLA. That means one month is a critical value to start heterogeneous hydrolysis, and due to this type of hydrolysis matrix becomes very small fragments and disappear with compost. This assumption was confirmed by conducting same type of experiment with PLACN prepared by using dimethyl dioctdecyl ammonium salt modified synthetic mica which has no terminal hydroxylated edge group, and the degradation tendency almost same with neat PLA [82].

They also conducted respirometric test to study degradation of the PLA matrix in compost environment at 58 ± 2°C. For this test the used compost was made from bean-curd refuse, food waste, and cattle feces. Unlike weight loss, which reflects the structural changes in the test sample, CO_2 evolution provides an indicator of the ultimate biodegradability of PLA in PLACN4 (prepared with (*N*(cocoalkyl)*N*,*N*-[bis(2-hydroxyethyl)]-*N*-methylammonium modified synthetic mica,), i.e. mineralization, of the samples. Figure 32 shows the time dependence of the degree of biodegradation of neat PLA and PLACN4, indicating that the biodegradability of PLA in PLACN4 is enhanced significantly. The presence of organoclay may thus cause a different mode of attack on the PLA component, which might be due to the presence of hydroxy groups. Details degrading the mechanism of biodegradability are presented in relevant literature [82,3].

K. Okamoto and M. Okamoto also investigated biodegradability of neat PBS before and after nanocomposites preparation with three different types of organoclay. They used alkylammonium or alkylphosphonium salts for the modification of pristine layered silicates, and these surfactants are toxic for microorganisms [43,86].

Figure 33a shows the real pictures of recovered samples of neat PBS and various nanocomposites from the compost after 35 days. From the figure it is clearly observed that many cracks are appeared in nanocomposite samples compared to that of neat PBS. This observation indicates the improved degradability of nanocomposites in compost. This

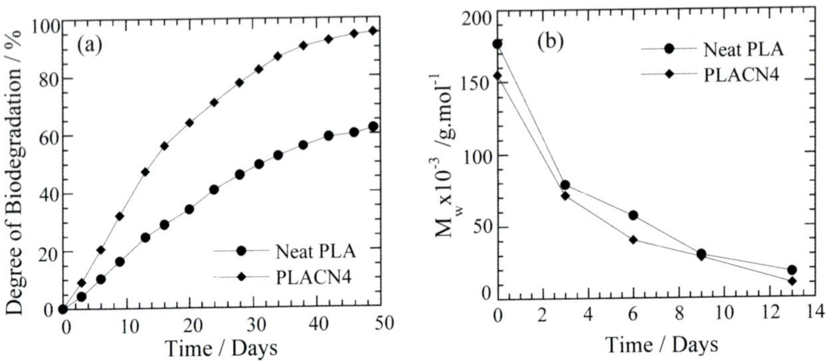

Fig. 32. Degree of biodegradation (i.e. CO_2 evolution), and (b) time-dependent change of matrix M_w of neat PLA and PLACN4 (MEE clay = 4 wt%) under compost at 58 ± 2°C. Reprinted from [82], S. Sinha Ray *et al.,* Macromol. Rapid Commun. 23, 943 (2002) © 2004 WILEY-VCH Verlag GmbH & Co.

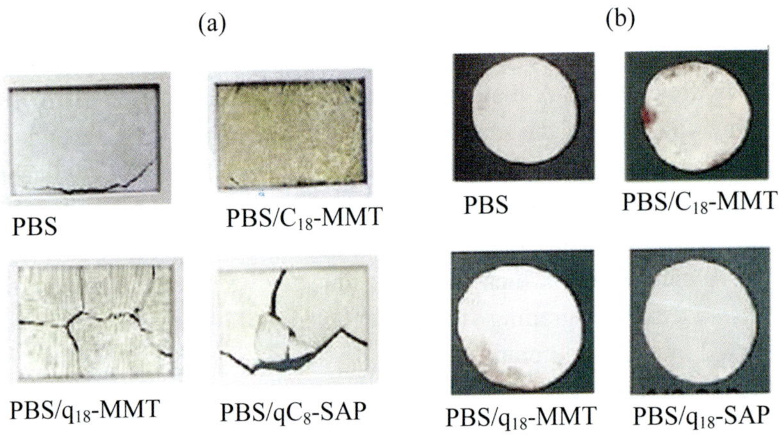

Fig. 33. Biodegradability of neat PBS and various nanocomposites sheets (a) under compost, and (b) under soil field. Reprinted from [86], K. Okamoto *et al.,* J Polym. Sci. Part B: Polym. Phys. 41, 3160 (2003) © 2003 John Wiley & Sons, Inc.

kind of fracture has an advantage for biodegradation because of easy to mix with compost and create much more surface area for further attack by microorganisms, and it should be noted here that the extent of ragmentation is directly related to the nature of organoclay used for nanocomposites preparation. They also conducted the GPC (gel permeation chromatography) measurement of recovered samples from compost, and we found that extent of molecular weights loss is almost the same for all samples (see Table 5). This result indicates that the extent of hydrolysis of PBS in pure state or OMLS filled systems is the same in compost.

Except for the PBS/qC_{16}-SAP (*n*-hexadecyl tri-*n*-butyl phosphonium cation modified saponite) system, the degree of degradation is not different for other samples. This observation indicates that MMT or alkylammonium cations, and at the same time other properties have no effect on the biodegradability of PBS. The accelerated degradation of PBS matrix in the presence of qC_{16}-SAP may be due to the presence of alkylphosphonium surfactant. This kind of behavior is also observed in the case of PLA/LS nanocomposite systems.

They also observed nature of degradation of PBS and various nanocomposites under soil field. This experiment was conducted for one, two, and six months. After one and two months, there is no change on the nature of samples surface, but after six months black or red spots are appeared on the surface of nanocomposites samples. Figure 33b represents results of degradation of neat PBS and various nanocomposites sheets recovered from soil field after six months. They reported these spots on the sample surface are due to the fungus attacked, because when we put these parts into the slurry and we observed clear growth of fungus. These results also indicate that nanocomposites exhibit same or higher level of biodegradability compared with PBS matrix.

Table 5. GPC results of various samples recovered from compost after 35 days.

Samples	$M_w \times 10^{-3}$/g/mol	$M_n \times 10^{-3}$/g/mol	$M_w^o \times 10^{-3}$/g/mol	M_w/M_w^o
PBS	16	3.8	101	0.16
PBS/C_{18}-MMT	17	6.6	104	0.16
PBS/qC_{18}-MMT	17	4.4	112	0.15
PBS/qC_{16}-SAP	8.7	1.2	91	0.096

M_w^o is molecular weight before composting

8.2. *Photodegradation*

Very recently, Hiroi and Okamoto *et al.* [87] first reported the photodegradability of neat PLA and corresponding PLA nanocomposite prepared by using organically modified layered titanate as a new nanofiller. One of the feature of this material is photocatalytic reactivity, such as titania (TiO_2). The photocatalytic reaction of anatase-TiO_2, such as evolution of hydrogen gas from water or oxidative degradation of organic compounds, have attracted intense research interest because of their possible application to the conversion of solar energy into chemical energy [88].

Figure 34 shows the UV/vis transmission spectra of pure PLA and Nanocomposite (PLANC1.7). The spectra show that the VIS region (>~ 400 nm) is changed with increasing absorbency by the presence of titanate layers compared with neat PLA. For UV wavelengths, there is strong absorption up to 320 nm, resulting in 0% transmittance. This significant change in the spectra may indicate that occurrence of the photodegradation of PLA matrix. To confirm this, they conducted some preliminary experiments on photodegradation of PLANCs under sunshine weathermeter at 60°C. After 300 h, there was no change in the nature of sample surfaces of neat PLA, however, the surface color of

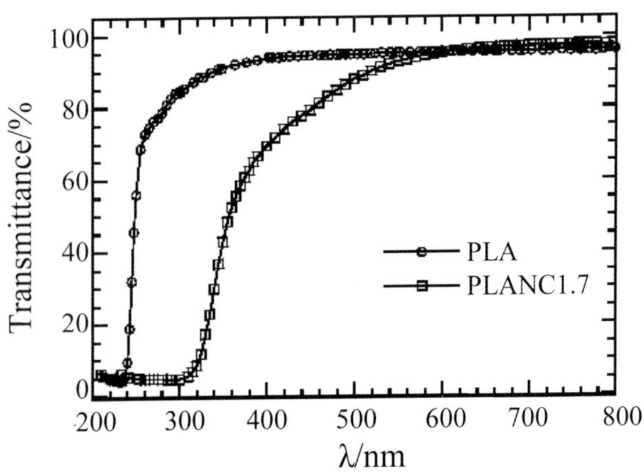

Fig. 34. UVvis transmission spectra of neat PLA and nanocomposite (PLANC1.7). Reprinted from [87], R. Hiroi *et al.*, Macromol. Rapid. Commun., 25, 1359 (2004) © 2004 WILEY-VCH Verlag GmbH & Co.

Table 6. GPC results of sample recovered from weathermeter after 300 h.

Samples	$M_w \times 10^{-3}$ (g/mol)	M_w/M_n	$M_w/M_w^{o\,a}$
PLA	198	1.53	0.94
PLANC1.7	93.7	1.89	0.68
PLANC3.9	86.3	1.86	0.76

[a] M_w^o is molecular weight before test

Source: Reprinted from [87], R. Hiroi *et al.*, Macromol. Rapid. Commun., 25, 1359 (2004) © 2004 WILEY-VCH Verlag GmbH & Co.

nanocomposite samples altered yellow and/or light brown. Table 6 shows the GPC measurement of recovered samples from the test. The drop in M_w accompanied with broadening of M_w/M_n indicates that the enhancement of degradation of PLA in the titanate-filled system has occurred.

8.3. *Pressure-volume-temperature (PVT) behavior*

Recently, in N6-based nanocomposites (N6CNs) the reduction of free volume by 12 to 17% was reported through a pioneering effort by Utracki *et al.* [89]. Figure 35 shows the temperature dependence of the free volume reduction by the addition of nanoclay particles. They speculated that the N6 chains are absorbed on the solid clay surface and the first few absorbed layers are immobilized. This loss of mobility translates in their model into a free volume loss. The effect was simulated by a large adhesion ratio $\varepsilon^*_{12}/\varepsilon^*_{11}$, (= 313.54 kJ/mol/ 32.09 kJ/mol) which are energetic interaction parameter between polymer-clay and polymer-polymer. They also reported the value of the free volume reduction for the polystyrene-based nanocomposite system [90]. In case of 4 wt% organoclay loading, the reducton is much lower (4 to 6%) than that of N6-based nanocomposites due to the weak interaction between polymer and silicate surfaces.

Recently, some literatures are also available [91] related to the PVT dependencies of commercial PP melt and its nanocomposites including small amount of a compatibilizer. Another recent approach is positron annihilation lifetime spectroscopy for evaluation of the free volume fraction in PLSNCs [92].

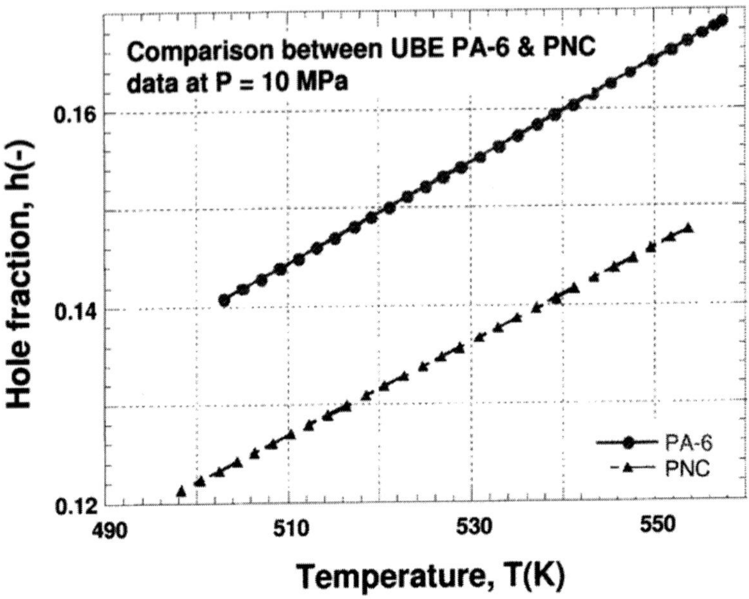

Fig. 35. Hole fraction versus temperature for N6-based nanocompopsite. Reprinted from [89], R. Simha *et al.*, Composite Interfaces, 8, 345 (2001) © 2001 VSP.

9. Computer Simulation

9.1. *Theoretical phase diagram of PLSNCs*

Recently, Ginzburg and Balazs developed simple models that describe the liquid crystalline ordering in the polymer-platelet systems [93–96]. They combined a density functional theory (DFT) with a self-consistent field model (SCF) to calculate the phase behavior of thin, oblate colloidal particles that are coated with surfactants and dispersed in a polymer melt. These coated particles represent organoclay sheets. By intergrafting the two methods, they investigated the effect of the surfactants characteristics (grafting density, ρ_{gr} and length, N_{gr}) and the polymer-intercalant interaction energy on the polymer/clay phase diagram (see Figure 36) [94]. Depending on the values of these critical parameters and clay volume fraction, ϕ, the system can be in an isotropic or nematic phase (exfoliated system). The system can also form a smectic, crystal, columnar, or house-of-cards plastic solid as well as a two-phase

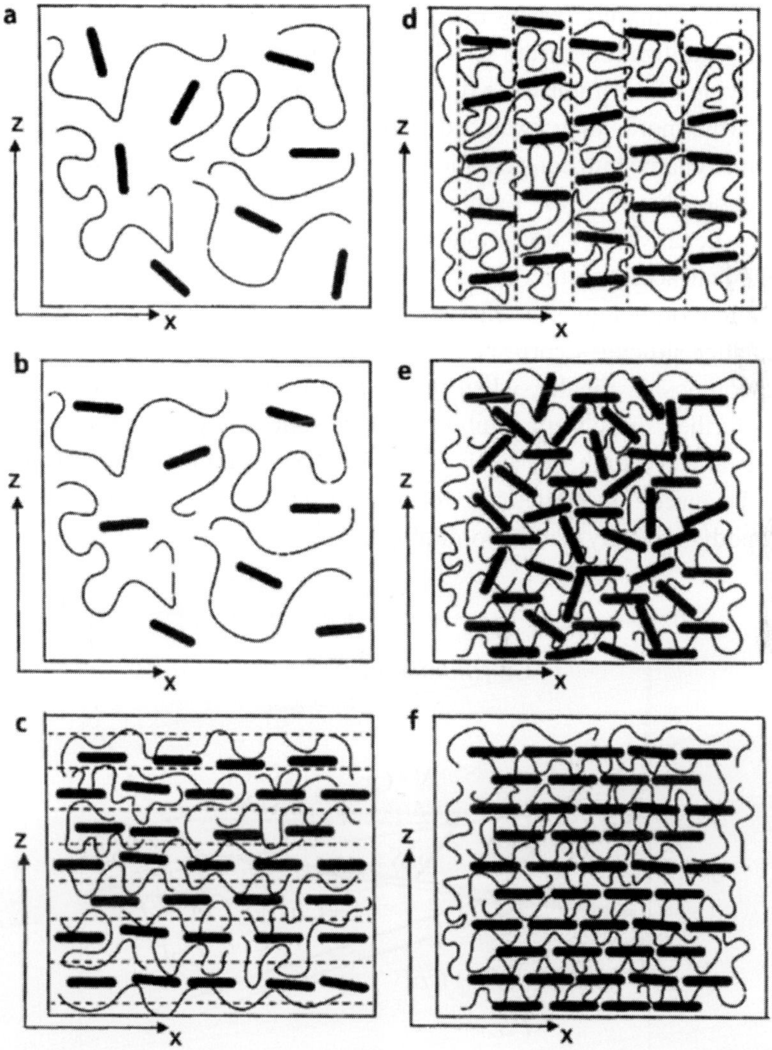

Fig. 36. Mesophases of oblate uniaxial particles dispersed in a polymer: (a) isotropic (I), (b) nematic (N), (c) smectic (S), (d) columnar (Col), (e) plastic solid (house-of-cards) (PS), and (f) crystal (Cr). The nematic director **n** in ordered phases is aligned along the Z axis, the disks lie in the XY plane. Dashed lines show smectic layers (c) and columns (d). Reprinted from [95], V V. Ginzburg *et al.*, Macromolecules, 33, 1089 (2000) © 2000 American Chemical Society.

immiscible mixture. Using this model they isolated conditions that lead to the stabilization of the exfoliated nanocomposite system and to the narrowing of the immiscible two-phase regions.

Furthermore, they extended their study to take end-functionalized polymer chains into consideration [95]. The parameter ε characterizes the interaction between the terminal functional group and the surface. For small negative ε, the system is completely immiscible As the sticker/surface attraction is increased, the system exhibits isotropic and nematic morphologies at relatively low clay volume fractions ($\phi < 0.12$). The latter structures correspond to thermodynamically stable, exfoliated nanocomposite. Finally, when the surface/sticker attraction is increased even further, the system exhibits a plastic solid at low ϕ, and a columnar phase at very high ϕ (see Figure 37). The resulting phase diagrams can be used as guidelines for the nanocomposite formation with thermodynamically stable morphologies.

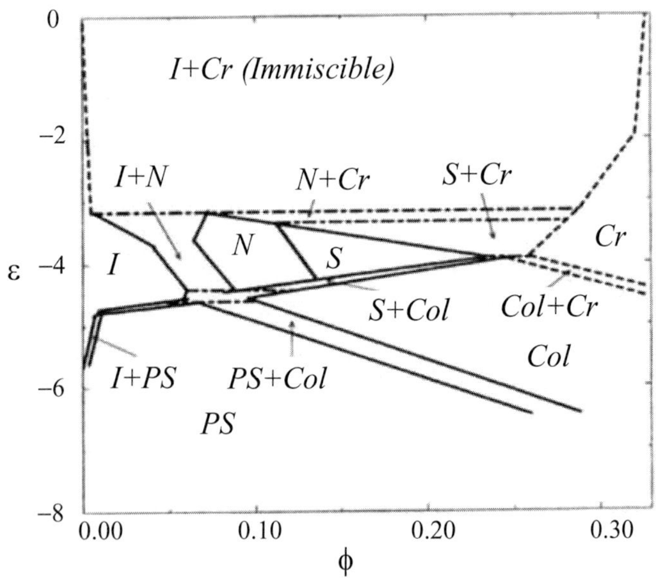

Fig. 37. Phase diagram for polymer/clay mixture. Here ε is the sticker-clay adhesion energy. Reprinted from [95], V V. Ginzburg *et al.*, Adv. Mater., 12, 1805 (2000) © 2000 WILEY-VCH Verlag GmbH & Co.

9.2. Atomic scale structure and binding energy in PLSNCs

Molecular simulation techniques are used to explore and characterize the atomic scale structure, and to predict binding energies and basal spacing of PLSNCs based on PP and maleated (MA) PP (PP-MA), MMT, and different alkylammonium ions as intercalants [97]. The basic machinery of the procedure consists in building a molecular model comprising PP, a given intercalant and MMT, refining and equilibrate it by molecular mechanics/molecular dynamics (MM/MD), and calculating the binding energies as guidelines for screening among different intercalants to make PPLSNCs characterized by strong interface between dispersed clay platelets and the PP matrix (see Figure 38). Figure 39 shows the

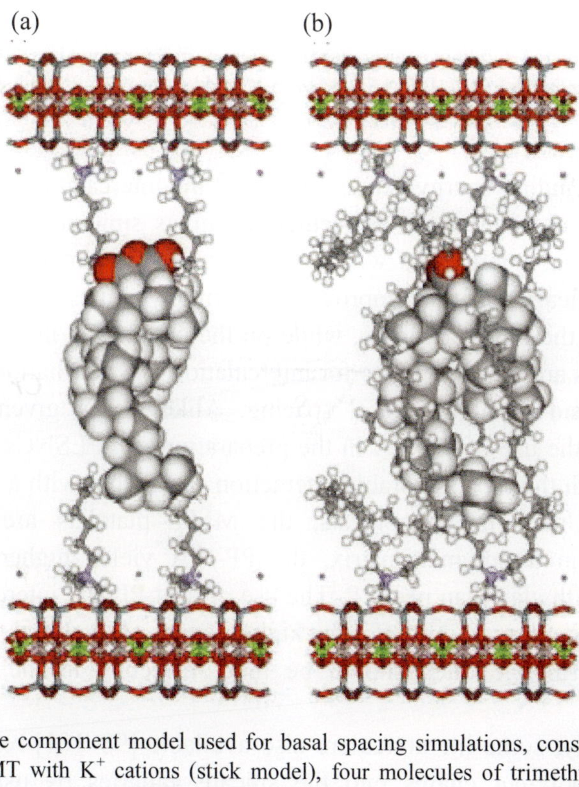

Fig. 38. Three component model used for basal spacing simulations, consisting of two layers of MMT with K^+ cations (stick model), four molecules of trimethylammonium cation (a) or dimethylstearylammonium cation (b) (stick and ball model), and one molecule of maleated PP (PP-MA) (ball model). Reprinted from [97], R. Toth *et al.*, Polymer 45, 8075 (2004) © 2004, Elsevier Science.

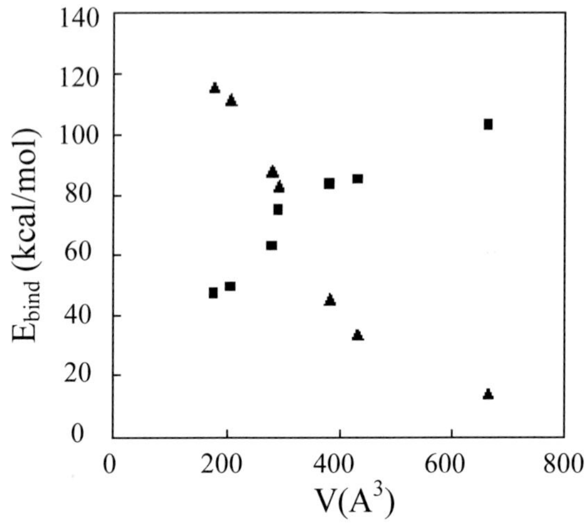

Fig. 39. Predicted binding energy versus volume of the intercalant. (■) E_{bind} (PP-MA/intercalant); (▲) E_{bind} (PP-MA/MMT). Reprinted from [97], R. Toth *et al.*, Polymer 45, 8075 (2004) © 2004, Elsevier Science.

predicted binding energy versus volume of the intercalant. From a global interpretation of all these molecular dynamics simulation results, they conclude that intercalants with smaller volume are more effective for clay modification as they improve the thermodynamics of the system by increasing the binding energy, while on the other hand intercalants with longer tails are more effective for intercalation and exfoliation processes, as they lead to higher basal spacing. Alike, for a given surfactant molecule, the use of PP-MA in the preparation of PPLSNCs results in a system with the most favorable interaction energy but with a lower basal spacing. Under the hypothesis, the MMT platelets are uniformly dispersed in a polymer matrix, the PP-MA yields higher interfacial strength with clay than neat PP. The use of neat PP and intercalants with higher molecular volume offer the higher values of the basal spacing and thus, in principle, they should be more effective in the exfoliation process.

Recently, some literatures are also available [98–101] related to the confined polymer chains into the silicate galleries by using coarse-grained MD simulation.

10. Future Prospects of PLSNCs

Development of the PLSNCs is one of the latest evolutionary step of the polymer technology. The PLSNCs offer attractive potential for diversification and application of conventional polymeric materials. Some of PLSNCs are already commercially available and applied in industrial products.

Biodegradable polymers based nanocomposites have a great deal of future promise for potential applications as high-performance biodegradable materials. These are entirely new types of materials based on plant and nature materials (organoclay). When disposed of in compost, these are safely decomposed into CO_2, water, and humus through the activity of microorganisms. The CO_2 and water will become corn or sugarcane again through plant photosynthesis. Undoubtedly, there unique properties originated from the controlled nanostructure paves the way to much broader range of applications (already commercially available through Unitika Ltd., Japan), and open a new dimension for plastics and composites.

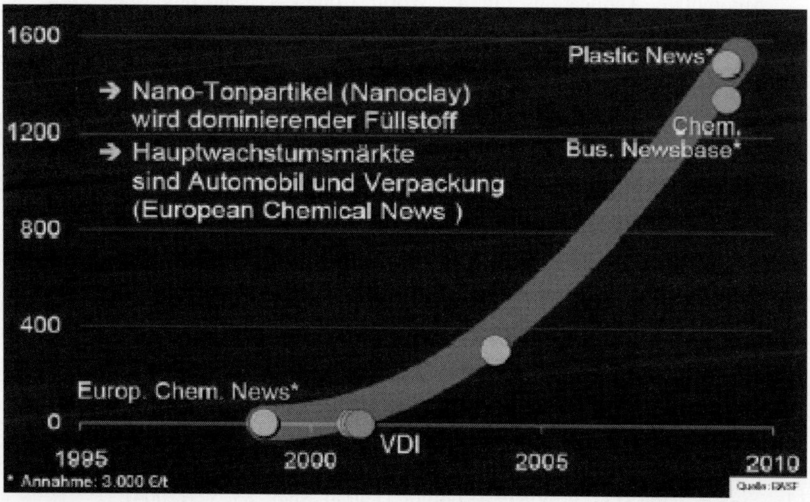

Fig. 40. Market forecasts (turnover in million Euro) for polymer nanocomposites by various sources. Through the courtesy of A. Ebenau [102].

According to several sources (Chemical Business Newsbase, Plastic News), a significant increase in turnover of about 100% p.y. leading to a value of about 1500 million Euro (ca. 500,000 t/y of polymer nanocomposites) in 2009 is expected (see Figure 40) [102]. Even if these numbers may be over-optimistic, they highlight the tremendous technological and economical potential associated with polymeric nanocomposites including not only clay but also other inorganic nanofillers, such as carbon nanotubes and SiO_2, SiC, and Si_3N_4. At present it is difficult to predict which, if any, market sector would not be able to benefit from this technology. Thus it may well be that polymeric nanocomposites in the mid- and longer-term will pervade all aspects of life, similar to the way plastics did in the last century. Clearly a diverse range of sectors such as aerospace, automotive, packaging (particularely food but also solar cells), electrical and electronic goods, household goods etc. will profit substantially from a new range of materials which would be offered by this technology.

References

1. Cox H, Dearlove T, Rodges W, Verbrugge M, "Nanocomposite Systems for Automotive Applications" in 4th World Congress Nanocomposites 2004, San Francisco, Sep. 1–3 (2004).
2. Okamoto M, "Polymer/Layered Silicate Nanocomposites", Rapra Review Report No 163, 166pp Rapra Technology Ltd., London, (2003).
3. Usuki A, Kojima Y, Okada A, Fukushima Y, Kurauchi T, Kamigaito O, J. Mater. Res. 8, 1174 (1993).
4. National Lead Co., U.S. Patent No. 2531396 (1950).
5. Union Oil Co., U.S. Patent No. 3084117 (1963).
6. Unitika Ltd., Japanese Kokai Patent Application No. 109998 (1976).
7. Fukushima Y, Inagaki S, J. Incl. Phen. 5, 473 (1987).
8. Kojima Y, Usuki A, Kawasumi M, Okada A. Fukushima Y. Karauchi T, Kamigaito O, J. Mater. Res. 8, 1185 (1993).
9. Jose-Yacaman M, Rendon L, Arenas J, Puche M.C.S. Science, 273, 223 (1996).
10. Hermosin M.C, Cornejo J, Clays Clay Miner. 34, 591 (1986).
11. Krishnamoorti, Vaia R.A, Giannelis E.P. Chem Mater. 8, 1728 (1996).
12. Sinha Ray S, Yamada K, Okamoto M, Ogami A, and Ueda K, Chem. Mater., 15, 1456 (2003).
13. Su S.P, Jiang D.D, Wilkie C.A, Polym. Degrad. Stab., 83, 321(2004).
14. Su S.P, Jiang D.D, Wilkie C.A, Polym. Degrad. Stab., 83, 333 (2004).
15. Vuillaume PY, Glinel K, Jonas AM, Laschewsky A, Chem. Mater., 15, 3625 (2003).

16. Acosta E.J, Deng Y, White G.N, Dixon J.B, McInnes K.J, Senseman SA, Frantzen AS, Simanek E.E, Chem. Mater., 15, 2903 (2003).
17. Fischer H, Mater. Sci. Eng., C23, 763 (2003).
18. Hasegawa N, Okamoto H, Kato M, Usuki A, Sato N, Polymer, 44, 2933 (2003).
19. van Olphen, H. An Introduction to Clay Colloid Chemistry, Wiley, New York (1977).
20. Zerda A.S, Caskey T.C, Lesser A, Macromolecules, 36, 1603 (2003).
21. Zhao Q, Samulski E.T, Macromolecules, 36, 6967 (2003).
22. Maiti P, Nam P.H, Okamoto M, Kotaka T, Hasegawa N, Usuki A, Macromolecules, 35, 2042 (2002).
23. Nam P.H, Maiti P, Okamoto M, Kotaka T, Hasegawa N, Usuki A, Polymer, 42, 9633 (2001).
24. Maiti P, Nam P.H, Okamoto M, Kotaka T, Hasegawa N, Usuki A, Polym. Eng. Sci., 42, 1864 (2002).
25. Khare R, Pablo J.J.de, Yethiraj A, Macromolecules 29, 7910 (1996).
26. Sheng N, Boyce M.C, Parks D.M, Rutledge G.C, Abes J.I, Cohen R.E, Polymer, 45, 487 (2004).
27. Carrado K.A, in Handbook of Layered Materials, Auerbach S.M, Carrado K.A, Dutta P.K, Eds., Marcel-Dekker: NY, pp.1–38 (2004).
28. Sato H, Yamagishi A, Kawamura K, J. Phys. Chem., B 105, 7990 (2001).
29. Seo Y-S, Ichikawa Y, Kawamura K, Mater. Sci. Res. Int. 5, 13 (1999).
30. Tamura K, Setsuda H, Taniguchi M, Yamagishi A, Langmuir, 15, 6915 (1999).
31. Medellin-Rodriguez F.J, Burger C, Hsiao B.S, Chu B, Vaia R.A, Phillips S, Polymer, 42, 9015 (2001).
32. Maiti P, Okamoto M, Macromole. Mater. Eng., 288, 440 (2003).
33. Landis W.J, Connect. Tissue Res., 35, 1 (1996).
34. Lincoln D.M, Vaia R.A, Wang Z.G, Hsiao B.S, Polymer, 42, 1621 (2001).
35. Lincoln D.M, Vaia R.A, Macromolecules, 37, 4554 (2004).
36. Kim G.M, Lee D.H, Hoffmann B, Kressler J, Stoppelmann G, Polymer, 42, 1095 (2001).
37. Stern T, Wachtel E, Marom G, J. Polym. Sci., Part B. Polym. Phys., 35, 2429 (1997).
38. Fratzl P, Gupta H.S, Paschalis E.P, Roschger P, J. Mater. Chem., 14, 2115 (2004).
39. Priya L, Jog JP, J. Polym. Sci. Part B: Polym. Phys., 40, 1682 (2002).
40. Shah D, Maiti P, Gunn E, Schmidt D.F, Jiang D.D, Batt C.A, Giannelis E.P, Adv. Mater., 16, 1173 (2004).
41. Sinha Ray S, Yamada K, Okamoto M, Ueda K, Polymer, 44, 857 (2003)
42. Chen J.S, Poliks M.D, Ober C.K, Zhang Y, Wiesner U, Giannelis E.P, Polymer, 43, 4895 (2002).
43. Sinha Ray S, Okamoto K, Okamoto M, Macromolecules, 36, 2355 (2003).
44. Okamoto K, Sinha Ray S, Okamoto M, J. Polym. Sci. Part B: Polym. Phys., 41, 3160 (2003).
45. Utracki, L.A, Polymer Alloys and Blends: Thermodynamics and Rheology, Hasser Publishers, NY 1990.
46. Cox W.P, Merz E.H, J Polym. Sci., 28, 619 (1958).

47. kamoto M, Sato H, Taguchi H, Kotaka T, Nippon Rheology Gakkaishi, 28, 201 (2000).
48. Okamoto M, Nam P.H, Maiti P, Kotaka T, Hasegawa N, Usuki A. Nano Lett., 1, 295 (2001).
49. Kotaka T, Kojima A, Okamoto M, Rheol. Acta, 36, 646 (1997).
50. Nam, P.H, Master Thesis, Toyota Technological Institute (2001).
51. Kojima Y, Usuki A, Kawasumi M, Okada A, Kurauchi T, Kamigaito O, Kaji K, J. Polym. Sci. Part B: Polym. Phys., 33, 1039 (1995).
52. Yalcin B, Cakmak M, Polymer, 45, 2691 (2004).
53. Medellin-Rodriguez F.J, Burger C, Hsiao B.S, Chu B, Vaia R.A, Phillips S, Polymer, 42, 9015 (2001).
54. Lele A, Mackley M, Galgali G, Ramesh C, J. Rheol., 46, 1091 (2002).
55. Bafna A, Beaucage G, Mirabella F, Mehta S, Polymer, 44, 1103 (2003).
56. Roe R, Methods of X-ray and neutron scattering in polymer science. New York: Oxford University Press, p.199 (2000).
57. Bafna A, Beaucage G, Mirabella F, Skillas G, Sukumaran S, J Polym. Sci., Part B: Polym. Phys. 39:2923 (2001).
58. Kim J.W, Kim S.G, Choi H.J, Jhon M.S, Macromol. Rapid Commun., 20, 450 (1999)
59. Kim J.W, Noh M.H, Choi H.J, Lee D.C, Jhon M.S, Polymer, 41, 1229 (2000).
60. Park J.H, Lim Y.T, Park O.O, Macromol. Rapid Commun., 22, 616 (2001).
61. Lim Y.T, Park J.H, Park O.O, J. Colloid Interface Sci., 245, 198 (2002).
62. Okamoto M, Nam P.H, Maiti M, Kotaka T, Nakayama T, Takada M, Ohshima M, Usuki A, Hasegawa N, Okamoto H. Nano Lett., 1, 503 (2001).
63. Fratzl P, Current Opinion in Colloid Interface Sci., 8, 32 (2003).
64. Nam P.H, Okamoto M, Maiti P, Kotaka T, Nakayama T, Takada M, Ohshima M, Hasegawa N, Usuki A, Polym. Eng. Sci., 42(9), 1907 (2002).
65. Mitsunaga M, Ito Y, Okamoto M, Sinha Ray S, Hironaka K, Macromol. Mater. Eng., 288, 543 (2003).
66. Klempner D, Frisch K.C, Handbook of Polymeric Foams and Foam Technology , Hanser Publishers, Munich Vienna (1991).
67. Colton J.S, Suh N.P, Polym. Eng. Sci., 27, 485 (1987).
68. Ito Y, Okamoto M, Polymer (2005) in press.
69. Goel S.K, Beckman E.J, Polym. Eng. Sci., 34, 1137 (1994).
70. Fujimoto Y, Sinha S. Ray, Okamoto M, Ogami A, Ueda K, Macromol. Rapid Commun., 24, 457 (2003).
71. Cao X, Lee LJ, Widya T, Macosko C, Polymer, 46, 775 (2005).
72. Chandra A, Gong S, Turng LS, Gramann P, Cordes H, Polym. Eng. Sci., 45, 52 (2005).
73. Taki K, Yanagimoto T, Funami E, Okamoto M, Ohshima M, Polym. Eng. Sci., 44, 1004 (2004).
74. Strauss W, D'Souza N.A, J. Cellular Plastics, 40, 229 (2004).
75. Fong H, Liu W, Wang C.S, Vaia R.A, Polymer, 43, 775 (2002).
76. Brown J.M, Curliss D.B, Vaia R.A, Proc. of PMSE, Spring Meeting, San Francisco, California, (2000), p278.

77. Sinha Ray S, Okamoto K, Yamada K, Okamoto M. Nano. Lett., 2, 423 (2002).
78. Gibson L.J, Ashby M.F, editors, Cellular Solids, P.8. Pergamon Press, N. Y, (1988).
79. Tetto J.A, Steeves D.M, Welsh E.A, Powell B.E, ANTEC'99, 1628 (1999).
80. Lee S.R, Park H.M, Lim H.L, Kang T, Li X, Cho W.J, Ha C.S, Polymer 43, 2495 (2002).
81. Sinha Ray S, Okamoto M, Yamada K, Ueda K, Nano. Lett., 2, 1093 (2002).
82. Sinha Ray S, Yamda K, Ogami A, Okamoto M, Ueda K, Macromol. Rapid Commun., 23, 943 (2002).
83. Sinha Ray S, Yamada K, Okamoto M, Ueda K, Macromol. Mater. Eng., 288, 936 (2003).
84. Sinha Ray S, Yamada K, Okamoto M, Ueda K., Polymer, 44, 857 (2003).
85. Liu J.W, Zhao Q, Wan C.X, Space Medicine and Medical Eng. 14, 308 (2001).
86. Okamoto K, Sinha Ray S, Okamoto M, J. Polym. Sci. Part B: Polym. Phys., 41, 3160 (2003).
87. Hiroi R, Sinha Ray S, Okamoto M, Shiroi T, Macromol. Rapid. Commun., 25, 1359 (2004).
88. Fujishima A, Honda K, Nature, 37, 238 (1972).
89. Simha R, Utracki LA, Garcia-Rejon A, Composite Interfaces, 8, 345 (2001).
90. Tanoue S, Utracki LA, Garcia-Rejon A, Tatibouet J, Cole KC, Kamal MR, Polym. Eng. Sci., 44, 1046 (2004).
91. Utracki LA, Simha R, Macromolecules, 37, 10123 (2004).
92. Wang Y, Wu Y, zhang H, Zhang L, Wang B, Wang Z, Macromol. Rapid Commun., 25, 1973 (2004).
93. Ginzburg V.V, Balazs A.C, Macromolecules, 32, 5681 (1999).
94. Ginzburg V.V, Singh C, Balazs A.C, Macromolecules, 33, 1089 (2000).
95. Ginzburg V.V, Balazs A.C, Adv. Mater., 12, 1805 (2000).
96. Ginzburg V.V, Gendelman O.V, Manevitch L.I, Phys. Rev. Lett., 86, 5073 (2001).
97. Totha R, Coslanicha A, Ferronea M, Fermeglia M, Pricl S, Miertus S, Chiellini E, Polymer, 45, 8075 (2004).
98. Sinsawat A, Anderson K.L, Vaia R.A, Farmer B.L, J. Polym. Sci. Part B: Polym. Phys., 41, 3272 (2003).
99. Kuppa V, Menakanit S, Krishnamoorti R, Manias E, J. Polym. Sci :Part B: Polym. Phys., 41, 3285 (2003).
100. Zeng Q.H, Yu A.B, Lu G.Q, Standish R.K, Chem. Mater., 15, 4732 (2003).
101. Sheng N, Boyce MC, Parks DM, Rutledge GC, Abes JI, Cohen RE, Polymer, 45, 487 (2004).
102. Ebenau A, "Wirtschaftliche Perspektiven der Nanotechnologie: Enorme Markte fur kleinste Teilchen" Journalisten und Wissenschaftler im Gesprach, "Nanotechnologie in der Chemie-Experience meets Vision" Mannheim, Oct. 28–29, (2002).

CHAPTER 7

Polymer-Clay Nanocomposites — A Review of Mechanical and Physical Properties

Zhong-Zhen Yu, Aravind Dasari and Yiu-Wing Mai*

Center for Advanced Materials Technology (CAMT)
School of Aerospace, Mechanical and Mechatronic Engineering J07
The University of Sydney, Sydney, NSW 2006, Australia
Tel: +61-2-93512290; Fax: +61-2-93513760
*E-mails: *mai@aeromech.usyd.edu.au*
zhongzhen.yu@aeromech.usyd.edu.au
aravind.dasari@aeromech.usyd.edu.au

1. Introduction

Different materials with synergistic properties are generally chosen to produce composites with tailored physical and mechanical properties; for example, high modulus and strength, superior fracture toughness, good barrier and flame retardant properties, enhanced electrical conductivity, and so forth. The great interest and demand generated for these macro- and micro-composites in scientific, technological and structural applications in recent years have finally led to the limits of optimizing composite properties, because the properties achieved usually involve compromises [1]: e.g. stiffness and/or strength are achieved at the expense of fracture toughness, or vice versa. So, to overcome these drawbacks from macro- and micro-scale reinforcements, nano-scale fillers provided a radical alternative - where the filler is < 100 nm in at least one dimension. In general, three major characteristics define and form the basis of performance of polymer nanocomposites [2]: nanoscopically confined matrix polymer chains, nanoscale inorganic constituents, and nanoscale arrangement of these constituents. The full exploitation of these fundamental characteristics of nano-reinforcements in polymers facilitates the achievement of enhanced properties in polymer nanocomposites.

Among the vast nano-reinforcements available for fabricating polymer nanocomposites, clays have been focused and studied the most because they are naturally occurring minerals that are commercially available, exhibit a layered morphology with high aspect ratio, large specific surface areas, and have substantial cation exchange capacities. Linus Pauling first elucidated the structures of clay minerals using early X-ray crystallographic methods and thereby provided the basis for subsequent extensive studies of these materials in different disciplines like polymer engineering and science, colloids, rheology, mineralogy, catalysis, tribology, civil and ceramic engineering, and so on. Polymer-clay nanocomposites have their origin in the pioneering research conducted at the Toyota Central Research Laboratories in the late 1980s, where the replacement of inorganic exchange cations in the galleries of native clay has been done using alkylammonium surfactants that compatibilized the surface chemistry of nano-clay and the hydrophobic polymer matrix [3–5]. Fully exfoliated nylon 6/clay nanocomposites were made by *in-situ* polymerization. The silicate layers were well separated to 1 nm thick individual layers with other dimensions in 100–500 nm range. This nanocomposite exhibited large increases in tensile strength, modulus, and heat distortion temperature without any loss in impact strength. Also, the nanocomposite possessed lower water sensitivity, permeability to gases and thermal expansion coefficient. This nylon-montmorillonite nanocomposite has been used as a timing belt cover on a Toyota Camry automobile. Since then, the fabrication and characterization of polymer/clay nanocomposites have generated much interest in scientific research and industrial applications.

These significant improvements in the properties of nanocomposites, as mentioned above, are not displayed by their 2-phase macro- and micro-composite counterparts. This stems from the fact that the fully exfoliated silicate layers (nano-scale reinforcement) in a polymer matrix results in large interfacial contacts between clay and polymer, high aspect ratio of clay layers, and high surface energy, as compared to the micro- or macro-scale reinforcements. To-date, the use of nano-clays as a dispersed phase to prepare nanocomposites has been extended to various thermo-plastic, thermosetting and elastomeric polymers, such as, epoxies [6–8], polystyrene [9], polyimide [10,11], polypropylene [12–14], polyurethane [15–18], and poly(vinylidene fluoride) [19].

Polymer-clay nanocomposites not only exhibit markedly improved mechanical properties but also significant enhancements in thermal, optical and physico-chemical properties when compared with the neat polymer or conventional composites. The potential to achieve these properties improvements indicates the importance of these materials and why it is necessary to understand how the addition of clay influences the fundamental relationships between processing, structures, and properties so as to tailor and/or manipulate the properties of interest depending on the end application. In the vast field of polymer-clay nanocomposites, in this chapter, we consider only a relatively small set of aspects relevant to the applications of these materials and review the fundamentals, positive and negative impacts, and recent advancements made. We will start with mechanical (stiffness, strength and fracture toughness: Section 2) and physical (barrier properties, flame retardancy and thermal stability: Section 3) property changes that result from the addition of nano-clay layers to different polymers (thermoplastics and thermosets). Later, in Section 4, attempts to optimize and complement the experimental results of different mechanical and physical properties of polymer-clay nanocomposites by theoretical studies and computer simulations (from conventional continuum mechanics-based models to novel molecular dynamic computer simulations) will be reviewed. Section 5 will shed light on the potential applications of these materials and Section 6 will briefly summarize what have been achieved in terms of mechanical and physical properties of these materials and provide an outlook of what should be done in the future for further improvements.

2. Mechanical Properties

For many engineering applications of polymer-clay nanocomposites, such as in the automotive and packaging industries, the properties of primary concern are stiffness, strength and fracture toughness. Other properties such as durability or tribological response of these materials also play a dominant role in determining their end use. However, here we will mainly focus on the direct changes in the mechanical properties that occur as a result of the addition of clay.

2.1. *Stiffness and strength*

Many previous studies on nano-clay reinforced polymeric materials have clearly revealed drastic improvements in the elastic modulus and tensile strength, even at low filler loadings of 1–5% by volume, compared to the pure polymers independent of the preparation methods (that is, whether solution blending, melt-compounding or in-situ polymerization) [4–25]. The high aspect ratio of clay layers and huge interfacial contact area between clay and polymer matrix make this possible. Examples illustrating the effect of nano-clay on elastic modulus and tensile strength are shown in Table 1 for nylon 6/clay nanocomposites [4]. Only 4.7 to 5.3 wt% of nano-clay has resulted in a nearly two-fold increase in Young's modulus and ~40 % increase in tensile strength of nylon 6. Also, the complete exfoliation of clay into individual layers optimizes the number of available reinforcing elements, which also explains why exfoliated morphologies show better improvement in elastic modulus and tensile strength than intercalated structures. An example of the different morphologies (unexfoliated clay tactoids, intercalated, and exfoliated) obtained in nylon 6 nanocomposites by varying the type of surfactant (hence changing the number of tails in the surfactant) used is shown in Figure 1 [26]. The clear differences in the tensile properties of the nanocomposites with different morphologies are also given in Figure 2 plotted as a function of organoclay loading. Even though the morphologies, particularly intercalated and exfoliated, are thermodynamically stable, still the differences that exist in the properties explain the effect of individual layers and the high aspect ratio.

Similarly, for intercalated PMMA [27] and polystyrene [28] nanocomposites, the increase in Young's modulus is relatively small. This confirms the inefficiency of intercalated structures to improve stiffness compared to exfoliated morphologies. Furthermore, Usuki *et al.* [29] studied the ionic interaction between nylon 6 and clay by measuring the end group contents of nylon 6 and its nanocomposite with about

Table 1. Mechanical properties of nylon 6-clay nanocomposites prepared by *in situ* polymerization (From Ref. 4).

	Nylon 6	Nylon 6-clay[a]	Nylon-clay[b]
Clay content (wt %)	0	4.7	5.3
Molecular weight ($\times 10^3$)	13.0	16.3	19.7
Crystallinity (%)	33.3	28.7	25.0
Young's modulus (GPa)	1.11	1.87	2.04
Tensile strength (MPa)	68.6	97.2	97.3

[a]Clay was modified by protonated 6-aminocaproic acid. [b]Clay was modified by protonated ε-caprolactam.

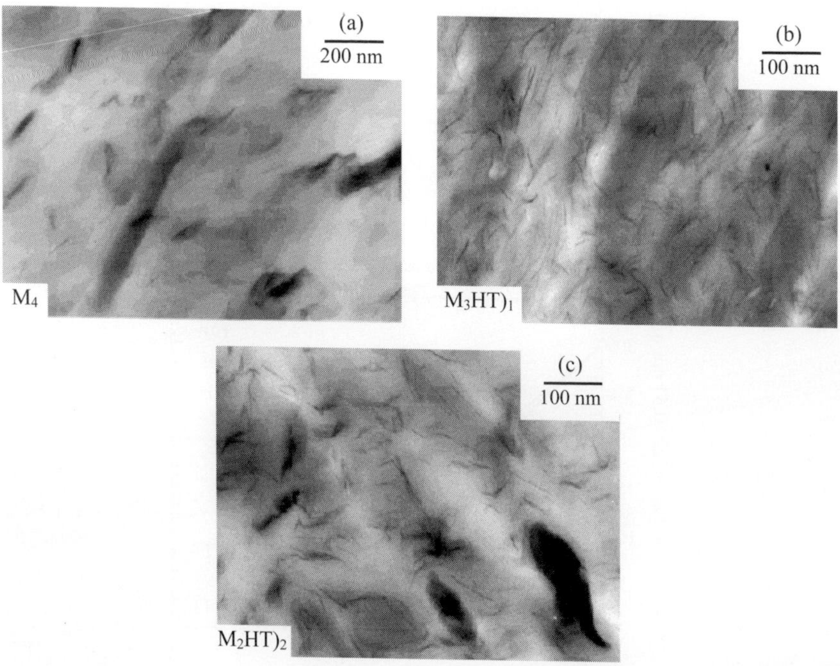

Fig. 1. TEM micrographs of high molecular weight nylon 6 composites illustrating the different morphology obtained by varying the type of surfactant (by changing the number of tails in the surfactant): (a) aggregated morphology, (b) exfoliated, and (c) intercalated. The symbols M and HT designate methyl and hydrogenated tallow groups attached to the nitrogen. (Reprinted with permission from Ref. 26. Copyright (2004) American Chemical Society)

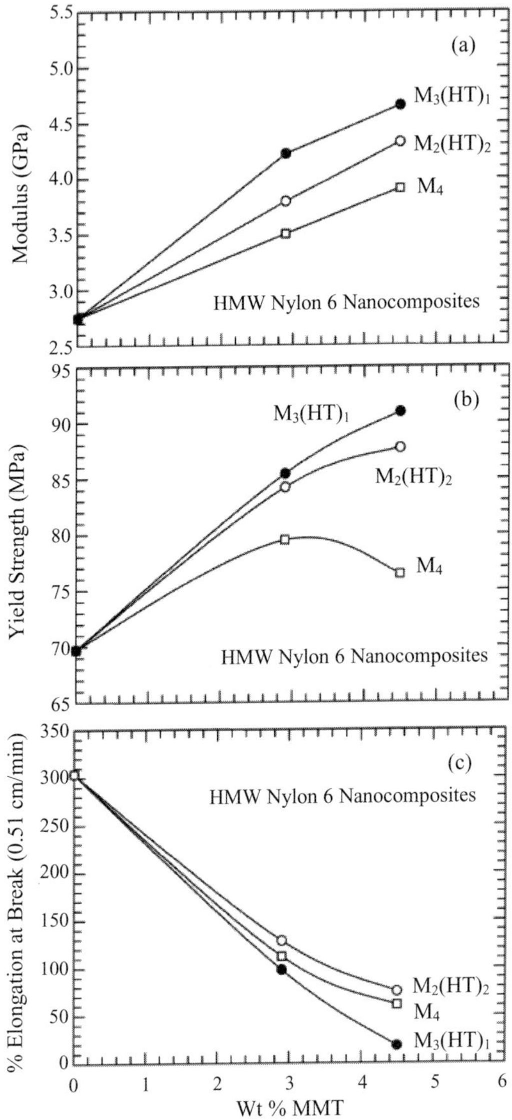

Fig. 2. Effect of different morphology (aggregated, exfoliated, and intercalated) on the tensile properties of high molecular weight nylon 6 composites as a function of organoclay loading. The symbols M and HT designate methyl and hydrogenated tallow groups attached to the nitrogen. (Reprinted with permission from Ref. 26. Copyright (2004) American Chemical Society)

Table 2. Terminal group analysis of nylon 6 and nylon 6-clay nanocomposite prepared by *in situ* polymerization (values are × 10^{-5} mol/g) (From Ref. 29).

	Nylon 6	Nylon 6-clay
COOH	5.14	9.46
NH_2	5.09	4.86
NH_3^+	–	4.6

5 wt% of clay. As shown in Table 2, the content of carboxyl terminal groups is equal to that of amino terminal groups in neat nylon 6. However, in nylon 6 nanocomposite, the content of carboxyl ends is higher than that of the amino ends, and their difference is the concentration of the cationic ammonium NH_3^+ ends. This NH_3^+ end group is believed to interact with the negative charge of the clay layers. This ionic interaction is also thought to be a reason why clay shows superior reinforcement.

Much work has been devoted to determine the parameters that may affect clay and its dispersion. Such factors include the preparation methods (melt compounding, solution blending, or *in-situ* polymerization), processing conditions, modification and compatibilizers used, shear stress developed during compounding, and matrix molecular weight. Additionally, the effect of the surfactant structure and its molecular weight also seems to play a dominant role in influencing the intra-gallery separation and mechanical properties of the resulting nanocomposites. In this regard, poly(styrene-*co*-acrylonitrile) (SAN) with montmorillonite organoclays were examined [30]. The composite that exhibited the greatest change in gallery height, the highest modulus, and greatest aspect ratio (~50) was produced from clay treated with the lowest molecular weight surfactant, dimethyl hydrogenated tallow ammonium. It was also concluded that the surfactant with a shorter tail length resulted in non-swollen stacks of clay in the composite.

Additional to the above-mentioned parameters, the orientation and spatial distribution of nano-clay layers also affect significantly the clay reinforcement efficiency. Kojima *et al.* [31] studied the preferred orientation of clay and nylon 6 crystallites in a 3-mm thick injection-molded bar of nylon 6/clay nanocomposites prepared by polymerizing

ε-caprolactum in an intercalated compound of montmorillonite and 12-aminolauric acid. Before any further discussion, it should be noted that the ammonium cations at the end of the nylon 6 molecules were bonded with the anionic sites on the clay monolayers, and if the nylon molecules were densely bonded, the chain segments near the clay monolayers might be oriented normal to its surface. Also, in the absence of an external stress field, the crystallites formed from a bundle of these bonded molecules would orient normal to the clay monolayer.

Now, returning to the preferred orientation of clay and nylon 6 crystallites, it was found that in the surface region of the injection-molded bar where the shear stresses would be high, both the clay monolayers and the nylon 6 crystallites were oriented parallel to the bar surface, in spite of large differences in their aspect ratios. In the intermediate region of the bar, as the shear stresses were reduced, the clay monolayers were still parallel to the bar surface, whereas the chain axes of nylon crystallites were perpendicular to clay monolayers. The preferred orientations between the clay monolayers and nylon crystallites were attributed to the large differences in their aspect ratios. Although the shear stresses were reduced in the intermediate regions of the bar, owing to their large aspect ratio, the clay monolayers remained still parallel to the bar surface. Conversely, due to their small aspect ratio, the nylon 6 crystallites could not maintain the parallel orientation and returned to their original stable configuration where their chain axes were perpendicular to the clay monolayers. In contrast, in the center region of the bar, the shear stresses were very low, so the clay monolayers were randomly oriented around the flow axis, while the nylon crystallites followed the orientations of the silicate monolayers and their chain axes remained perpendicular to the planes of clay layers. The preferred orientation in a 60 μm-thick film was found to be even more serious. All the clay monolayers and the chain axes of the nylon 6 crystallites were parallel to the bar surface [32].

The preferred orientation of clay along the flow direction has also been found in polymer/clay nanocomposites prepared via melt extrusion followed by injection molding [33–36]. Such an anisotropic microstructure was confirmed to influence the physical and mechanical properties of polymer/clay nanocomposites in 3-directions. Yoon *et al.*

[34] measured linear thermal expansion behavior in flow direction (FD), transverse direction (TD), and normal direction (ND) of ~3.2 mm-thick, 6.3 mm-wide and 12.7 mm-long rectangular bars of nylon 6/clay nanocomposites. As shown in Figure 3, the addition of clay reduced the thermal expansion coefficient in both FD and TD but increased that in ND. Since the nylon 6 in the FD and TD was highly constrained, this created a "squeezing" effect in ND, thus increasing the expansion in this direction. Masenelli-Varlot *et al.* [35] in a similar study found that the compression modulus and flow stress of nylon 6/clay nanocomposites were different in all three directions as shown in Table 3.

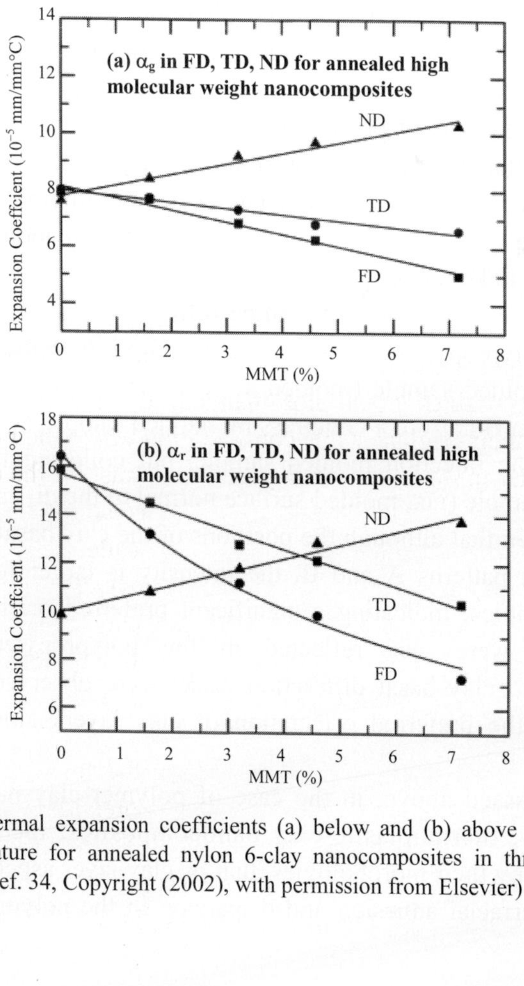

Fig. 3. Linear thermal expansion coefficients (a) below and (b) above nylon 6 glass transition temperature for annealed nylon 6-clay nanocomposites in three directions. (Reprinted from Ref. 34, Copyright (2002), with permission from Elsevier)

Table 3. Compression modulus and flow strength of nylon 6 and nylon 6-organoclay nanocomposite prepared by melt extrusion followed by injection-molding process (From ref. 35).

Directions	Modulus (GPa)		Flow strength (MPa)	
	Nylon 6	Nylon 6-clay[a]	Nylon 6	Nylon-clay[a]
Flow direction	1.22	2.08	85	141
Transverse direction	1.18	1.69	84	104
Normal direction	1.18	1.43	79	96

[a]The organoclay content is 6.5 wt%. The surfactant was methyl, octadecyl, bis-2-hydroxyethyl ammonium methylsulfate.

Besides the influence of spatial distribution and orientation of clay layers on the reinforcement ability, it is important to consider these factors like preferred orientation, clay dilution, peak broadening, etc, when assessing clay dispersion using XRD in a polymer nanocomposite. Simple dilution of clay can contribute to peak broadening and lead to false conclusion that exfoliation has occurred. Alternatively, preferred orientation effects may result in the false conclusion about exfoliation. In this regard, Eckel *et al.* [37] studied the preferred orientation effect on the XRD data. Figure 4 shows three diffraction patterns obtained from the same polypropylene/organoclay nanocomposite. Pattern A is from an injection molded sample (molded surface parallel to the diffractometer axis), pattern B is from a compression molded sample, while pattern C is also from the injection molded sample, but collected from the cross-sectioned sample (i.e., molded surface normal to the diffractometer axis). It can be seen that although the positions of the clay basal reflections are the same in patterns A and B, the intensity is larger for the injection molded samples, indicating a significant preferred orientation. Similar differences were also reflected in the polypropylene reflections. However, no clay basal diffraction peaks were observed in pattern C, supporting the preferred orientation of clay layers along the molded surface.

As discussed above, in the case of polymer-clay nanocomposites, particularly, thermoplastic-clay nanocomposites, the properties are determined by their morphologies, that is, clay layer aspect ratio, surface energy, interfacial adhesion and dispersion in the polymer. In contrast,

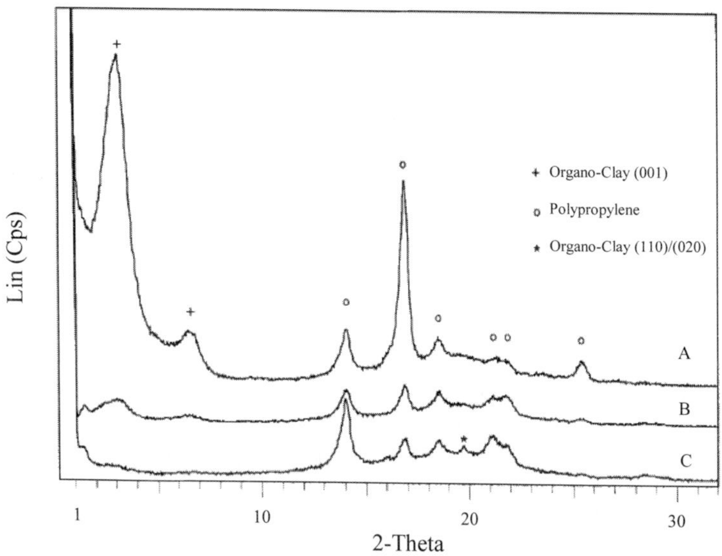

Fig. 4. X-ray diffraction patterns for the different polypropylene-clay nanocomposites: (A) injection molded and aligned parallel to the diffractometer axis; (B) compression molded and aligned parallel to the diffractometer axis; and (C) injection molded and aligned normal to the diffractometer axis. (Reprinted with permission from Ref. 37. Copyright (2004) Wiley Periodicals, Inc.)

the major problem in correlating the mechanical properties of thermosets with their morphology is the inability to characterize accurately their microstructure. The only structural parameter that can be measured directly is the average molecular weight between crosslinks calculated from tests. However, the existence of other microstructural features, such as inhomogeneous crosslinking and heterogeneous structure called nodules, affects significantly the performance of thermosets [38] and so requires an in-depth understanding of a broad range of thermoset resins. Also, in this regard, Zilg *et al.* [39] reported the correlations between polymer morphology, silicate structure, stiffness, and toughness of thermoset nanocomposites as a function of layered silicate type and content. They showed that the main factor for the matrix stiffness improvement resided in the formation of supramolecular assemblies obtained by the presence of the dispersed anisotropic laminated nano-particles.

In epoxy resin/clay nanocomposites, it is also necessary to consider the effect of dangling chains that are formed as a result of interaction between the alkylammonium ions of the surfactant used to treat clay and the epoxy resin during polymerization. This problem was studied by Pinnavaia and co-workers [40] by comparing the effect of tensile modulus of epoxy resin filled with three different layered materials (montmorillonite modified with octadecylammonium cation, magadiite modified with the same alkylammonium, and magadiite modified with methyl-octadecylammonium cation). It was shown that the increase of tensile modulus was higher in the montmorillonite-based nanocomposites. This behavior was attributed to differences in the layer charge density for magadiite and montmorillonite. Organomagadiites with a higher layer charge density had higher alkylammonium content than organo-montmorillonites and resulted in more dangling chains (dangling chains were formed as the alkylammonium ions interacted with epoxy resin while polymerizing). These dangling chains thus weakened the polymer matrix by reducing the degree of network crosslinking and compromising the reinforcement effect of the silicate layer exfoliation.

Besides all those factors affecting the correlation of mechanical properties of thermoset resins reinforced with nano-clay, nevertheless, significant improvement in mechanical properties were reported. Lan and Pinnavaia [7] prepared exfoliated epoxy nanocomposites with $CH_3(CH_2)_{17}NH_3^+$ modified clay and polyether amine (Jeffamine D2000) having a high molecular weight of 2000 as the curing agent. The final nanocomposites had sub-ambient glass transition temperatures. The modulus and strength of epoxy nanocomposites showed a more than 10-fold improvement as compared to neat epoxy (Figure 5). This improvement was attributed to a possible strain-induced alignment of the silicate layers in the rubbery state. In another study, a nanocomposite containing 4% by volume of clay showed 60% increase in storage modulus in the glassy region compared to unfilled epoxy [41]. The equivalent increase in the rubbery region was ~450%. This increase in modulus is remarkable, as conventionally filled epoxies with micron-size or larger particles do not show much of an improvement in modulus at loading levels of less than 10 vol %. Similar kinds of improvements in

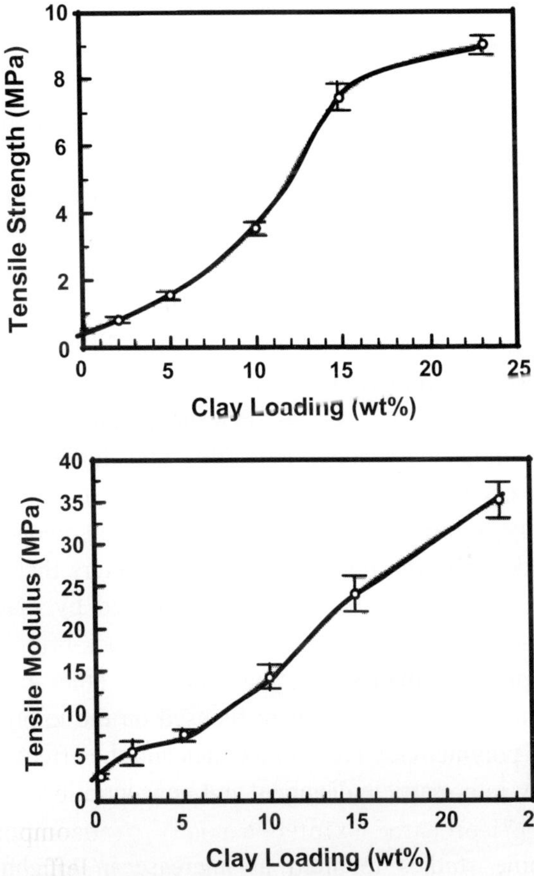

Fig. 5. Plots of tensile strength and modulus on clay loading for epoxy-$CH_3(CH_2)_{17}NH_3^+$ modified clay nanocomposites. (Reprinted with permission from Ref. 7. Copyright (1994) American Chemical Society)

modulus and strength in thermoset/clay nano-composites have also been reported by many other researchers [42–45].

2.2. *Fracture toughness*

Polymer-silicate nanocomposites, as described earlier, generally show improved mechanical properties, even at low volume fraction loadings (1 to 5%), which may be attributed to the interactions of the nano-clay layers with the polymer at the molecular scale that affects both the physical and material parameters at scales inaccessible to conventional fillers. Despite the large improvement in elastic modulus, a vast range of potential engineering applications of polymeric silicate nanocomposites are limited due to their reduced fracture toughness.

In many polymeric materials, such as the polyolefins and polyamides, it is generally accepted that the major contribution to toughness comes from the plastic deformation. Thus, it is important to understand these complex deformation mechanisms. For un-reinforced semi-crystalline polymers like polypropylene and polyethylene, different generalized models for plastic deformation during uniaxial tensile testing were proposed in the literature, such as Peterlin's micro-fibrillar model [46,47], Nitta's molecular model [48, 49], and Ichihara and Iida's entangled chain crystal model [50]. But, for polymer nanocomposites, there are currently no universally accepted deformation mechanisms. It is important to understand the microstructural parameters that control the deformation and failure processes, the roles played by matrix, nano-particle, and nano-particle/matrix interface on crack-bridging, crack-deflection and pinning, matrix yielding, and any other processes on toughness. Recently, many efforts were focused on the improvement of the toughness of polymer/clay nanocomposites and the effects of particle size, aspect ratio, inter-particle distance, and concentration were studied [51–53].

Although some studies reported an increase in toughness along with modulus and strength in exfoliated polymer/clay nanocomposites [54–58], there are debates on toughening mechanisms in polymer-silicate nanocomposites using length-scale arguments. That is, toughening occurs over a specific size range, and effective toughening may not be energetically favorable at the nano-scale. This generally requires a filler size greater than ~100 nm [59]. It is also thought that the size range of individual clay layers is too small to provide toughening via mechanisms like crack bridging, deflection and pinning. Thus, an intercalated structure, where polymer chains have entered into the galleries between silicates but have not fully delaminated them is considered to provide better toughening efficiency rather than a well-exfoliated system. Based on this concept some studies [58,60] have reported an increase of toughness due to voiding and debonding of clay particles. However, a negative consequence to this approach, as described in Section 2.1, is that the elastic modulus of layered silicate nanocomposites is dependent on the extent of exfoliation in thermoplastics and thermoset nanocomposites [39,61]. Also, the large aggregates of clay may serve as stress concentrators that lead to premature brittle failure. Besides the length-scale argument, it has been shown recently, for nylon 6 nanocomposites, that clay induces less stable γ-form crystals with lamellae growth transverse to the clay sheets, which in turn does not favor toughening [62].

Hence, many approaches and efforts have been attempted to achieve exfoliated nanocomposites with enhanced mechanical and physical properties. These include the addition of an elastomeric phase to the polymer/clay nanocomposites [63–65], inclusion of a compatibilizer that improves the interfacial interaction [66], usage of organoclay as a compatibilizer for immiscible polymer blends [67], further modification of organically treated clay with epoxy monomer before blending to produce a nanocomposite [68], etc. Amongst all these methods, adding an elastomeric phase to polymer/clay nanocomposites has been a most common method to enhance toughness. For rubber toughened polymer/clay materials, the key is the cavitation ability of the dispersed rubbery phase in the matrix. However, it also depends on the mechanical properties of the matrix, the location of clay (i.e., whether located in the matrix or in the soft rubbery phase), compatibilizer, dispersion of rubbery phase, and interfacial interaction of rubbery phase and matrix. Though cavitation does not consume a very large amount of energy, it does enable plastic dilatation of the surrounding material to undergo extensive plastic deformation. The most important point of the toughening mechanism is that the deformation zone ahead of the crack-tip is under plane-strain condition and thus the material is subjected to high plastic constraint. Without a relief of constraint mechanism, the material under the triaxial tension tends to fail in a brittle mode with low toughness. Because of the low tear-strength of the rubber particles, cavitation occurs under triaxial tension and thus releases the high plastic constraint and enables large scale plastic deformation in the surrounding material [69,70].

Using this approach, polyamide 6/polypropylene blends containing organically modified clay were compatibilized with a maleic anhydride-grafted ethylene-propylene rubber (EPR-g-MA) and a balance in stiffness, strength, and toughness were achieved [65]. However, the organoclay was mostly in the polyamide 6 phase of the blends in both exfoliated and intercalated forms. Li *et al.* [71] also prepared toughened poly(butylene terephthalate) (PBT) with higher tensile strength and modulus by using a similar method where maleated poly(ethylene-co-vinylacetate) (EVA-g-MA) elastomer was used as a toughener and organoclay a strengthening agent. Although, in this case, organoclay was not completely exfoliated in either binary or ternary PBT system, blending of PBT/organoclay nanocomposite with EVA-g-MA was found to have improved mechanical properties. Tjong and Meng [72] toughened polypropylene (PP)/clay nanocomposites with maleated styrene-ethylene/butylene-styrene triblock copolymer (SEBS-g-MA). 4.5 times increase in notched impact strength of PP/clay nanocomposite was achieved at 15 wt% of SEBS-g-MA, however, at the expense of stiffness and tensile strength. A more in-depth analysis on the effect of adding a rubbery phase (SEBS-g-MA) to nylon 66/organoclay nanocomposite was recently reported where four different blending sequences were adopted in fabricating the ternary nanocomposites (nylon 66/organoclay/SEBS-g-MA) [64]. The four ternary nanocomposites showed distinct microstructure and mechanical properties with different deformation mechanisms. It was concluded that the best microstructure for toughness and other mechanical properties was to have the maximum amount of the exfoliated organoclay in the continuous nylon 66 matrix rather than to have it in the dispersed SEBS-g-MA phase. The presence of organoclay in the SEBS-g-MA phase reduces the latter's ability to cavitate, making it rigid, and finally resulting in reduced toughening efficiency. The study has clearly indicated that addition of an elastomeric phase to the binary polymer/clay nanocomposites is beneficial to achieve a balance in mechanical properties. However, it is also important to consider the blending sequence, location and dispersion of the clay, and size distribution and cross-linking ability of the dispersed soft domains.

In contrast, in thermoset resin-clay nanocomposites, it was shown that higher cure temperatures improved clay delamination (Figure 6), and simultaneously increased toughness and modulus (Table 4) [73]. The improved organoclay exfoliation was related to the improvement in mass transfer into the clay galleries due to the enhanced molecular mobility or lower viscosity at higher cure temperatures. In addition, at higher cure temperatures, the catalytic effect of the organoclay on both resin-amine cure and homopolymerization may also shift the equilibrium between intra- and extra-gallery reactions towards a higher reaction rate within the clay galleries. Therefore, it was pointed out that to achieve exfoliation in thermoset-clay nanocomposites or to overcome the electrostatic attractive forces between the clay layers, control of intra- and extra-gallery reaction is critical [73,74]. In another study, it was theoretically shown that completely exfoliated clay layers would only promote stiffness; but formation of anisotropic laminated nano-particles was important to improve the toughness [39].

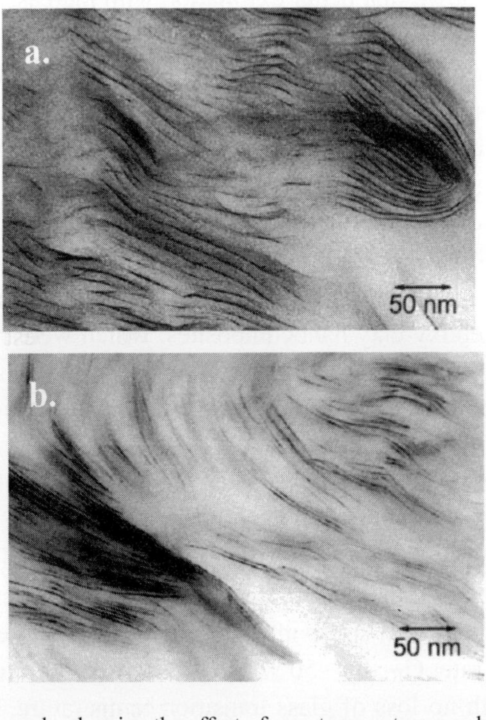

Fig. 6. TEM micrographs showing the effect of cure temperature on clay delamination in DGEBA nanocomposites at (a) 100°C and (b) 160°C. (Reprinted with permission from Ref. 73 Copyright (2003) American Chemical Society)

Table 4. Mechanical property variations of epoxy nanocomposites cured at different temperatures (From Ref. 73).

DGEBA curing temperature (°C)	Clay content (wt%)	Young's modulus (MPa)	K_{IC} ($MNm^{1.5}$)
100	0	2408 ± 184	0.50 ± 0.05
160		2871 ± 66	0.91 ± 0.26
100	2.5	2536 ± 91	0.71 ± 0.16
160		3071 ± 115	1.19 ± 0.24
100	5	2646 ± 166	0.80 ± 0.11
160		3211 ± 51	1.32 ± 0.17
100	7.5	2883 ± 99	0.86 ± 0.25
160		3144 ± 41	1.13 ± 0.23

It was also shown that better exfoliation with increased toughness and modulus could be achieved in bi-functional DGEBA resin than other resins of higher functionalities due to better catalysis of the intra-gallery reaction by the organo-ions that reside within the galleries [73]. In another method to obtain epoxy/exfoliated clay nanocomposites with improved toughness, stiffness, and strength, a slurry-compounding process was used, where sodium montmorillonite was first suspended in water and then treated with acetone to form slurry followed by chemical modification using silane [45]. This was further mixed extensively with epoxy to form epoxy-clay nanocomposites, which were reported to have both improved Young's modulus and fracture toughness. The initiation and development of a large number of micro-cracks (Figure 7) and the increase of fracture surface area due to crack deflections were suggested as the major toughening mechanisms in these materials. In a similar study, Fröhlich *et al.* [75] synthesised anhydride cured epoxy nanocomposites in which two types of phenolic imidazolineamides were used for clay modification. A steady increase in fracture toughness, measured by the critical stress intensity factor, K_{IC}, and critical potential energy release rate, G_{IC}, by ~50 % was observed with increasing silicate content but with no loss of glass transition temperature. The creation of additional surface area due to crack growth is assumed to be the primary cause for the improved fracture toughness.

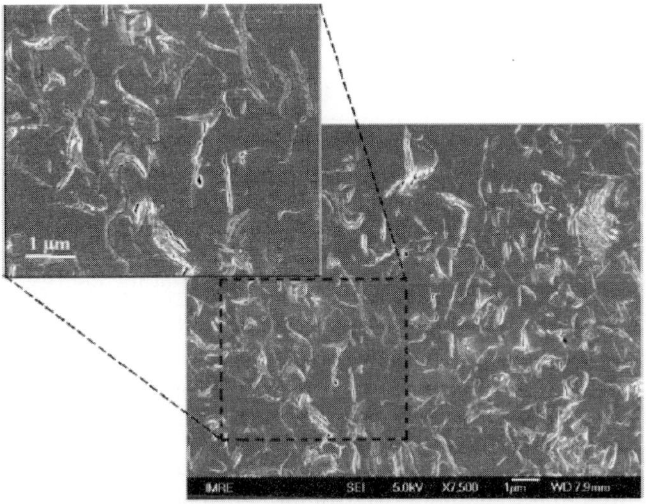

Fig. 7. SEM micrographs of the damaged zone showing the initiation and development of a large number of micro-cracks between clay layers. (Reprinted with permission from Ref. 140. Copyright (2005) American Chemical Society)

To further support the length-scale argument discussed above regarding toughness improvement in intercalated morphology compared to exfoliated structure, some more recent arguments have been made why intercalated nanocomposites are capable of property enhancements, particularly fracture toughness. It was proposed that facilitating the individual silicate layers to interact allows cooperative mechanisms of energy dissipation to occur in the polymer over much larger length-scales relative to the individual silicates, and thereby enhancing the toughness [76]. Hence, the combination of the spatial configuration produced in the intercalated phase, the modulus mismatch between polymer and silicate, and the high aspect ratio of the silicates produce high stress concentrations at the silicate edges that mechanically interact with other silicates within their proximity. Based on this concept, Zerda and Lesser [77] dealing with intercalated diglycidyl-ether of bisphenol A (DGEBA)/Jeffamine D230 layered silicate nanocomposites have shown significant improvements in K_{IC} of these materials. K_{IC} increased from 0.9 to 1.5 $MPam^{3/2}$ with an addition of 3.5 wt% of organically-modified clay.

However, analogous to thermoplastic nanocomposites, to obtain an improved balance between elastic modulus, strength, and toughness, impact modifiers are also added to thermoset resin-clay nanocomposites. Yet, it is important to note that in epoxy polymers the impact modifier should be initially miscible in the resin, not to degrade the resin viscosity, and is phase-separated during cure for easy processing and yield enhanced mechanical properties. The phase-separated rubber particles should act as stress concentrators initiating energy absorbing 'toughening' processes due to multiple plastic deformation processes, in a manner similar to the case of thermoplastic/rubber/clay systems. Generally, for epoxy resins, two types of impact modifiers were established: flexibilizers and tougheners. Both are miscible with uncured epoxy resin. However, in contrast to flexibilizers, toughening agents improve toughness without sacrificing stiffness and glass transition temperature. Prominent impact modifiers are based upon liquid rubbers such as nitrile rubbers [78,79] (low molecular weight butadiene acrylonitrile copolymers bearing carboxyl (CTBN), amine (ATBN) or epoxy (ETBN) reactive end groups), which dissolve in the epoxy resin and phase separate during cure. Other agents include poly(propylene oxide), poly(tetra-hydrofurane), and recently, a new type of impact modifier based upon hyper-branched nanometer-scaled polymers, especially polyesters, were also being used [80–84].

Furthermore, when dealing with the addition of impact modifiers to thermoset resin-clay nano-composites, different parameters should be considered, such as, reactivity of the end groups of modifiers and epoxy resin during the curing reaction, low viscosity and no large viscosity build-up of the uncured system, complete phase separation and formation of dispersed rubber phases with reasonable particle size distribution to prevent plasticization of the matrix, and good interfacial adhesion between dispersed rubber phase and epoxy matrix. In addition, generally, these impact modifiers lack heat resistance due to the double bond and so the final epoxy-rubber-clay system may have poor thermal stability [81]. To overcome this problem, silicone rubber was used to modify epoxy instead of normal nitrile rubber. However, silicone rubber does not have compatibility with epoxy resin so that upon direct blending it can easily separate from the matrix.

3. Physical Properties

3.1. *Barrier properties*

Barrier properties are of prime importance in bottling, food packaging, packaging hydrocarbon solvents, and protective coating industries. As such, many different techniques have been developed to improve the barrier properties of neat polymers, which include surface treatment by fluorination or sulfonation, laminar-blend blow molding process, etc [85]. However, significant improvements in barrier properties were not achieved that satisfied the need for all different end applications. Hence, if polymer nanocomposites are to be used in these sorts of applications, they should have decreased permeability to different gases and liquids. Also, in biodegradable polymer composites where the hydrolysis of matrix polymer is likely to depend on the transport of water from the surface into the bulk of the material, reduced extent of water permeability will beneficially modify the degradation rates [86]. Studies on the barrier properties of polymer nano-composites have revealed that a dramatic decrease in permeability can be achieved compared to conventional polymer micro-composites having much higher filler contents.

The simple underlying concept for the decreased permeability is that the transport cross-section is reduced and the tortuous paths for gas or liquid molecules are increased. This is particularly true for polymer clay nanocomposites owing to the presence of well-dispersed and exfoliated silicate layers with large aspect ratios in the polymer matrix, which are impermeable to the gas or liquid molecules thus forcing them to traverse a tortuous path through the polymer matrix surrounding these silicate particles (Figure 8), thereby increasing substantially the effective path length for diffusion. Recently, Bharadhwaj [87] studied the effect of clay layer orientation on the relative permeability in exfoliated polymer layered nanocomposites. The presented idea is purely based on geometric arguments. As shown in Figure 9, a range of orientations of the clay layers relative to the flow direction of the gases or liquid across the film are considered. Obviously, permeability is highest when the clay layers are aligned in the flow direction; vice versa, permeability is lowest when they are across the flow direction. A sheet-like morphology (with high

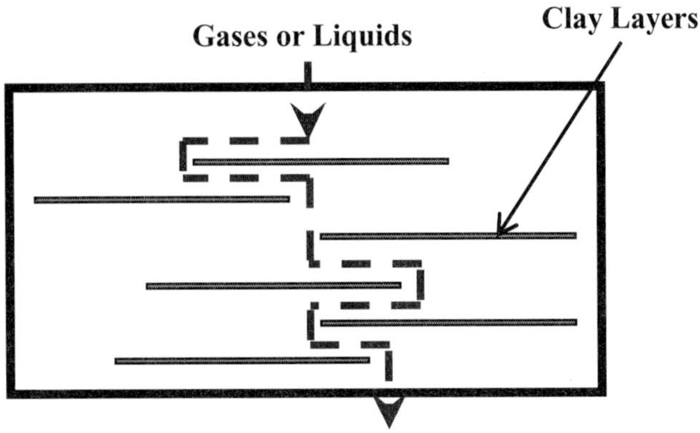

Fig. 8. Schematic representation of exfoliated large aspect ratio silicate layers in a polymer matrix forcing the liquids or gases to traverse a tortuous path surrounding the silicate particles. (Adopted from Ref. 11).

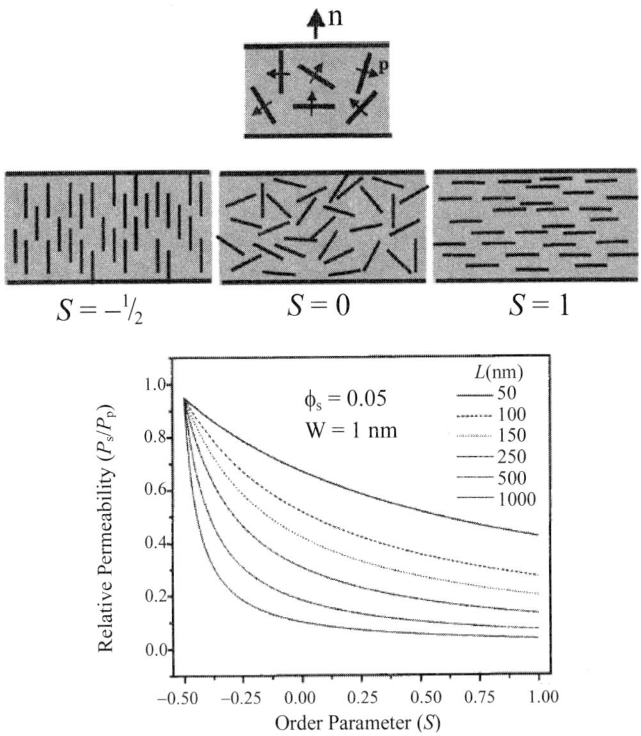

Fig. 9. Effect of clay layer orientation on relative permeability in exfoliated polymer layered nanocomposites. (Reprinted with permission from Ref. 87. Copyright (2001) American Chemical Society)

aspect ratios) is also more impermeable than other shapes like spheres or cubes (with low aspect ratios) for all clay layers orientations.

3.1.1. *Gases*

Yano *et al.* [11] firstly reported the permeability coefficients of various gases (O_2, H_2O, He, CO_2, and ethyl-acetate vapors) for polyimide (PI)/clay nanocomposites, where the clay layers were homogeneously dispersed and oriented parallel to the film surface (planar homogeneity). It was found that only 2 wt% of the clay reduces the permeability coefficients of various gases to less than half of those shown by neat PI. In exfoliated polyethylene terephthalate (PET)-clay nanocomposites, a large barrier improvement to oxygen, carbon dioxide, and water vapor was found with a clay loading of ~5 wt% [88]. In another study on the oxygen transmission properties of injection-molded biodegradable polyesteramide-organoclay nanocomposites, a substantial reduction in oxygen permeability was observed despite TEM and XRD observations, which indicated an intercalated structure [89]. The beneficial effect on the oxygen permeability even with an intercalated system was probably due to the high degree of clay layer orientation as observed by TEM. Similarly, the gas barrier performance of a thermoset acrylonitrile butadiene copolymer that is commonly used as inner-tubes for tires, air springs, and cure bladders was improved considerably by reinforcing it with organoclay [90,91].

Recently, permeability to oxygen in organoclay reinforced polyurethane nanocomposites was reported, where three types of commercially available organoclay were used (Nanofil 804: treated with bis(2-hydroxyethyl) hydrogenated tallow ammonium, Nanofil 32: treated with alkylbenzyl-dimethylammonium (benzalkonium), and Nanofil 15: treated with dimethyl dihydrogenated tallow ammonium) [92]. TEM micrographs of these three nanocomposites indicated that the nano-clay was well exfoliated in the polyurethane matrix irrespective of the organic treatment (Figure 10). As shown in Table 5, the oxygen transmission rate through the polyurethane nanocomposites clearly revealed that Nanofil 32 and 804 reinforced nanocomposites exhibited a better permeation barrier when compared to pristine polyurethane. However, in contrast to

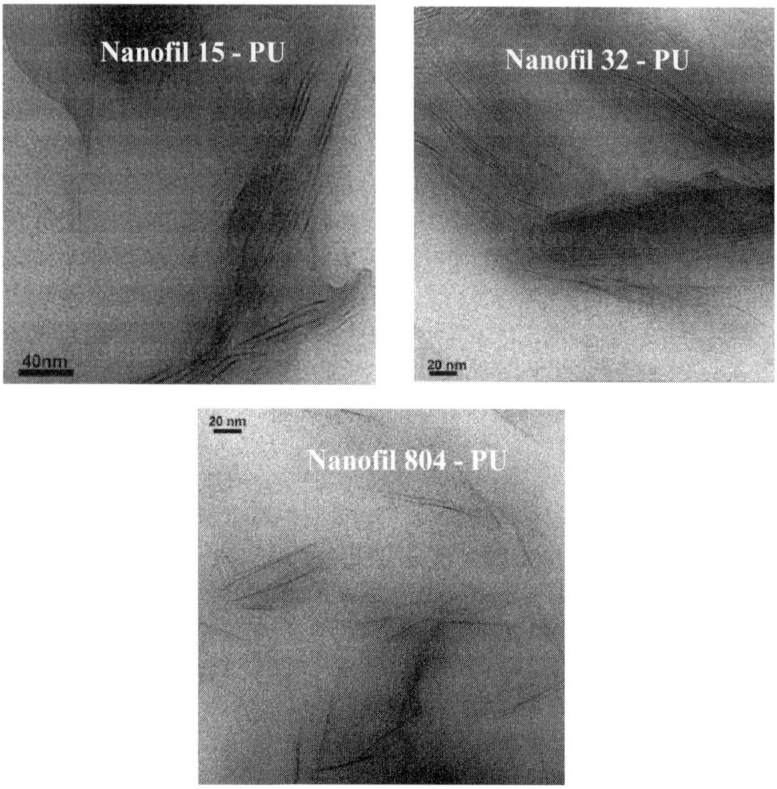

Fig. 10. TEM micrographs of three polyurethane nanocomposites prepared with three different types of organoclay indicating that the nano-clay was well-exfoliated in polyurethane matrix irrespective of organic treatment. (Reprinted with permission from Ref. 92. Copyright (2003) American Chemical Society)

Table 5. Oxygen transmission rates in PU/organoclay nanocomposites (23°C, 0% RH) (from Ref. 92).

Inorganic content (vol. %)	Oxygen transmission rate ($cm^3 \cdot mm/m^2 \cdot day$)		
	Nanofil 804	Nanofil 32	Nanofil 15
0	6.3	6.3	6.3
1.41	5.0	5.2	7.4
2.13	4.8	4.7	8.6
2.85	4.6	4.3	8.6
3.58	4.6	4.5	9.0
4.31	4.5	4.3	9.2

Nanofil 32 and 804, for Nanofil 15, even though the TEM micrographs indicated no differences in the microstructure or dispersion of clay at identical clay loadings, surprisingly, a negative effect on barrier properties was found and also the oxygen transmission rate increased with further increase in filler volume fraction. This increase in transmission rate was attributed to a probable change in morphology resulting from phase separation at the interface between the non-polar pure hydrocarbon coating on clay and the relatively polar polyurethane. This leads to a decrease in density of the organic material (higher free volume) at the interface around the platelets and consequently to an increase in the permeation rate through the composite. In addition to the coating of the clay platelets, it was also predicted that only exfoliation of clay layers did not contribute to the improvement in barrier properties [93]. Poorly oriented platelets were shown to be only one-third as effective as aligned platelets in reducing the permeability, which suggested that orientation of clay layers is a crucial factor in determining the barrier properties.

3.1.2. *Liquids*

Liquid barrier properties are particularly important to the use of polymer-matrix composites in aerospace, electronics and electrical engineering. Messersmith and Giannelis [94] found that thick films of degradable poly(ε-caprolactone)/clay nanocomposites showed a large reduction in water vapor permeability. The permeability of a nanocomposite containing as low as 4.8 vol.% clay was reduced by nearly an order of magnitude compared to neat poly(ε-caprolactone). Later, studies on barrier applications (e.g. coating, film and container) based on polyester, polyamide and other thermoplastic nanocomposites [95–97] also showed positive results.

In another report, permeation barrier properties of organoclay filled nylon 6 nanocomposites for two different solvents (non-polar solvent: toluene; polar solvent: ethanol) were studied [85]. It was shown that the toluene and ethanol permeation resistance of all nanocomposites were better than those of pure nylon 6. They attributed this improvement to the tortuous path created by the nano-clay layers in addition to the increased crystallinity in nanocomposites, which reduced the free volumes in the nanocomposites. Also, the enhancement of barrier properties of these nano-composites reached a maximum as the amount of organoclay contained in the nylon 6-orgnaoclay nanocomposites approached 0.1 phr, which was ~3 and ~4 times slower than that of pure nylon 6 to toluene and ethanol, respectively. After that, the permeation rate of the solvent in these nanocomposites increased. This was ascribed to the poor compatibility between organoclay and nylon 6.

Like the study on the effect of clay pre-treatment on the gas barrier properties of polyurethane nanocomposites filled with different organoclays in Section 3.1.1, water vapor transmission rates of high density polyethylene filled with different types of surface modified clays was reported. The clays and organic surfactants used here were montmorillonite treated with octadecylamine, montmorillonite treated with dodecylpyrrodidone, and a high aspect ratio fluoromica treated with octadecylamine. The results showed that dodecylpyrrodidone treated montmorillonite performed better than octadecylamine treated montmorillonite and fluoromica [98].

In summary, aspect ratio, concentration, orientation, surface treatment and degree of exfoliation of nano-clay layers all have significant effects on the barrier properties of polymer-clay nano-composites. However, based on a simple tortuous model, it was concluded that dispersing longer sheets ($L > 500$ nm) than the normal length of clay platelets in a polymer matrix not only increased the tortuosity, but also reduced the dependence of the relative permeability on the orientation of the layers (87). Additional parameters like the degree of crystallinity of the polymer, its chain configuration, extent of crosslink and packing density also affect the barrier properties of the nanocomposites.

3.2. *Flame retardancy*

In 1976, in a patent application on polyamide 6-montmorillonite nanocomposite, Unitika Ltd. first mentioned about the improved flame retardant properties of these nanocomposites [99]. Although, many issues are unresolved as to the exact mechanisms of this improvement, however, since then, the flammability properties for many different polymer-clay nanocomposite systems have been characterized using Cone Calorimeter, Limiting Oxygen Index (LOI), Vertical Burning Test (UL94), among others. By studying the combustion residues using electron microscopy and X-ray diffraction, a possible mechanism of enhanced flame retardancy of polymer-clay nano-composites have been identified [100–102]. The structure of the nanocomposite collapses during combustion. The multilayered carbonaceous-silicate structure seems to improve the performance of the char through structural reinforcement. That is, the silicate-rich char acts as an excellent insulator and mass transport barrier, thereby reducing the heat release rate (HRR) and slowing down the escape of the volatile products released as the polymer decomposes.

Another mechanism proposed by Wilkie and colleagues is the radical trapping by paramagnetic iron within the clay [103]. They showed that even when the clay was as low as 0.1% by mass fraction, the peak heat release rate of the polystyrene-clay nanocomposite was lowered by 40%, a value not much different from that observed with higher amounts of clay. This in turn improved the flame-retardant properties of polystyrene-clay nanocomposite compared to those of neat polystyrene (PS) (Table 6). This was particularly significant for PS, which normally produced little or no char when burnt alone. In polystyrene-clay nanocomposite, char was indeed formed and lower HRRs obtained. It was also shown that the clay-reinforced carbonaceous char was responsible for the reduced mass loss rates (and hence low HRRs). Another factor that was observed to lower the heat release rate of the nanocomposites was the polymer viscosity. Higher viscosity material exhibited a 30 % lower HRR relative to a low viscosity material.

Recently, an investigation was performed to study how the initially well-dispersed clay particles in a polymer nanocomposite accumulated on the surface during burning/gasification to form the protective silicate-rich char [101]. Figure 11 clearly reveals this phenomenon of re-aggregation of the originally well-dispersed clay layers. Two probable mechanisms for this process have been suggested: (a) Recession of polymer resin from the surface by pyrolysis leaving the clay particles behind. It is expected that at high temperature the amine grafted organic layer will degrade from the clay particle surface [104], resulting in more

Table 6. TGA and cone calorimetry data of PS and PS/clay nanocomposites (From Ref. 103).

Parameter	PS	PS-clay nanocomposite (3 wt.%)
$T_{10\%}$ (°C)	351	396
$T_{50\%}$ (°C)	404	435
Char (%)	0	4
Time to ignition (s)	35	40
Peak heat release rate (kWm^{-2})	1024	472
Time to peak heat release rate (s)	165	185
Time to burnout (s)	190	252
Energy released through 190s	981	529
Average mass loss rate (mgs^{-1})	127	108
Mass loss at 190s (%)	86	61
Specific extinction area (m^2kg^{-1})	1572	1321

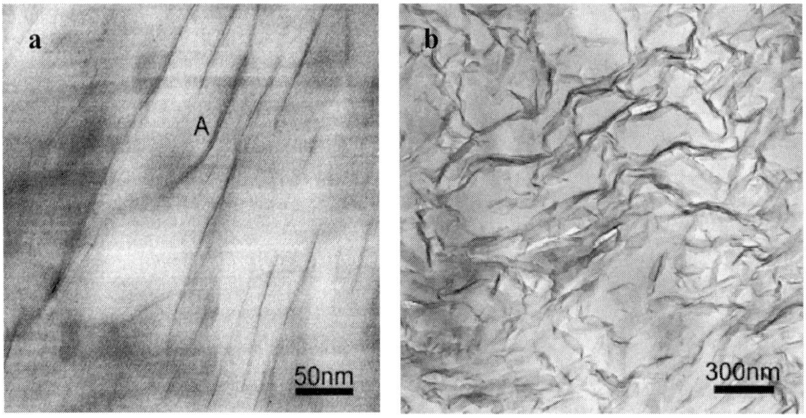

Fig. 11. TEM micrographs indicating the re-aggregation of the originally well-dispersed clay layers (a) on the surface during burning/gasification to form protective silicate-rich char (b). (Reprinted with permission from Ref. 101. Copyright (2004), with permission from Elsevier)

hydrophilic and less compatible clay particles with the resin. Thus, the hydrophilic clay particles separate from the resin and re-aggregate. (b) Another mechanism is the transportation of clay particles pushed by numerous rising bubbles of degradation products and associated convection flow in the melt from the interior of the sample towards the surface. However, too many rising bubbles may have another effect on the transport of clay particles. Bursting of the bubbles at the sample surface pushes any accumulated clay particles on the melt surface outwards from the bursting area promoting agglomeration to island-like floccules rather than formation of a continuous net-like structure of a clay-filled protective layer.

Thus, put simply, factors that identify the resistance to combustion of polymer-layered silicate nanocomposites include: (a) increase in the peak heat release temperature; (b) change in char structure and char permeability; (c) decrease in the rate of mass loss; and (d) tendency to self-extinguish. Many studies (100,105–107) have reiterated the main mechanism of improved flame retardancy in polymer-clay nanocomposites is the formation of clay-catalysed carbonaceous char, and the reinforcement of char by clay. This high-performance carbonaceous silicate char builds up on the surface during burning, insulates the underlying material and slows down the escape of the volatile products generated during decomposition. Additionally, as indicated by the X-ray diffraction of the recovered char, the same interlayer spacing (about 13Å) is found for whatever nature of the matrix (thermoplastics or thermosets) and whatever structure of the nanocomposite (exfoliated or intercalated), inferring the same nature of the residue.

Some examples showing improved flame retardant properties in different polymer-clay nano-composites (108–112) are listed in Table 7. Moreover, it is important to note here that prior to the introduction of Cone Calorimeter, flame retardancy was evaluated by the conventional tests like Limiting Oxygen Index and the Vertical Burning test. In the absence of other flame retardants (example, antimony trioxide), in most cases, polymer nanocomposites do not pass these tests. Therefore, in this respect, Zanetti *et al.* [113] and Hu *et al.* [114] have recently studied the flame retardancy of polypropylene/clay and nylon 6/clay nanocomposites, respectively, with and without conventional flame retardant additives (combination of decabromodiphenyl oxide (DB) and antimony trioxide (AO)). The idea being that the nanocomposite

Table 7. Influence of clay on the flame retardancy of polymers.

Polymers	Clay content (wt%)	Dispersion of clay	Reduced flame retardancy (%)	Ref.
Nylon 6	5	Exfoliated	40	108
PE	5	Intercalated	32	109
PP	3	Intercalated/Exfoliated	34	110
PS	5.9	Exfoliated	46	111
PMMA	5.5	Exfoliated	31	111
ABS	5	Intercalated/Exfoliated	28	112

provides the base reduction flammability, and the secondary flame retardant provides the ignition resistance. The results clearly indicated that in the presence of DB-AO, the nanocomposites satisfied both the UL-94 and Cone Calorimeter tests; whilst without the DB-AO, they satisfied only the Cone Calorimeter test and failed the UL-94 test. Similarly, Alcan International Co. Ltd. (Canada) provided a polymer composition that comprises a polymer and a combination of a common flame retardant (aluminium trihydroxide) and nano-clay [115]. The presence of this flame retardant additive combination in polymers increases the strength of the char that is formed during combustion, thus creating a barrier to ignition of the underlying material.

More recently, nylon 6 nanocomposite with 2 wt% of nano-clay was used as an a carbonization agent together with ammonium phosphate (APP) to replace conventionally used pentaerythritol, mannitol or sorbitol in an intumescent flame retardant formulation for ethylene-vinyl acetate copolymers (EVA) as these additives normally lead to exudation and water solubility problems [116]. Also, these additives are not compatible with the polymeric matrix, and the mechanical properties of the formulations become very poor. It was shown that the clay allowed the thermal stabilization of a phosphorocarbonaceous structure in the intumescent (swollen) char which increased the efficiency of the shield and, in addition, the formation of a ceramic: which could act as a protective barrier, similar to other studies described earlier in the section.

3.3. *Thermal stability*

Since processing of most polymer-clay nanocomposites need high temperatures irrespective of the fabrication route, thermal stability of organoclay has a significant role on the performance and application of nanocomposites. In general, as discussed in Sections 3.1 and 3.2, nano-clay layers possess substantial barrier properties (including thermal and mass transport) that protect the polymer from fire and make it difficult for the degraded products to leave. Molecular dynamics simulation of thermal degradation of nano-confined polypropylene also supports this mechanistic hypothesis [117]. The presence of alkylammonium cations on the clay surface (organoclay) may result in decomposition following Hofmann's elimination reaction that depends on the basicity of the anion, the steric environment around the ammonium, temperature, and its product, in addition to clay itself, which can catalyze the degradation of the polymer matrix [118,119]. The multiple pathways are attributed to a fraction of excess (un-exchanged) surfactant and the chemically heterogeneous morphology of the layered silicate. This will reduce the thermal stability of the polymer-clay nanocomposites. It has been shown that with the addition of low fractions of nano-clay that is well-dispersed into fine layers in the polymer, the barrier effect is predominant. But with increasing clay loading, the catalyzing effect rapidly increases so that the thermal stability of the nanocomposites decreases [103]. The positive and negative aspects of adding nano-clay to different polymers with regard to their thermal stability is discussed in detail below.

Xie *et al.* [120] studied thermal stability of organoclay using thermogravimetric analysis (TGA), infrared spectroscopy (FTIR) and mass spectrometry (MS). The onset decomposition temperature of organoclay was found to be ~155°C using TGA and ~180°C with TGA-MS. The low thermal stability of alkyl ammonium pre-treated clay makes it unsuitable as filler for thermoplastic and thermoset nanocomposites that require either high melt processing or cure temperatures. For example, melt extrusion and injection molding temperatures of engineering plastics, such as nylon, PET, PBT, polycarbonate, are usually in the range of 250–320°C, which are well above the onset decomposition temperature of the organoclay. Thus, alkyl quaternary ammonium modified clay will inevitably degrade. Cho *et al.* [121] and Qin *et al.* [122] in their studies on the thermal stability of nylon 6-organoclay and nylon 66-organoclay nanocomposites, respectively, found that the addition of organoclay accelerated the thermal decomposition processes even if the clay was fully exfoliated. In another study, Vanderhart *et al.* [123] investigated the chemical stability of a dimethyl-di (hydrogenated tallow) ammonium clay during melt blending with nylon 6 at 240°C. It was found that when such clay surface was exposed to nylon 6, most of the ammonium surfactant on the surface was decomposed, releasing a free amine with one methyl and two tallow substituents, which was caused by the combination of temperature and shear stress during melt blending. Fornes *et al.* [124] studied the molecular weight degradation mechanism of nylon during melt processing of nylon 6-clay nanocomposite and they attributed this to the reactions between the surfactant of the organoclay and the polyamide chains. Different types of organoclay lead to different levels of polymer degradation and coloring, depending on the unsaturation level in the organic surfactant; the higher the number of double bonds the greater the degradation and the deeper the color formation. In addition to unsaturation effects, the presence of hydroxyl-ethyl groups, as opposed to methyl groups in the organic surfactant, also results in more coloring.

In contrast, many previous studies also showed that organoclay filled polymer nanocomposites could enhance thermal stability compared to the pristine polymer. The improved thermal stability observed in these nanocomposites was generally attributed to the hindered diffusion of volatile decomposition products (such as small cyclic siloxanes for polydimethylsiloxane decomposition) as a direct result of their decreased permeability. Hence, the improvement in thermal stability is related to barrier properties and the radical-trapping effect of clay platelets. Clays can act as free radical scavengers and traps by reacting with the propagating or initiating radicals [125].

Pramoda *et al.* [126] investigated the thermal degradation behavior of nylon 6 nanocomposites at different loadings of organoclay (2.5, 5, 7.5, and 10 wt%) by analyzing the major evolved gas products (cyclic monomers, hydrocarbons, CO_2, CO, NH_3 and H_2O) under nitrogen.

They found that the onset temperature for degradation of nylon 6 with 2.5 wt% of organoclay was 12°C higher than that for neat nylon 6. Also, they showed that for higher clay loadings (5, 7.5, and 10 wt%), there was no change in the onset temperature for degradation compared to neat nylon 6. They attributed these differences to the dispersion of organoclay. A well exfoliated structure of the organoclay was obtained at 2.5 wt% where the improvement in thermal stability was observed. At other loading conditions, agglomerated clay particles were found and so were thought not to contribute to the thermal stability of the polymer matrix. However, the mechanism of improvement in thermal stability with exfoliated structure is not understood. In another study by Parija *et al.* [127] on polypropylene-organically modified layered silicate nanocomposites, enhanced thermal stability was obtained relative to the pristine polypropylene. This was thought to be related to the strong interaction of the base polymer and clay surface via chemical linkage between compatibilizer (maleic anhydride grafted polypropylene) and nano-clay, which in turn mediated the surface polarity of clay and the host polymer at the clay-polymer interface [127]. Giannelis *et al.* [128] reported that poly(methylmethacrylate) (PMMA)-clay (montmorillonite) nanocomposites had better thermal stability than those from fluorohectorite. This was due to the higher amount of aluminosilicates in montmorillonite. Zhu *et al.* [103] proposed an effect of iron on thermal stability improvement in polystyrene nanocomposite. Xu *et al.* [129] in their work on PMMA-clay nanocomposites showed that both iron and aluminum contributed to the improved thermal stability, but iron was more effective. The effect of magnesium on thermal stability improvement was found to be very small. Blumstein [130] also reported enhanced thermal stability in intercalated PMMA-clay nanocomposites. The enhanced thermal stability was attributed to both the difference of chemical structure and restricted thermal motion of polymer chains within the clay interlayers.

Further, various alternatives have been proposed to process conventional organoclay modified polymer nanocomposites when thermal limitations are a concern. These include emulsion and suspension polymerization in which unmodified (alkali-metal-containing) montmorillonite is used [131]; sol-gel technology consisting of direct crystallization of organically modified layered silicates by hydrothermal treatment with a gel containing organics and organometallics [132]; partial exchanged systems that decrease the amount of surfactant needed [133]; utilization of alkyl imidazolium as a surfactant to increase the initial decomposition temperature [134]; and use of phosphonium-based surfactants to improve thermal stability [135]. In particular, the treatment of clay with alkyl imidazolium salts showed tremendous improvement in thermal stability of polyamide 6. An improvement of ~100°C was obtained in nitrogen in comparison to that of common alkyl ammonium clays [136]. In another similar study, Liang *et al.* [137] prepared polyimide/montmorillonite nanocomposites based on rigid-rod aromatic amine modifiers. They obtained 97°C higher onset decomposition temperature than the commonly used surfactants. Recently, a new processing methodology was developed in which water was used to replace the organic alkyl ammonium surfactants to fabricate thermoplastic- and thermoset-exfoliated clay nanocomposites [138–144]. The underlying concept is that water is a powerful swelling agent for pristine clay and thus would assist intercalation/exfoliation of the clay layers.

In summary, the role of clay in the nanocomposite structure and the type of surfactant used may be the main reasons responsible for the difference in thermal stability of different polymer-clay nanocomposite systems. Clay can act as a heat barrier (that enhances the overall thermal stability of the system) and to assist the char formation after thermal decomposition. That is, in the initial stages of thermal decomposition, clay would shift the decomposition to higher temperature. After that, this heat barrier effect would result in a reverse thermal stability. Therefore, the stacked silicate layers could hold up the accumulated heat that might be used as a heat source to accelerate the decomposition process, in concert with the heat flow supplied by the outside heat source [24]. In addition, the variation of the surface polarity of clay and host polymer at their interface owing to the use of different compatibilizers and the radical-trapping effect of clay platelets also affect the thermal stability of polymer-layered silicate nanocomposites.

4. Modeling of Polymer Nanocomposites

Further development and optimization of different mechanical and physical properties of polymer nanocomposites requires that experimental results be complemented by theoretical studies and computer simulations as they have the advantage of providing insights into the role of individual factors, and to rationally elucidate their significance. From the conventional continuum mechanics-based models to the novel molecular dynamic computer simulations, many attempts have been made to predict the properties of these nanocomposites, such as stiffness, strength, fracture toughness, barrier properties, etc [145–168]. In this section, a brief overview of some of the most extensively used models for predicting some of the mechanical and physical properties of polymer nanocomposites are presented pointing out their positive and negative aspects.

Although many different continuum mechanics models to predict the tensile modulus of polymer nanocomposites have been proposed, however, the Halpin-Tsai [145] and Mori-Tanaka [146] methods are most successful. These models generally depend on parameters like particle/matrix stiffness ratio, particle volume fraction, particle aspect ratio, and orientation. A number of assumptions are inherent to both approaches, such as, the filler and matrix are linearly elastic, isotropic, and fully bonded. The filler is perfectly aligned, uniform in shape, size, and dispersion state. Particle–particle interactions are not explicitly considered. As with all composite theories the properties of matrix and filler are considered to be identical to those of the pure components.

Halpin-Tsai's composite theory for predicting the stiffness is based on the early micromechanics works of Hermans [147] and Hill [148]. Halpin-Tsai reduced Hermans' results into a simpler analytical form adapted for a variety of reinforcement geometries, including discontinuous filler reinforcement. The Mori-Tanaka average field theory was derived on the principles of Eshelby's equivalent inclusion method for prediction of the elastic stress state in and around an ellipsoidal particle in an infinite matrix [149]. To account for a finite filler concentration, Mori-Tanaka, however, considered a non-dilute composite consisting of many identical ellipsoidal particles that cause the matrix to experience an average stress different from that of the applied stress due to the inclusions; but through equilibrium conditions the average stress over the entire composite equals the applied stress.

Tucker and Liang [150] considered the application of several classes of micromechanical models to aligned short-fiber composites. They noted that of the existing models the widely used Halpin-Tsai model gave reasonable estimate for the effective stiffness, but the Mori-Tanaka type model gave better results for the entire range of component parameters considered and the complete set of elastic constants for the composite. They [150] also mentioned that the shear lag models following the classical work of Cox [151] have enjoyed rather strong popularity in short-fiber composites and recently in polymer/carbon nanotube composites despite some serious theoretical flaws due to their algebraic simplicity. Further, modified versions of shear lag model considering the platelet-like reinforcement geometry [152] and platelet-platelet interactions at high filler contents [153] were also proposed. Since then, many different models have been developed. Fornes and Paul (154) emphasized the importance of aspect ratio and exfoliation ratio of clay in modeling stiffness. They quantified those parameters and used the composite theories of Halpin-Tsai and Mori-Tanaka to predict the stiffness of high molecular weight nylon 6/clay nano-composites and obtained good agreement with experimental data (Figure 12). In Figure 12a, experimental modulus data was compared with model predictions for aligned layered alumino-silicate nanocomposites having an aspect ratio identical to the experimental value of 57 (number average aspect ratio). The ratio of 97 corresponds to perfectly exfoliated morphology, i.e. the number average particle length, 91 nm, divided by the thickness of an individual platelet, 0.94 nm. It is clear from Figure 12a that Halpin-Tsai equations slightly over-predict the experimental data, while the Mori-Tanaka theory under-predicts the experimental data. However, both theories demonstrate that even higher levels of reinforcement might be possible with higher levels of exfoliation, larger platelet diameters, improved orientation, etc. Figures 12b and 12c compare the experimental data to model predictions based on an intercalated morphology (stacks of clay intercalated with polymer having one or more platelets per stack). The stack properties

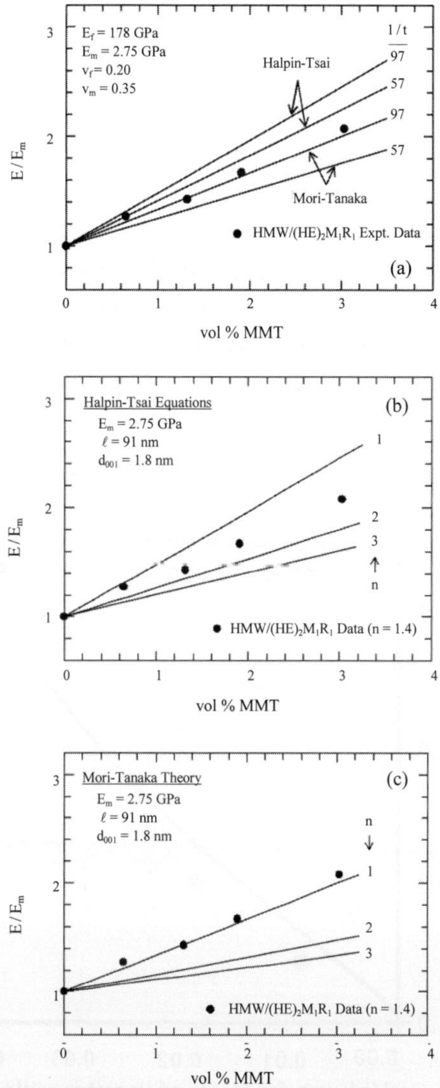

Fig. 12. Comparison of experimental and theoretical (based on unidirectional reinforcement) stiffness for high molecular weight nylon 6 nanocomposites (a) pure montmorillonite having a filler modulus of 178 GPa and aspect ratio of 57 (experimentally determined number average value) and 97, corresponding to complete exfoliation, and (b, c) intercalated morphology having one or more platelets per stack. (Reprinted with permission from Ref. 154. Copyright (2003), with permission from Elsevier)

were based on experimental data, i.e. the stacks are 91 nm in length, have a repeat spacing of 1.8 nm, and each individual disk has a modulus of 178 GPa. It can be seen that the experimental data lie between the Halpin-Tsai curves corresponding to 1 and 2 platelets per stack, which is very close to the experimental determined value of 1.4. However, when Mori-Tanaka theory is used, the experimental data matches a completely exfoliated morphology, i.e. n = 1.

Using Mori-Tanaka method, recently Wang and Pyrz [155] predicted the modulus of different polymer/clay nanocomposites by treating the intercalated silicate layers as transversely isotropic spheroids. For the chosen elastic constants, the predicted values agreed well with

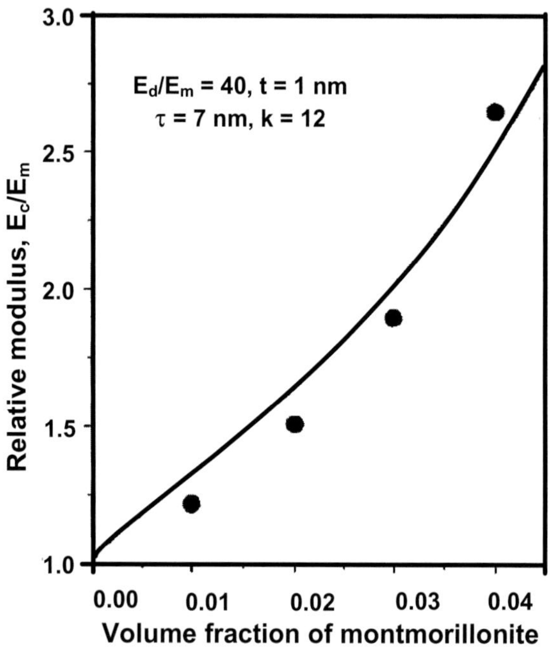

Fig. 13. Experimental results and theoretical curve predicted based on three-phase model for nylon 6/montmorillonite nanocomposite. E_d, E_m are the modulus of the dispersed phase and matrix, respectively, t is the thickness of clay layer, τ is the thickness of interfacial region, and k is a linear gradient change in modulus between the matrix and the surface of particle. (Reprinted with permission from Ref. 158. Copyright (2002), with permission from Wiley)

experimental data. Based upon Halpin-Tsai equations, Brune and Bicerano [156] addressed three important aspects of modulus prediction in thin platelet-reinforced nanocomposites: (a) buckling of platelets under compression; (b) reinforcement deficiency as a result of incomplete exfoliation; and (c) reinforcement deficiency due to deviation of the platelet orientation from being perfectly biaxial in-plane. Tandon and Weng [157] using the Mori-Tanaka approach predicted the stiffness for composites with aligned spheroidal isotropic inclusions. The Tandon-Weng approach was further used by Masenelli-Varlot *et al.* [35] to consider the filler geometry in nylon-6/montmorillonite nanocomposites. Recently, three-phase models were introduced to explain better the improved stiffness by considering the interphase between nano-filler and polymer. Ji *et al.* [158] developed such a model considering the dispersed particle size, shape, thickness of the interfacial region, particulate-to-matrix modulus ratio, and a parameter describing a linear gradient change in the modulus between the matrix and the surface of particle on the modulus of nanocomposites. Comparison of test data to the theoretical curve based on this model for nylon 6/montmorillonite nanocomposites is shown Figure 13, which show that they agree well with each other.

Even though many different composites models have been introduced to predict the stiffness of polymer nanocomposites, still disparities exist between theory and experimental results. Fornes and Paul [154] summarized some of the issues that limit the ability to model stiffness in polymer-layered silicate nanocomposites, which is shown in Table 8.

Relatively, there is much less work on analytical modeling of strength and toughness of polymer nanocomposites. Especially, for the latter property, confusing toughness data (both improvements and reductions have been reported) make it difficult to compare with any models quantitatively. It seems that accurate and validated toughness results should be first obtained and then appropriate toughening models developed for those failure mechanisms actually observed.

Among the physical properties that have received considerable interest from the modeling point-of-view is the barrier property. The influence of the platelet orientation imperfections on the overall barrier properties (gas permeability) of polymer nanocomposites was studied

Table 8. Some of the important issues that has limited the modeling of stiffness prediction in polymer-layered nanocomposites. (Reprinted with permission from Ref. 154. Copyright (2003), with permission from Elsevier).

Issue	Theory	Experimental
Filler shape and size	- Uniform shape - Constant dimensions	- Non-uniform shape - Distribution of lengths and thickness - Imperfect exfoliation of LAS
Filler orientation	- Unidirectional	- Some degree of misalignment
Filler interface	- Filler and matrix are well bonded	- Imperfect bonding between the filler and matrix
Filler modulus	- Assumes filler modulus is the same in all directions	- Filler is anisotropic
Matrix considerations	- Assumes matrix is isotropic	- Polymer chain orientation - Presence of polymer crystallites
Filler concentration effects	- No particle-particle interactions - Ignores changes in viscosity - No particle agglomeration	- Particle-particle interactions and agglomeration - Changes in viscosity can alter morphology during injection molding - Changes in crystalline morphology (e.g. type, crystallite size and amount)

Among the physical properties that have received considerable interest from the modeling point-of-view is the barrier property. The influence of the platelet orientation imperfections on the overall barrier properties (gas permeability) of polymer nanocomposites was studied recently using the continuum level approach [159]. A wide range of platelet aspect ratios and concentrations was considered. For this purpose, the microstructures representative of three different regimes (dilute, semi-dilute and concentration regimes) were selected, which were firstly introduced by Fredrickson and Bicerano [93] depending on the platelet aspect ratio, a, and the platelet volume fraction, f. The dilute regime, corresponds to $x = af \ll 1$, the semi-dilute regime, with $x \gg 1$ but $f \ll 1$, and the concentrated regime, with f no longer negligible compared with unity. In the dilute regime, where the platelets are separated by an average distance that is much larger than the platelet diameters, the platelets contribute practically independently to the overall permeability reduction, and the results obtained confirmed the validity of the asymptotic theoretical predictions of Fredrickson and Bicerano [93], i.e. $\chi \rightarrow 1/3$ as $x \rightarrow 0$, where χ is defined as:

$$\chi = \frac{P_0 - P_{Random}}{P_0 - P_{Aligned}}$$

where P_0 is the permeability coefficient of the pure matrix, and P_{Random} and P_{Aligned} denote the effective, overall permeability coefficients of the morphologies with randomly disoriented and fully aligned platelets, respectively. For morphologies typical of real nanocomposites (semi-dilute and concentrated regimes), the numerical results suggested that, as a rule of thumb, one can take $\chi \approx 0.5$ to estimate the improvement in the overall barrier reduction when changing from morphologies with fully disoriented platelets to those with fully aligned ones, provided that all the other factors stay unchanged.

In another study, the effects of clay sheet length, concentration, orientation, and degree of delamination were considered by Bharadwaj [87] on the relative permeability by modifying the simple tortuosity model. It was concluded that dispersing longer sheets ($L > 500$ nm) in a polymer matrix is particularly beneficial because of increased tortuosity, reduced dependence of relative permeability on the orientation order of the sheets, and slowing the degradation in barrier property with decreasing state of delamination (Figure 8).

Finite element calculations of gas permeability in clay reinforced polymers were also reported [160]. For simplicity of analysis, it was assumed that the clay layers were randomly dispersed, non-overlapped, and were impermeable in an isotropic matrix. However, in reality, such idealized morphologies are hardly achievable. The role of geometric factor was studied, assuming perfectly aligned platelets and no molecular level transformations in the matrix. It was found that a simple stretched exponential function controlled the permeability reduction upon increasing the platelet aspect ratio and platelet volume fraction. Further, it was the product of the platelet aspect ratio and the volume fraction that determined the permeability reduction.

Despite the development and application of different continuum mechanics models to predict the improved mechanical and physical properties, however, a critical question is whether continuum mechanics is applicable to polymer-clay nanocomposites having nanoscale microstructures. Most analytic expressions for predicting elastic properties have several disadvantages as they are based on idealized assumptions based on morphology. As nanocomposites have a vast interfacial area per unit volume, they are completely different from micro-composites. Although some three phase models were proposed by considering this interfacial area, there are still many other parameters that should be considered when predicting the mechanical properties of polymer-clay nanocomposites. For example, the polymer chains in a polymer-clay nanocomposite are often confined between the platelets that are closer to each other than the radius of gyration of a chain [161]. This was proposed to modify the thermodynamics of polymer chain conformations and kinetics of chain motions, which will potentially affect the mechanical properties of the polymer nanocomposites.

More recently, to provide better appreciation of the underpinning physics for the observed phenomena, molecular simulation methods, such as molecular mechanics (MM) based on molecular dynamic (MD) simulations, were used to study the diverse properties of polymeric systems in terms of energetics and kinetics [162,163]. It is believed that MD simulations can fill the gap between theory and experimental results in the mechanical behavior of polymer nanocomposites. They generally involve tracking the motion of atoms and molecules as a function of time [164,165]. MD simulations of polymers filled with nanoscopic particles recently revealed the mechanisms by which 'nano' particles improve the toughness of the material. It was shown that the mobility of the particle, rather than its surface area, controls its ability to dissipate energy [165]. These studies also showed that the strength of the polymer-particle interaction, particle specific surface area, surface structure, and proximity to the glass transition were the most important factors which controlled the properties of the interfacial region [166,167]. These simulations also resolved some major problems in complex systems that are characterized by a large number of degrees of freedom, lack of symmetry, non-linearity and complicated interactions [168]. Additionally, these methods can also be used in conjunction with different length-scale methods via multi-scale modeling [154].

Apart from these advantages of using MD simulations, there are also many drawbacks. One of which is that they are limited on the space and time scales, since they deal with relatively few molecules in the system [169]. As an alternative, coarse-grained models [170] and lattice models [171] have been developed. Coarse-grained model group a number of molecules or mers into a single grain. In the lattice models based on statistical mechanics, polymer chains are represented by random walks on numerical lattices, and thus, many states can be generated and equilibrated. However, the lattice models do not consider the detailed molecular structure, which is their main limitation. Despite this, lattice modeling can be advantageous because the infinite number of conformations of real chains in continuum space is reduced to a finite number, which makes computing feasible.

Additionally, all of these advanced models, such as MD simulations, lattice models, etc, require a lot of computing power and time. Thus, even though there are deficiencies of analytical models based on continuum mechanics for prediction of mechanical and physical properties of polymer nanocomposites, the ease and convenience of using such models make them applicable whenever a first estimate is required.

5. Potential Applications

Improved reinforcement, corrosion resistance, noise dampening, dimensional stability, scratch and wear resistance, and superior barrier properties all make polymer nanocomposites prime candidates for automotive and packaging applications. For auto-manufacturers, polymer nano-composite parts are lightweight with an average 25% weight-saving over highly filled-plastics and 80% over steel [172], which means large fuel cost reductions. Owing to their good thermal stability, flame retardancy and heat distortion temperature, polymer nanocomposites can also be used for under-the-hood and interior parts. Superior gas barrier properties have been employed in different applications, including packaging for processed meats, cheese, confectionery, cereals and boil-in-the-bag foods, also extrusion-coating applications in association with paperboard for fruit juice and dairy products, together with co-extrusion processes for the manufacture of beer and carbonated drinks bottles.

The barrier properties of clay layers have been utilized in developing novel nanoscopic-pigment particles that exhibit improved oxygen, temperature and UV stability of organic molecules in addition to the normal leaching by a combination of soluble dyestuff and clay particles via ionic interactions [173]. In addition, the resulting colored products are transparent since the primary particle size of the nano-pigments is lower than the wavelength of light. Nanocomposites can be processed as thin films for packaging (and in bulk form for beverage bottles) exploiting their gas barrier properties and optical transparency. Recently, nanocomposites have also been utilized as alternatives to polymer electrolytes. In poly(ethylene oxide) (PEO) electrolytes, a serious drawback has been the precipitous decrease of conductivity below their melting temperature. This decrease is caused by the formation of crystallites that severely impede the ionic mobility. As intercalation impedes polymer crystallization, it potentially enhances the electrolyte conductivity. Since the counter anions in the nanocomposites are the comparatively immobile silicate layers, single-ion conductor behavior is anticipated [41]. Furthermore, layered silicate nanocomposite coatings are being used as a fast drying, high-quality, image-receiving layer in a typical inkjet printer [174]. The presence of filler incorporation at nano-levels has also been shown to have significant effects on the transparency and haze characteristics of films. In comparison to conventionally filled polymers, nano-clay incorporation has been shown to significantly enhance transparency and reduce haze.

In recent years, there have also been considerable commercial activities concerning polymer-clay nanocomposite. For example, commercial grades of polyamide 6 nanocomposites for packaging films are available from Ube and Bayer companies. Montell has partnered with General Motors (GM) to develop a PP nanocomposite for automotive exterior applications [175]. Additionally, sports balls and tires need to hold air pressure, have elastomeric flexibility, and minimize the energy losses that occur when the product is deformed [176]. Exxon has patented in the area of elastomer-clay system citing improvements in the resistance to air permeability. The low air permeability of rubber modified by exfoliated clay was used to develop high performance rubber tires [176]. Thus, InMat LLC recently developed elastomeric nanocomposite coatings to help these products hold air better [176]. Wilson Sporting Goods has commercialized a new tennis ball (Double Core™) using this technology, which maintains its original air pressure and bounce two times longer than other tennis balls. The inner core of the tennis ball was coated with a butyl-based rubber filled with nano-clay particles, which acted as a barrier and restricted air flow from escaping the core. This new technology utilizing a coated inner core was shown to inhibit air permeation by 200%. Eastman Chemical pursued patents on multi-layer and single layer nanocomposite barrier packages [177]. One of the commercial targets appears to be the disposal/recyclable beer bottle market where the current plastics product is a multi-layer construction. Tetra Laval has been awarded patents that cite the use of nanocomposites of both single layer and multi-layer construction [178]. Moreover, environmental durable epoxy-clay nanocomposites adhesives were being used in aerospace due to their high heat- and moisture-resistance.

6. Summary and Outlook

From the above discussions, it is evident that in the case of mechanical properties, in the presence of clay, in almost all materials systems, a dramatic increase in elastic modulus of the nano-composites is seen; though in most cases this is at the expense of fracture toughness. Despite many efforts to obtain balanced mechanical properties, it is essential to develop an in-depth basic understanding of how the changes in these properties occur with addition of clay from different viewpoints and length scales. For example, the role of the interphase/interface between the nano-additive and polymer matrix on the ultimate mechanical properties should be studied and clarified with care, perhaps first with well-defined materials systems. Further, theoretical models should also be developed to predict and verify the deformation, strength and toughness of these polymer nanocomposites, whether based on continuum mechanics or MD simulations. A lot of confusions in the literature are caused by (a) poor experiments, especially those on toughness evaluations; and (b) inconsistent materials of apparently same compositions.

For physical properties, the opportunity to obtain excellent flame retardant and thermally stable polymer-clay nanocomposites exists, but the exact recognition of the role of the main factors controlling their mechanisms is still under debate. There is now a drive to develop a new class of highly flame-retardant and environmentally benign polymer nanocomposites without the addition of commercial flame retardants that are eco-unfriendly. On permeability, for critical applications, it is useful to develop a means of predicting the percolation threshold. This has recently been done based on the group re-normalization theory for several polymer/clay nanocomposites. It is shown that the theoretical results compare well with available experimental data [179]. The tortuousity model discussed in this chapter is purely geometrical. Modifications are required in cases where there is physico- or chemico-absorption with the polymer or the clay layers when diffusion of gaseous or liquid molecules occurs through the film thickness.

It is very important to ensure and emphasize that it is not what is being achieved, but it is basic understanding of the nature and structures of these nanocomposites that must continue. For only this will enable us to design novel materials with enhanced properties avoiding the trade-offs of conventional systems. It may also be concluded that to realize the most exciting prospects and appropriate utilization of this class of new materials needs creative imaginations, which can only be made possible by an interdisciplinary approach (i.e. combining fields of physics, chemistry, engineering, materials and molecular science). This was elegantly expressed by Francis Jacob, a Nobel Laureate in Medicine in 1965: "The part of imagination in scientific work is the same as in the work of painter or writer. It consists of cutting the real and recombining the pieces in order to create something new."

Acknowledgements

The authors thank the Australian Research Council (ARC) for the continuing support of the project on: "Polymer Nanocomposites". YWM and ZZY are, respectively, Australian Federation Fellow and Australian Postdoctoral Fellow, supported by the ARC and tenable at the CAMT, University of Sydney. AD thanks the Australian Government for an International Postgraduate Research Scholarship (IPRS) award and the University of Sydney for an International Postgraduate Award (IPA) to undertake a PhD program at the CAMT. We also wish to acknowledge the contributions of various members of the Polymer Nanocomposites Group in the CAMT for useful discussions and constructive comments.

References

1. Schadler LS. in 'Polymer-based and Polymer-filled Nanocomposites', Nanocomposite Science and Technology, Ed. Pulickel M. Ajayan, Linda S. Schadler, Paul V. Braun, Wiley-VCH Verlag GmbH & Co. KGaA, Weinheim, 2003.
2. Vaia RA, Krishnamoorti R. in 'Polymer Nanocomposites: Synthesis, Characterization, and Modeling,' ACS Symposium Series 804, Ed. Krishnamoorty R, Vaia RA. American Chemical Society, Washington, DC, USA, p. 1, 2002.
3. Fukushima Y, Inagaki S. J. Incls. Phen. 5 (1987) 473.
4. Kojima Y, Usuki A, Kawasumi M, Okada A, Kurauchi T, Kamigaito O. J. Polym. Sci. Part A: Polym. Chem. 31 (1993) 983.
5. Kojima Y, Usuki A, Kawasumi M, Okada A, Kurauchi T, Kamigaito O. J. Polym. Sci. Part A: Polym. Chem. 31 (1993) 1755.
6. Messersmith PB, Giannelis EP. Chem. Mater. 6 (1994) 1719.
7. Lan T, Pinnavaia TJ. Chem. Mater. 6 (1994) 2216.
8. Kornmann X, Lindberg H, Berglund LA. Polymer 42 (2001) 1303.
9. Fu X, Qutubuddin S. Polymer 42 (2001) 807.
10. Agag T, Koga T, Takeichi T. Polymer 42 (2001) 3399.
11. Yano K, Usuki A, Okada A, Kurauchi T, Kamigaito O. J. Polym. Sci. Part A: Polym. Chem. 31 (1993) 2493.
12. Nam PH, Maiti P, Okamoto M, Kotaka T, Hasegawa N, Usuki A. Polymer 42 (2001) 9633.
13. Maiti P, Nam PH, Okamoto M, Hasegawa N, Usuki A. Macromolecules 35 (2002) 2042.

14. Kato M, Usuki A, Okada A. J. Appl. Polym. Sci. 66 (1997) 1781.
15. Chen TK, Tien YI, Wei KH. Polymer 41 (2000) 1345.
16. Wang Z, Pinnavaia TJ. Chem. Mater. 10 (1998) 3769.
17. Rhoney I, Brown S, Hudson NE, Pethrick RA. J. Appl. Polym. Sci. 91 (2004) 1335.
18. Song L, Hu Y, Li BG, Wang SF, Fan WC, Chen ZY. Intl. J. Polym. Anal. Charac. 8 (2003) 317.
19. Priya L, Jog JP. J. Polym. Sci. Part B: Polym. Phys. 41 (2003) 31.
20. LeBaron, P. C.; Wang, Z.; Pinnavaia, T. J. Appl. Clay. Sci. 15 (1999) 11.
21. Giannelis EP, Krishnamoorti R, Manias E. Adv. Polym. Sci. 138 (1999) 107.
22. Jeon HG, Jung HT, Lee SW, Hudson SD. Polym. Bull. 41 (1998) 107.
23. Peter C. LeBaron, Zhen Wang, Thomas J. Pinnavaia. Appl. Clay Sci. 15 (1999) 11.
24. Ray SS, Okamoto M. Prog. Polym. Sci. 28 (2003) 1539.
25. Alexandre M, Dubois P. Mater. Sci. Engng. 28 (2000) 1.
26. Fornes TD, Hunter DL, Paul DR. Macromolecules 37 (2004) 1793.
27. Lee DC, Jang LW. J. Appl. Polym. Sci. 61 (1996) 1117.
28. Noh MW, Lee DC. Polym. Bull. 42 (1999) 619.
29. Usuki A, Kojima Y, Kawasumi M, Okada A, Fukushima Y, Kurauchi T, Kamigaito O. J. Appl. Polym. Sci. 55 (1995) 119.
30. Stretz HA, Paul DR, Li R, Keskkula H, Cassidy PE. Polymer 46 (2005) 2621.
31. Kojima Y, Usuki A, Kawasumi M, Okada A, Kurauchi T, Kamigaito O, Kaji K. J. Polym. Sci. Part B: Polym. Phys. 33 (1995) 1039.
32. Kojima Y, Usuki A, Kawasumi M, Okada A, Kurauchi T, Kamigaito O, Kaji K. J. Polym. Sci. Part B: Polym. Phys. 32 (1994) 625.
33. Varlot K, Reynaud E, Kloppfer MH, Vigier G, Varlet J. J. Polym. Sci. Part B: Polym. Phys. 39 (2001) 1360.
34. Yoon PJ, Fornes TD, Paul DR. Polymer 43 (2002) 6727.
35. Masenelli-Varlot K, Reynaud E, Vigier G, Varlet J. J. Polym. Sci. Part B: Polym. Phys. 40 (2002) 272.
36. Yu ZZ, Yang MS, Zhang QX, Zhao CG and Mai Y.-W. J. Polym. Sci. Part B: Polym. Phys. 41 (2003) 1234.
37. Eckel DF, Balogh MP, Fasulo PD, Rodgers WR. J. Appl. Polym. Sci. 93 (2004) 1110.
38. Wu W-L, Hu J-T, Hunston DL. Polym. Engng. Sci. 30 (1990) 835.
39. Zilg C, Mulhaupt R, Finter J. Macromol. Chem. Phys. 200 (1999) 661.
40. Wang Z, Lan T, Pinnavaia TJ. Chem. Mater. 8 (1996) 2200.
41. Giannelis EP. Adv. Mater. 8 (1996) 29.
42. Lan T, Kaviratna PD, Pinnavaia TJ. Chem. Mater. 7 (1995) 2144.
43. Wang Z, Pinnavaia TJ. Chem. Mater. 10 (1998) 1820.
44. Becker O, Varley R, Simon G. Polymer 43 (2002) 4365.
45. Wang K, Wang L, Wu J, Chen L, He C. Langmuir 21 (2005) 3613.
46. Peterlin A. J. Mater. Sci. 6 (1971) 490.
47. Peterlin A. in 'The Strength and Stiffness of Polymers', M. Dekker, New York, 1983.
48. Nitta K. Comp. Theor. Simul. 9 (1999) 19.
49. Nitta K. Macromol. Symp. 170 (2001) 311.

50. Ichihara S, Iida S. in 'The Strength and Stiffness of Polymers'; M. Dekker, New York, 1983, p 129.
51. Sheng N, Boyce MC, Parks DM, Rutledge GC, Abes JI, Cohen RE. Polymer 45 (2004) 487.
52. Wu S. J. Appl. Polym. Sci. 35 (1988) 549.
53. Dijkstra K, Wevers HH, Gaymans RJ. Polymer 35 (1994) 323.
54. Liu W, Hoa SV, Pugh M. Comp. Sci. Technol. 65 (2005) 307.
55. Liu W, Hoa SV, Pugh M. Polym. Engng. Sci. 44 (2004) 1178.
56. Nair SV, Goettler LA, Lysek BA. Polym. Engng. Sci. 42 (2002) 1872.
57. Lu HJ, Liang GZ, Ma XY, Zhang BY, Chen XB. Polym. Intnl. 53 (2004) 1545.
58. Liu T, Tjiu WC, Tong Y, He C, Goh SS, Chung TS. J. Appl. Polym. Sci. 94 (2004) 1236.
59. Bucknall CB, Karpodinis A, Zhang XC. J. Mater. Sci. 29 (1994) 3377.
60. Zerda AS, Lesser AJ. J. Polym. Sci: Part B: Polym. Phys. 39 (2001) 1137.
61. Reichert P, Kressler J, Thomann R, Mulhaupt R, Stoppelmann G. Acta Polym. 49 (1998) 116.
62. Kelnar I, Kotek J, Kapralkova L, Munteanu BS. J. Appl. Polym. Sci. 96 (2005) 288.
63. Khatua BB, Lee DJ, Kim HY, Kim JK. Macromolecules 37 (2004) 2454.
64. Dasari A, Yu ZZ, Yang M, Zhang QX, Xie XL, Mai Y.-W. Comp. Sci. Technol. in press (2006).
65. Chow WS, Baker AB, Ishak ZAM, Karger-Kocsis J, Ishiaku, US Eur. Polym. J. 41 (2005) 687.
66. Liu XH, Wu QJ, Berglund LA, Fan JQ, Qi ZN. Polymer 42 (2001) 8235.
67. Wang Y, Zhang Q, Fu Q. Macromol. Rapid Commun. 24 (2003) 231.
68. Liu XH, Wu QJ. Macromol. Mater. Engng. 287 (2002) 180.
69. Yee AF, Li DM, Li XW. J. Mater. Sci. 28 (1993) 6392.
70. Chan CM, Wu JS, Li JX, Cheung YK. Polymer 43 (2002) 2981.
71. Li XC, Park HM, Lee JO, Ha CS. Polym. Engng. Sci. 42 (2002) 2156.
72. Tjong SC, Meng YZ. J. Polym. Sci. Part B: Polym. Phys. 41 (2003) 2332.
73. Becker O, Cheng YB, Varley RJ, Simon GP. Macromolecules 36 (2003) 1616.
74. Ke Y, Lu J, Yi X, Zhao J, Qi Z. J. Appl. Polym. Sci. 78 (2000) 808.
75. Fröhlich J, Golombowski D, Thomann R, Mulhaup R. Macromol. Mater. Engng. 289 (2004) 13.
76. Zerda AS, Caskey TC, Lesser AJ. Macromolecules 36 (2003) 1603.
77. Zerda AS, Lesser AJ. Polym. Engng. Sci. 44 (2004) 2125.
78. McGarry FJ, Willner AM. ACS Div. Org. Coat. Plast. Chem. 28 (1968) 512.
79. McGarry FJ, Sultan NJ. ACS Div. Org. Coat. Plast. Chem. 28 (1968) 526.
80. Aizpurua B, Franco M, Corcuera MA, Riccardi CC, Mondragon I. J. Appl. Polym. Sci. 76 (2000) 1269.
81. He S, Shi K, Bai J, Zhang Z, Li L, Du Z, Zhang B. Polymer 42 (2001) 9641.
82. Albert P, Lauger J, Kressler J, Mulhaupt R. Acta Polym. 46 (1995) 68.
83. Albert P, Lauger J, Kressler J, Mulhaupt R. Acta Polym. 46 (1995) 74.
84. Boogh L, Pettersson B, Manson JAE. Polymer 40 (1999) 2249.
85. Jiang T, Wang YH, Yeh JT, Fan ZQ. Eur. Polym. J. 41 (2005) 459.

86. Messersmith PB, Giannelis EP. J. Polym Sci: Part A: Polym. Chem. 33 (1995) 1047.
87. Bharadwaj RK. Macromolecules 34 (2001) 9189.
88. Ku B-C, Froio D, Steeves D, Kim DW, Ahn H, Ratto JA, Blumstein A, Kumar J, Samuelson LA. J. Macromol. Sci: Part A: Pure Appl. Chem. 41 (2004) 1401.
89. Krook M, Morgan G, Hedenqvist MS. Polym. Engng. Sci. 45 (2005) 135.
90. Kojima Y, Fukumori K, Usuki A, Okada A, Kurauchi T. J. Mater. Sci. Lett. 12 (1993) 889.
91. Nah C, Ryu HJ, Kim WD, Choi SS. Polym. Adv. Technol. 13 (2002) 649.
92. Osman MA, Mittal V, Morbidelli M, Suter UW. Macromolecules 36 (2003) 9851.
93. Fredrickson GH, Bicerano J. J. Chem. Phys. 110 (1999) 2181.
94. Messersmith PB, Giannelis EP. J. Polym. Sci: Part A: Polym. Chem. 33 (1995) 1047.
95. Frisk P. US Pat. 5,916,685. March 14, 1999.
96. Frisk P, Laurent J. US Patent 1999; 5, 876, 812. March 2.
97. Barbee R. B., US Pat. 6,034,163. March 7, 2000.
98. G.W. Beall in 'Polymer-Clay Nanocomposites,' Wiley Series in Polymer Science, Ed. Pinnavaia TJ, Beall GW. John Wiley & Sons, Ltd., Chichester, England, p. 267, 2000.
99. Fujiwara S, Sakamoto T. Kokai Patent Application no. SHO 511976–109998 (1976).
100. Gilman JW. Appl Clay Sci. 15 (1999) 31.
101. Kashiwagia T, Harris Jr RH, Zhang X, Briberb RM, Ciprianoc BH, Raghavan SR, Awada WH, Shields JR. Polymer 45 (2004) 881.
102. Zanetti M, Camino G, Mulhaupt R. Polym. Degrad. Stab. 74 (2001) 413.
103. Zhu J, Uhl FM, Morgan AB, Wilkie CA. Chem. Mater. 13 (2001) 4649.
104. Davis RD, Gilman JW, VanderHart DL. Polym. Degrad. Stab. 79 (2003) 111.
105. Bourbigot S, Le Bras M, Dabrowski F, Gilman JW, Kashiwagi T. Fire Mater. 24 (2000) 201.
106. Vaia RA, Price G, Ruth PN, Nguyen HT, Lichtenhan J. Appl. Clay Sci. 15 (1999) 67.
107. Wang S, Hu Y, Li Z, Wang Z, Zhuang Y, Chen Z, Fan W. Coll. Polym. Sci. 281 (2003) 951.
108. Bourbigot S, Devaux E, Flambard X. Polym. Degrad. Stab. 75 (2002) 397.
109. Wang SF, Hu Y, Qu ZK, Wang ZZ, Chen ZY, Fan WC. Mater. Lett. 57 (2003) 2675.
110. Wang DY, Wilkie CA. Polym. Degrad. Stab. 80 (2003) 171.
111. Su SP, Wilkie CA. J. Polym. Sci: Part A: Polym. Chem. 41 (2003) 1124.
112. Wang SF, Hu Y, Lin ZH, Gui Z, Wang ZZ, Chen ZY, Fan WC. Polym. Internl. 52 (2003) 1045.
113. Zanetti M, Camino G., Canavese D, Morgan AB, Lamelas FJ, Wilkie CA. Chem. Mater. 14 (2002) 189.
114. Hu Y, Wang S, Ling Z, Zhuang Y, Chen Z, Fan W. Macromol. Mater. Engng. 288 (2003) 272.

115. World Intellectual Property Organization Patent: WO 2000066657A1: Fire Retardant Compositions (April 2000).
116. Bourbigot S, Le Bras M, Dabrowski F, Gilman JW, Kashiwagi T. Fire Mater. 24 (2000) 201.
117. Nyden MR, Gilman JW. Comp. Theor. Polym. Sci. 7 (1997) 191.
118. Xie W, Pan WP, Hunter D, Vaia R. Chem. Mater. 13 (2001) 2979.
119. Xie W, Xie R, Pan WP, Hunter D, Koene B, Tan LS, Vaia R. Chem. Mater. 14 (2002) 4837.
120. Xie W, Gao ZM, Liu KL, Pan WP, Vaia R, Hunter D, Singh A. Thermochimica Acta 367–368 (2001) 339.
121. Cho JW, Paul DR. Polymer 42 (2001) 1083.
122. Qin HL, Su QS, Zhang SM, Zhao B, Yang MS. Polymer 44 (2003) 7533.
123. VanderHart DL, Asano A, Gilman JW, Chem. Mater. 13 (2001) 3796.
124. Fornes TD, Yoon PJ, Paul DR. Polymer 44 (2003) 7545.
125. Solomon DH, Swift JD. J. Appl. Polym. Sci. 11 (1967) 2567.
126. Pramoda KP, Liu T, Liu Z, He C, Sue H.-J. Polym. Degrad. Stab. 81 (2003) 47.
127. Parija S, Nayak SK, Verma SK, Tripathy SS. Polym. Comp. 25 (2004) 646.
128. Bandyopadhyay S, Giannelis EP. Polym. Mater. Sci. Engng. 82 (2000) 208.
129. Xu Y, Brittain WJ, Xue C, Eby RK. Polymer 45 (2004) 3735.
130. Blumstein A. J. Polym. Sci: Part A: Polym. Chem. 3 (1965) 2665.
131. Rossi GB, Beaucage G, Dang TD, Vaia RA. Nano Lett. 2 (2002) 319.
132. Carrado KA, Xu L. Chem. Mater. 10 (1998) 1440.
133. Lan T, Kaviratna PD, Pinnavaia TJ. J. Phys. Chem. Sol. 57 (1996) 1005.
134. Gilman JW, Morgan AB, Harris RH, Trulove PC, DeLong HC, Sutto TE. Polym. Mater. Sci. Engng. 83 (2000) 59.
135. Yeh JM, Liou SJ, Lin CY, Cheng CY, Chang YW, Lee KR. Chem. Mater. 14 (2002) 154.
136. Gilman JW, Awad WH, Davis RD, Shields J, Harris Jr RH, Davis C, Morgan AB, Sutto TE, Callaham J, Trulove PC, DeLong HC. Chem. Mater. 14 (2002) 3776.
137. Liang ZW, Yin J, Xu HJ. Polymer 44 (2003) 1391.
138. Hasegawa N, Okamoto H, Kato M, Usuki A, Sato N. Polymer 44 (2003) 2933.
139. Chen B, Liu J, Chen H, Wu J. Chem. Mater. 16 (2004) 4864.
140. Wang K, Chen L, Wu J, Toh ML, He C, Yee AF. Macromolecules 38 (2005) 788.
141. Yu ZZ, Hu G-H, Varlet J, Dasari A, Mai Y.-W. J. Polym. Sci: Part B: Polym. Phys. 43 (2005) 1100.
142. Kato M, Matsushita M, Fukumori K, Polym. Engng. Sci. 44 (2004) 1205.
143. Strawhecker KE, Manias E. Chem. Mater. 12 (2000) 2943.
144. Lee DC, Jang LW. J. Appl. Polym. Sci. 61 (1996) 1117.
145. Halpin JC. J. Compos. Mater. 3 (1969) 732.
146. Mori T, Tanaka K. Acta Metall. 21 (1973) 571.
147. Hermans J. Proc. Kon. Ned. Akad v Wetensch B: 65 (1967) 1.
148. Hill R. J. Mech. Phys. Solids 12 (1964) 119.
149. Eshelby JD. Proc. Roy. Soc. Lond. A241 (1957) 376.
150. Tucker III CL, Liang E. Compos. Sci. Technol. 59 (1999) 655.
151. Cox HL. Brit. J. Appl. Phys. 3 (1952) 72.

152. Padawer GE, Beecher N. Polym. Engng. Sci. 10 (1970) 185.
153. Lusis J, Woodhams RT, Xhantos M. Polym. Engng. Sci. 13 (1973) 139.
154. Fornes TD, Paul DR. Polymer 44 (2003) 4993.
155. Wang J, Pyrz R. Compos. Sci. Technol. 64 (2004) 925.
156. Brune DA, Bicerano J. Polym. 43 (2002) 369.
157. Tandon GP, Weng GJ. Polym. Comp. 5 (1984) 327.
158. Ji XL, Jing JK, Jiang W, Jiang BZ. Polym. Engng. Sci. 42 (2002) 983.
159. Lusti HR, Gusev AA, Guseva O. Modelling Simul. Mater. Sci. Eng. 12 (2004) 1201.
160. Gusev AA, Lusti HR. Adv. Mater. 13 (2001) 1641.
161. Balazs AC, Singh C, Zhulina E, Lyatskaya Y. Acc. Chem. Res. 32 (1999) 651.
162. Binder K, in 'Monte Carlo and Molecular Dynamics Simulations in Polymer Science,' Oxford University Press, New York, 1995.
163. McQuarrie DA, Simon JD, in 'Physical Chemistry: A Molecular Approach,' University Science Books, California, 1997.
164. Bhushan B. 'Handbook of Micro/Nanotribology', CRC Press, Chapter 4, p.187, 1999.
165. Gersappe D. Phys. Rev. Lett. 89 (2002) 0583011.
166. Borodin O, Bedrov D, Smith GD, Nairn J, Bardenhagen S. J. Polym. Sci. Part B: Polym. Phys. 43 (2005) 1005.
167. Kuppa V, Foley TMD, Manias E. Eur. Phys. J. E12 (2003) 159.
168. Yang ACM, Wu TW. J. Poly. Sci. Part B: Poly. Phys. 35 (1997) 1295.
169. Kim K, Utracki LA, Kamal MR. J. Chem. Phys. 121 (2004) 10766.
170. Olaj OF, Petrik T, Zifferer G. J. Chem. Phys. 108 (1998) 8214.
171. Utracki LA, in 'Polymer Blends Handbook,' Ed. Utracki LA, Kluwer Academic Publishers, Dordrecht, Chap. 2, 2002.
172. Garces JM, Moll DJ, Bicerano J, Fibiger R, McLeod DG. Adv. Mater. 12 (2000) 1835.
173. Fischer H. Mater. Sci. Engng. C23 (2003) 763.
174. Majumdara D, Blantona TN, Schwark DW. Appl. Clay Sci. 23 (2003) 265.
175 Hishinuma M, Yamamoto T. J. Mater. Sci. Lett. 3 (1984) 799.
176. Goldberg HA, Feeney CA, Karim DP, Farrell M. Tech. Papers-American Chemical Society, Rubber Division, Spring Technical Program, 161st, Savannah, GA, United States, Apr. 29-May 1, 2002, 254.
177. Kanbara T, Yamamoto T, Tokuda K, Aoki K. Chem. Lett. 11 (1987) 2173.
178. Gangopadhyay R, De A. Chem. Mater. 12 (2000) 608.
179. Lu C, Mai Y-W, Phys. Rev. Lett. 95 (2005) 088303.

CHAPTER 8

Preparation Methods and Properties of Polypropylene/Layered Silicate Nanocomposites

Joong-Hee Lee[1,2,*], Prashantha Kalappa[1], Chang Eui Hong[1], Nam Hoon Kim[1] and Gye-Hyoung Yoo[2]

[1]*Department of Advanced Materials Engineering*
Chonbuk National University
Duckjin-dong 1Ga 664-14, Jeonju, Jeonbuk, 561-756, Korea
[2]*Research Center, Korea Composite Research Co.*
Hyojeong-ri 526-4, Yongjimyun
Kimje, Jeonbuk, 576-972, Korea

1. Introduction

Polymer nanocomposites containing layered silicates are gaining a great attention because of their academic and industrial importance [1–7]. They have shown a significant improvement in material properties comparing with virgin polymer or conventional composites. These improvements can include enhanced mechanical properties [8–13], increased thermal stability [14–18], gas barrier properties [19–24] and inflammability [25–33] with low filler loading. These improved properties are resulted from the homogeneous dispersion of nano clay as filler and even with fewer demerits compared to conventional particulate filled composites [33]. Moreover, there has been a considerable progress in theory and simulations concerning the preparation and properties of these materials [34–42], and they are also considered to be unique model

* **Corresponding Author**: E-mail: jhl@chonbuk.ac.kr; Tel: 82-63-270-2342; Fax: 82-63-270-2341.

systems in order to study the structure and dynamics of polymers in confined environments [43–52].

Polymer–layered silicates nanocomposites were first reported in 1961, when Blumstein [53] obtained a polymer inserted in the structure of montmorillonite (MMT) by polymerizing a vinyl monomer. Two years later, Greenland [54] used a poly(vinyl alcohol)/MMT system to show that polymer could be directly inserted into clay layers in an aqueous solution. In 1975, Tanihara and Nakagawa [55] reported similar results by intercalating clay into poly (acryl amide) and poly(ethylene oxide) from an aqueous solution. The developments in clay-polymer system involved water soluble polymers, date back to 1960's and 1970's, were reviewed by Theng [56]. Later, Vaia, *et al.* [57] demonstrated the possibility of melt mixing of polymers with layered silicates without the use of organic solvents. In these days, efforts are being made globally by using almost all types of polymer matrices.

The first industrial application of polymer/layered silicate nanocomposite was provided by Okada, *et al.* [58] at Toyota's central research laboratories in Japan. In this case, a nylon 6 nanocomposite was formed by polymerization in the presence of the inserted monomer. This material exhibits an excellent mechanical property [59, 60]. This material was used in making the timing belt cover of Toyota's car engines and for the production of packaging film. However, this application was stopped because of the high cost. The high cost was caused by the time-intensive preparation process and the high price of materials. These facts lead to develop nanocomposites using polypropylene, which is one of the most widely used thermoplastics, not only because of its balance of physical and mechanical properties, but also due to its environmental friendliness (e.g. recyclability) and low cost. The most common type is isotactic PP, while syndiotactic PP is a minor contributor to the PP market. Polypropylene has an attractive combination of low density, high stiffness and toughness, and high heat distortion temperature above 100°C, which endow it with extraordinary versatility of properties and applications ranging from molded parts to films and fibers. Innovative catalyst and process technologies have significantly simplified its production, resulting in the minimization of catalyst residues, waxy by-products and low stereo regularity components by the use of high activity and highly stereo selective catalysts such as Ziegler-Natta and metallocene catalysts. The need for solvents has been minimized by high energy efficient processes, such as gas-phase and liquid-pool processes. Polypropylene offers ready means for recycling by remolding or feedstock recovery, or by recovery of its energy content in power plants and steel mills. It has also continued to make roads into the engineering plastics markets by replacing more expensive and less environmental friendly resins. The addition of mineral fillers to polypropylenes has widened its sphere of applications by enhancing its stiffness properties. However, the concomitant effect is typically a decrease in toughness properties, since the inorganic filler particles act as stress concentrators.

In recent years, there has been much interest in the developments of polypropylene (PP) nanocomposites to improve the properties of this low cost, commodity polymer by layered silicate reinforcement. Mixtures of PP and layered silicates have been prepared with a significant improvement in mechanical [61,62], flame retardant and barrier properties compared with the pure PP [63–67]. These enhancements have been obtained with only low clay loading, typically in the range of 1–10 wt%. Therefore, the polypropylene/clay nano hybrids so-called polypropylene/clay nanocomposites (PPCNs) are regarded as promising new materials in many industrial fields.

There are two classes of hybrids between a polymer and nanoscopically dispersed layered silicates:

(1) Intercalated hybrids
(2) Exfoliated (delaminated) hybrids

In an intercalated hybrid, the polymer chains are interacted between swollen silicate layers with their regular alteration of galleries and laminae. The distance between the layers, that is, the space occupied by the polymers is typically of the order of a few nanometers. As its name suggests, the second class of silicate is totally delaminated and dispersed in the polymer matrix. Its ordered structure is lost and the distance between the layers is the order of radius of gyration of the polymer. Both classes of hybrids are schematically compared with an immiscible system, namely a conventionally filled polymer or microcomposite as shown in Figure 1. It can be said that an intercalated hybrid displays

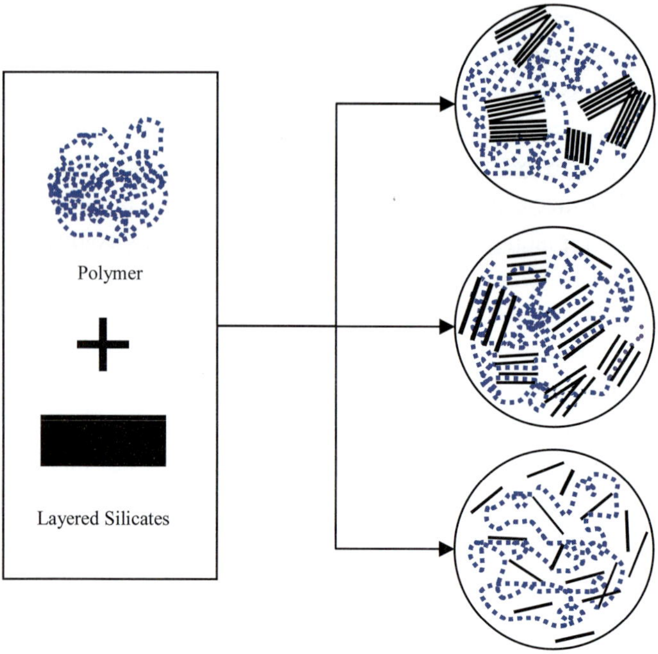

Fig. 1. Schematic representation of composite structure obtained using layered silicates.

limited miscibility and that of delaminated hybrid is totally miscible. The diagram illustrates two extremes and both of these hybrid structures can co-exist in the polymer matrix.

Montmorillonite is the most common clay mineral used in polymer/clay hybrids, and is composed of an octahedral alumina sheet sandwiched between two tetrahedral silica sheets. The anionically charged sheets, or clay platelets, are strongly held together by cations, such as Na^+, Li^+, Ca^{2+}, Fe^{2+} and Mg^{2+}. The tightly bound stacks of clay platelets may be readily separated by simple dissolution in a polar solvent, such as water. Also, the natural clay is miscible with polar polymers in which the platelets readily disperse. However, the separation of the clay platelets in a non-polar polymer such as PP and PE is difficult. The thickness of the individual clay platelets is about 1 nm. and the gallery spacing (spacing is defined in terms of d-spacing from x-ray diffraction and small angle x-ray scattering, which is the distance from one clay platelet center to the next) of the platelets in natural clay is about 1 nm. Organic modification of the clay with alkyl ammonium chlorides, in which the alkyl chain is typically composed of 6 to 25 carbons, expands the gallery spacing to about 2.2 to 2.4 nm in typical commercial organo-clays. Addition of maleic anhydride-modified PP often results in intercalated clay with gallery spacing of greater than 3 nm. Thus, the preparation of PPCNs can be achieved by mixing the organophilic clay with polymer and, optionally, compatibilizing additives as shown in Figure 2. The typically desired state is, upon addition of PP, the complete dispersion, i.e. exfoliation and dispersion of the clay platelets, so that there is no correlation between platelets and therefore no peak in the x-ray patterns. Considering that clay is composed of platelets that are about 1 nm in thickness and around 1000 nm in lateral dimensions and assuming a density of 2.65 g/cm^3, yields an upper limit of about 750 m^2/g surface area per gram for fully exfoliated clay. This is 5 to 10 times the surface area per gram, which is achievable with ordinary fillers. Exfoliated clay in polymers was shown to bestow unique properties to the system, such as depressed Tg, increased stiffness, maintenance of impact strength, increased barrier to gases and liquids, increased flame-retardancy, and etc.

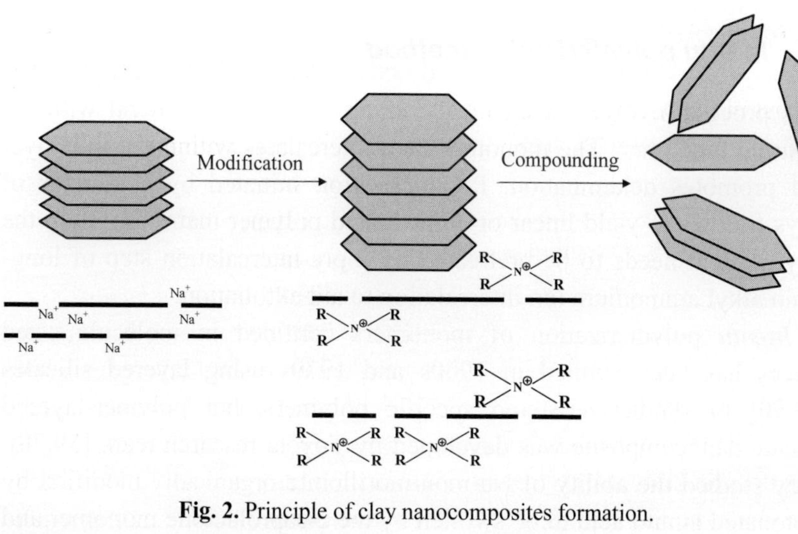

Fig. 2. Principle of clay nanocomposites formation.

This chapter is intended to highlight the developments in polypropylene/Clay nanocomposites up to now. Various processing techniques used to prepare PPCNs, their physicomechanical characterization methods, material properties, and potential applications of PPCN nanocomposites are reviewed in details. The current challenges and opportunities in development of PPCNs are also discussed.

2. Preparation Methods

Polymer-layered silicates are typically synthesized by three methods; (1) *In-situ* polymerization method, i.e. intercalation of a suitable monomer followed by polymerization [22], (2) Solution method, i.e. intercalation of dissolved polymer from a solution [68], and (3) Direct melt mixing method, i.e. mixing the layered silicates with the polymer matrix above its softening point in either static or flow conditions [9]. In this case, the modified silicates are usually employed to promote the intercalation. The polymer chains spread from the molten mass into the silicate galleries to form either intercalated or delaminated hybrids according to the degree of penetration.

2.1. *In-situ polymerization method*

This process involves mechanical mixing of the clay mineral with the required monomer. The monomer then intercalates within the interlayer and promotes delamination. Polymerization initiated by a number of ways follows to yield linear or cross-linked polymer matrices. Often the clay mineral needs to be pretreated by a pre-intercalation step of long-chain alkyl ammonium ion intercalation to aid exfoliation.

In-situ polymerization of monomers confined in molecule sized spaces has been studied in 1960s and 1970s using layered silicates [69,70] to synthesize stereo specific polymers, but polymer-layered silicate nanocomposite was developed by Toyota research team [59,70]. They studied the ability of Na-montmorillonite organically modified by protonated amino acid to be swollen by the ξ-caprolactone monomer and subsequently initiated its ring opening polymerization to obtain nylon 6 based nanocomposites [24,70]. The resulting composite does not show any diffraction peak in X-RD patterns, which indicate a fully exfoliation of layers. This fact can be also supported by TEM images, which show the molecular dispersion of the silicate sheets. The studies have proved that the nature of the amino acid used is sensitive to promote the intercalation of the ξ-caprolactone. Later, *in-situ* intercalative polymerization has been largely studied for producing poly(styrene), polyamide, poly(methylmethacrylate), poly(benzoxazole), polyurethanes, and polyolefin nanocomposites [71–76]. The process of in-situ formation of polymer-layered silicate nanocomposite is schematically illustrated in Figure 3. Polypropylene is one of the most extensively used polymers. It has no polar groups in its backbone, so the well-dispersion of the silicate layers in PP cannot be realized easily. Therefore, dispersing silicate layers of MMT at the nanometer level in non-polar polymers is difficult. Literature survey reveals that reports on the in-situ formation of polypropylene/layered silicate nanocomposites are very limited. Tudor, *et al.* [77] demonstrated the ability of soluble metallocene catalysts to intercalate inside silicate layers and to promote the coordination polymerization of propylene. Accordingly, a synthetic hectorite was initially treated with methylaluminoxane (MAO) to remove all the acidic protons and to prepare the interlayer spacing to obtain the catalyst. However, this treatment failed to show a noticeable increase in the interlayer spacing. But, upon the addition of the metallocene catalyst, a cation exchange occurred with sodium cations and the catalyst was incorporated in the hectorite galleries, yielding an increase in the interlayer spacing. This metallocene-immobilized catalyst exhibited high activity in the polymerization of propylene when it was in contact with an excess of MAO, producing PP oligomers.

In another publication, Bergman, *et al.* [78] reported the *in-situ* polymerization approach by an aliphatic 1-tetradecylammonium cations modified fluoro-hectorite intercalated by a well-defined cationic palladium complex resulted in an exfoliated polyolefin–silicate nanocomposite material when exposed to olefinic monomer. In order to make an organically modified fluoro-hectorite, the exchanged fluoro-hectorite was mixed in toluene with a Brookhart-type palladium catalyst $\{[2,6\text{-}^iPr_2C_6H_3N{=}C(Me)C(Me){=}NC_6H_3{}^iPr_2]\ [Pd(CH_2)_3CO_2\text{-}Me]\ [B(C_6H_3(CF_3)_2\text{-}3,5)_4]\}$. After the thorough mixing, the white organoclay

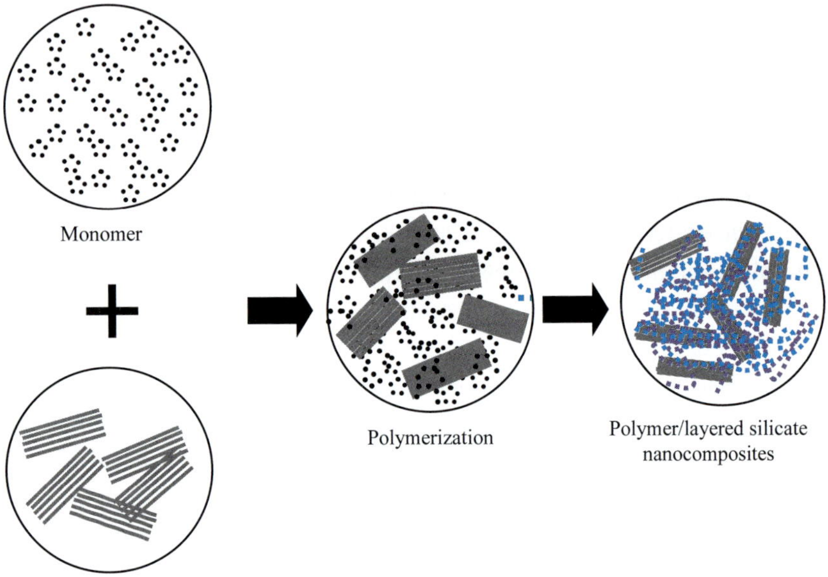

Fig. 3. Schematic representation of polymer/layered silicate nanocomposite obtained by *in-situ* polymerization.

was turned into an orange brown product indicating the formation of fluoro-hectorite. Perfluoroborates were then used as an activator to initiate polymerization with ethylene at 80 psi of the monomer pressure and at 22°C temperature. The recovered dry solid was a rubbery clay-ethylene nanocomposite. The catalyst had a turnover frequency of 162 h^{-1}. The X-ray powder diffraction (X-RD) patterns of the nanocomposite showed disappearance of ordering in the clay structure during the polymerization indicating that the clay was exfoliated or became totally disordered. Using a MAO activator and quarternized clay fillers in an approach similar to Bergman's, Heinemann, *et al.* [79] prepared nanocomposites of polyolefins as a result of the polymerization of ethylene and long chain α-olefins by various catalytic systems in the presence of dispersed organo-modified silicates. Intercalation and exfoliation were assessed by X-RD patterns and TEM images. *In-situ* polymerization catalyzed with MAO-activated zirconocene, nickel and palladium catalysts, proved more effective in nanocomposite formation, as evidenced by larger interlayer spacing and formation of exfoliated anisotropic nanosilicates with high aspect ratio.

Further Sun and Garcos [80] reported the highly advanced novel synthesis approach to make high performance PPCN by *in-situ* polymerization under mild polymerization conditions, which amazingly improved catalytic efficiency, mechanical performance and processability compared to the previous reports. In this novel approach, no external activators such as methylaluminoxane or perfluoroarylborates needed for initiating the propylene polymerization. High-pressure or high-temperature process conditions were also not required. The organoclay precursors are made by ion exchanging clays with selected amine complexes, which are capable of activating metallocene olefin polymerization catalysts. The organically modified clay powder is suspended in toluene, which acts as a solvent, by mechanical agitation, and is then mixed with hydrolytic scavengers and a highly isotactic metallocene propylene polymerization catalyst. The homogenized slurry is exposed to a propylene gas stream at ambient (about 25°C) conditions to produce PPCN. It was further found that polymerization did not occur in the absence of the organoclay nanofillers. Therefore, a stop-flow polymerization was performed to achieve the desired nanofiller loading. The high catalyst efficiency and simplicity of this process, with no addition of external activators, make this approach a potentially valuable route to prepare high performance polyolefin nanocomposites. X-RD patterns and microscopic images were used to confirm changes in the clay/polymer nanocomposite at various nanofiller loading levels during the polymerization process. X-RD patterns revealed that the organoclay precursor had an increased d_{001}-spacing compared to the original clay. Microscopic image analysis of the nanocomposite revealed the state of the organoclay dispersion and/or exfoliation in the PPCN. The prepared PPCN showed a dramatically improved Young's modulus up to 600 kpsi, which was much higher than the modulus of highly crystalline PP. This significant improvement in Young's modulus may help these in-situ PPCN to compete costly engineering plastics. Other improved properties of the PPCN observed in their studies are the reduced thermal expansion coefficient, improved melt flow strength, increased heat-distortion temperature, and lowered gas diffusion coefficient.

Recently, Hwu, *et al.* [81] have prepared the PPCN by using rac-$Et(Ind)_2ZrCl_2$ catalyst supported on montmorillonite. In their research, an in-situ method was adopted in preparing PPCN by selecting stearyltrimethyl ammonium chloride (SAC) to increase the d_{001}-spacing of MMT. They also used TEM images to show that each silicate sheet of MMT is randomly dispersed into the PP matrix following polymerization by using a supported catalyst. The synthetic procedure suggested is schematically represented in Figure 4. Through this suggested method, they also obtained the improved thermal, mechanical, and barrier characteristics of the nanocomposites in comparison with those of pure PP.

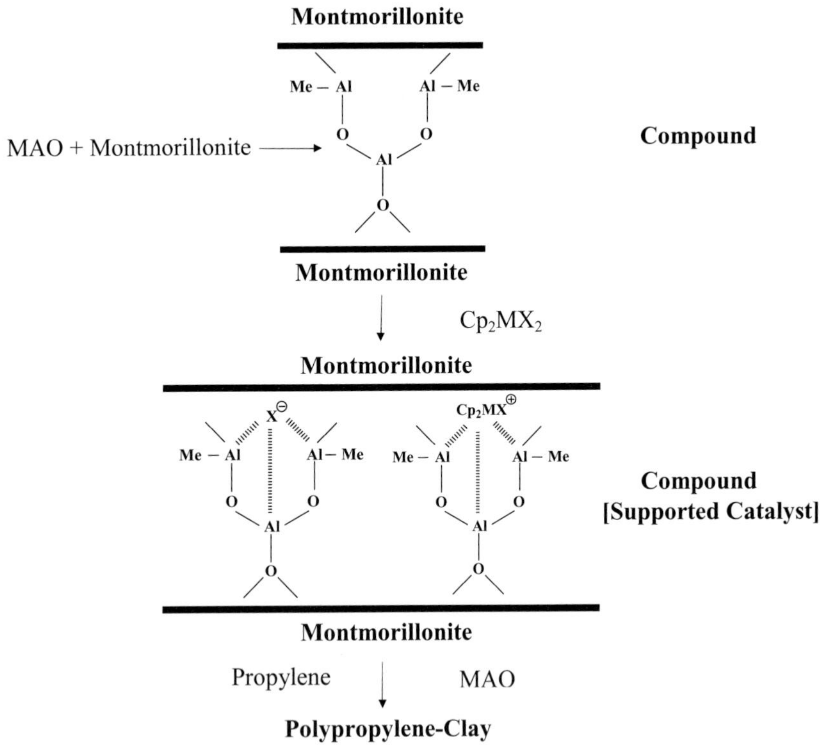

Fig. 4. Schematic representation of *in-situ* formation of PP-MMT nanocomposites [81]. Reproduced from reference [81] by Jyh-Ming Hwu and George-J. Jing, *Journal of Applied Polymer Science*, 95,1228 (2005).

2.2. *Solution method*

Intercalation of the polymer from the solution [82–87] involves a solvent system in which polymers are soluble and the silicate layers are swellable. When the polymer and layered silicate solutions are mixed, the polymer chains dissolved in the solvent intercalate into the interlayer of the silicates with the solvent. Upon solvent removal, the intercalated structure remains, resulting in nanocomposites (Figure 5). For the overall process, a negative variation in the Gibbs free energy is required in order to exchange polymer with the previously intercalated solvent in the gallery. The driving force for the polymer intercalation into interlayers of the silicates from solution is the entropy gained by desorption of solvent molecules, which compensates for the decreased entropy of the confined intercalated polymer chains.

The drawback of this method is the requirement of a suitable solvent. It has, in fact, been shown that intercalation only occurs for certain polymer/solvent or monomer/solvent pairs. Application of this method in

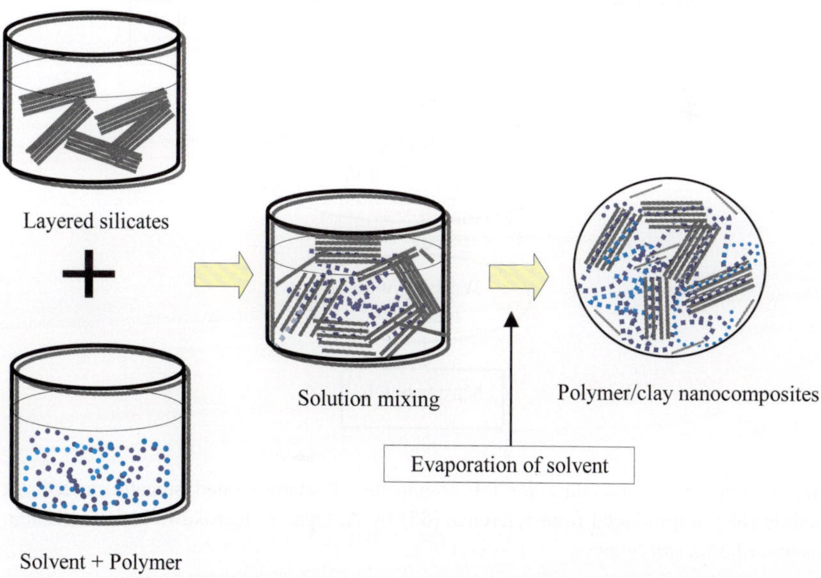

Fig. 5. Schematic representation of polymer/layered silicate nanocomposite obtained by solution method.

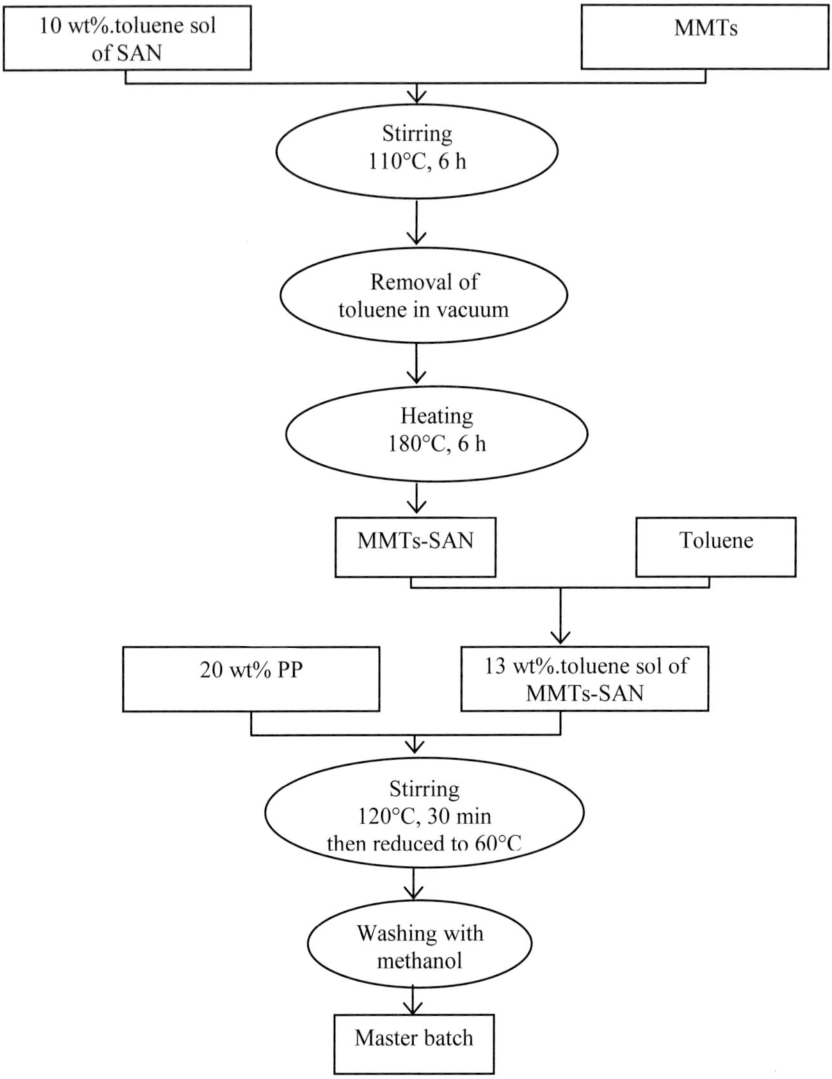

Fig. 6. Preparation procedure for the composite PP/silane treated silicate via solvent method [88]. Reproduced from reference [88] by A. Oya, Y. Kurokawa and H. Yasuda, *Journal of Material Science*, 35, 1045 (2000).

the industrial scale production may thus be impracticable, especially in a point of view of the high costs associated with solvent themselves and their disposal and environmental impact.

Oya and Kurokawa [88] reported a novel and somewhat complex procedure for the formation of nanocomposites via solvent method to prepare PPCN. In their report, the hydrophobic hectorite clay was dried at 60°C, then clay was dissolved in toluene (10 wt%) to give a transparent sol. Methyl trimethoxysilane (MTMS) was added to the sol at one-third of the sol by weight, followed by stirring at 110°C for 6 hours for MTMS treatment. After removal of toluene under reduced pressure, the silane surface treated clay (MTMS-clay) was heated at 180°C for 1 hour to remove unreacted chemical entities on the clay surface. The MTMS-clay was again dissolved in toluene to give a sol, mixed with 20 wt% of conventional PP and stirred at 120°C for 30 minutes to homogenize the mixture. The solution was dropped into methanol after cooling at 60°C with stirring and finally washed with methanol to remove unreacted MTMS. The MTMS-clay/PP composite thus prepared was used as a master batch. The master batch after drying was kneaded and shaped in a mold. The preparation procedure is schematically shown in Figure 6. X-RD profiles of the clay in PPCN are broader than those of the corresponding clays indicating that the stacked layer structures of clay were separated into thinner platelets through the composite formation. The properties of the prepared PPCN are strongly controlled by the stiffness of the clay. This fact can be proven by the increased bending modulus of the resulting material. As a result, a nanocomposite with excellent mechanical properties is prepared through preferable separation of very stiff clay layers throughout the matrix. Moreover, it is well known that a clay filler with a high aspect ratio exhibits preferably better reinforcement effects on the resulting composite.

2.3. *Direct melt mixing method*

Direct intercalation of the molten polymer (Figure 7) into layered silicates is a new method that does not involve a solvent. A polymer is heated above the glass transition temperature in either static or flow condition and mixed with the layered silicates. Modified layered silicates

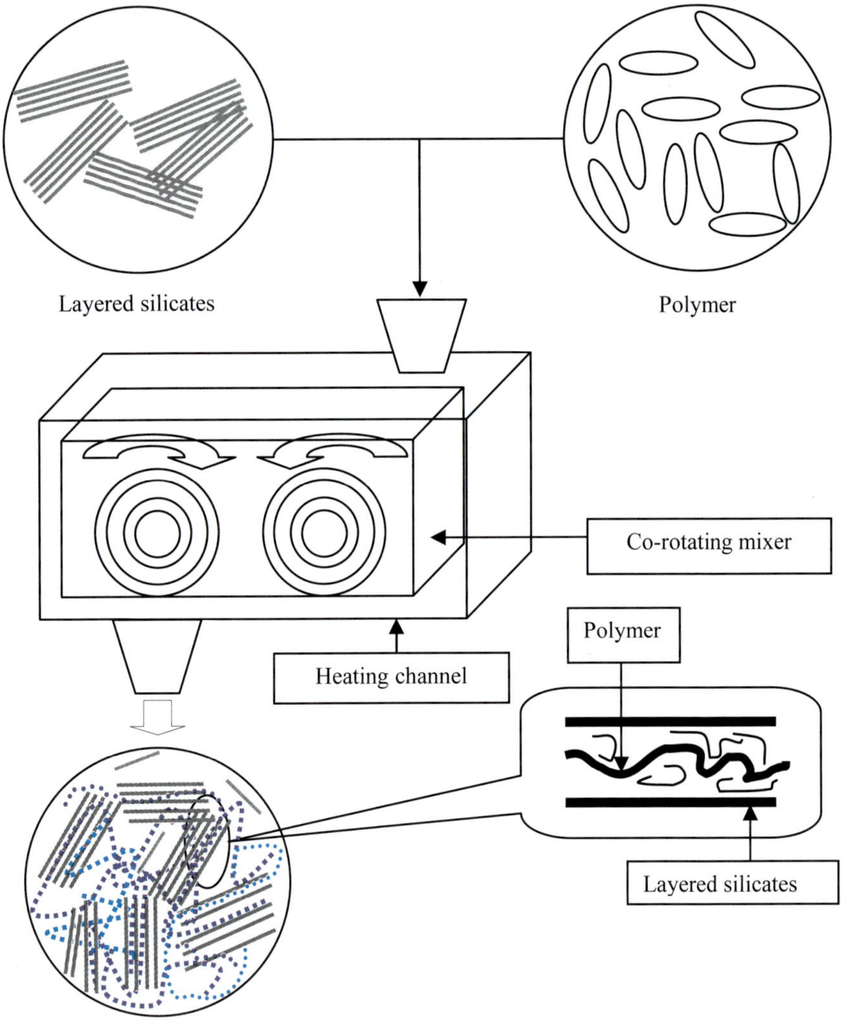

Fig. 7. Schematic representation of polymer/layered silicate nanocomposite obtained by direct melt mixing method.

are usually employed to promote intercalation. The polymer chains from the molten mass spread into the silicate galleries to form either intercalated or delaminated hybrids according to the degree of penetration. The critical factor that determines the degree of exfoliation is probably linked to thermodynamic factors.

A mean-field lattice-based model of polymer intercalation in organically-modified silicates was presented by Vaia and Giannelis [89,90]. The thermodynamic factors which controlled polymer intercalation into the clay galleries were outlined in terms of interplay between entropic and energetic factors. The entropic penalty for confining the polymer inside the clay gallery may be partially compensated by an increase in entropy of the surfactant chains (alkyl chains on organic-modifier) as the clay layers separate. Although the tethered alkyl chains cannot increase their entropy by an increase in translational freedom, the expanded gallery spacing permits increased conformational entropy. Intercalation will be thermodynamically possible when favorable energetic interactions between polymer and organically modified silicates overcome any unfavorable entropic factors. Further separation, i.e. driving exfoliation, depends on maximization of favorable polymer-surface interactions with the organically-modified silicate, as well as minimization of unfavorable interactions with the tethered alkyl chains of the organic-modifier. This explains the use of polymer functionalization with polar moieties. The polar functionalities on polypropylene participate in polar-polar interactions with the silicate surface, decreasing the free energy of the system. If the total free energy of the system is decreased by the combined entropic and energetic factors then intercalation and exfoliation may occur.

Maintenance of spontaneity requires sufficient enthalpy as a driving force, i.e. a negative change in the enthalpy due to the augmented energy of the interaction between the polymer and silicate. This condition is reached by the establishment of weak bonds such as hydrogen bond, dipole-dipole, and van-der waals interactions. The change in enthalpy associated with these interactions is small and more interactions are needed to bring about a substantial total variation.

In general, interplay of entropic and enthalpy factors determine the outcome of whether an organically modified montmorillonite will be dispersed, intercalated or exfoliated in a polymer [91–93]. Dispersion of MMT in a polymer requires sufficiently favorable enthalpy contributions to overcome any entropic penalties. For most polar or polarizable polymers, an alkylammonium surfactant (the most commonly used organic modification) is adequate to offer a sufficient excess enthalpy and thus to promote the nanocomposite formation. However, polypropylene is one of the most widely used polyolefin polymers. Since it has no polar groups in the chain, direct intercalation of PP into the silicate galleries is difficult [4]. Thus, the challenge with polypropylene is to design the system where the polymer-layered silicate interactions are more favourable than the surfactant-layered silicate interactions. There were new approaches to synthesis polypropylene/layered silicate nanocomposites [94–100]. Maleicanhydride grafted polypropylene (PP-g-MA) or PP oligomers modified with hydroxyl groups was mixed with octadecylammonium exchanged MMT, creating a master batch, and then it was subsequently blended with neat polypropylene by using an extruder or mixer. In this way, the PP-g-MA and pretreated silicate layers are effective to promote the swelling of the inter-distance of clays in nanocomposites, and the extrusion is also promoting mixing due to the effect of the mechanical shear. As a result, the structure and the properties of the resulting hybrid materials depended strongly on the processing conditions [95–99]. Obviously, PP-g-MA pre intercalation with very low maleicanhydride content made the master batch so robust that silicate layer did not mix further with neat PP [95,96]. The authors believed that the driving force of the interaction originated from the maleicanhydride group and the oxygen groups of the silicate through hydrogen bonding.

Reichert, *et al.* [99] reported the influence of silicate modification and compatibalizer addition on morphology development and mechanical properties of melt extruded *isotactic* polypropylene (*i*PP) nanocomposites. Synthetic sodium fluoro mica was used as water swellable layered silicate, which was rendered to organophilic by means of cation exchange with various protonated n-alkyl amines. Exfoliation and self assembly of individual silicate layers within the polypropylene matrix was achieved by using amine modifiers in conjugation with PP-g-MA as a compatibilizer. The nature of the interfacial coupling at the interfaces between filler and polypropylene was not explored to substantiate the stress transfer during processing in order to achieve improved exfoliation and stabilization of uniformly dispersed anisotropic nanoparticle.

Okamato, *et al.* [101] prepared PPCN using the method similar to initial efforts of Usuki, *et al.* [94]. In this work, a blend of PP-g-MA and clay was melt extruded at 200°C in a twin screw extruder. PPCNs were prepared with three different amounts of clay content (2, 4 and 7.5 wt% clay content), abbreviated as PPCN2, PPCN4 and PPCN7.5, respectively. They suggested that the contents of clays in the nanocomposites would changed their nano-structure formations such as nearly exfoliated structures with PPCN2, disordered intercalated structures for PPCN4, and the ordered intercalated structures for PPCN7.5.

Liu and Wu [102] reported the preparation of PPCN via grafting-melt compounding by using a new type of co-intercalation organophilic clay which had a larger interlayer spacing than the ordinary organophilic clay modified only by alkylammonium cations and one of the co-intercalation monomer to be attached on the PP backbone by virtue of a grafting reaction. The larger interlayer spacing caused by grafting enhanced the dispersion effect of silicates in PP matrix. The incorporation of clay showed the considerable enhancement in properties demonstrating the reinforcement effects on PP matrix.

Manias, *et al.* [103] carried out the extensive research work on the preparation of PPCN by melt intercalation technique with both intercalated and exfoliation structures. In order to achieve nanocomposite formation, two methods were introduced; (a) nanocomposite formation through PP functionalization and (b) nanocomposite formation through fluoro-organic modified silicates. Random co-polymers of PP with typically 0.5 mol% of functionalized co-monomers with 1 mol% of non-PP block co-polymers [50] were used for the functionalization of PP. Functional groups that promote nonanocomposite formation with modified silicates include p-methylstyrene, maleic anhydride, styrene and hydroxyl containing styrene. For fluoro-organic modification of silicates, a semi fluorinated alkyltrichloro silane was used to re-modify the silicates, rendering it miscible with neat PP [20].

Kaempfer, *et al.* [104] reported the formation of PPCN via melt compounding of *syndiotactic* polypropylene (sPP) with organohectorite obtained via cation exchange of fluorohectorite with octadecylammonium cations, by using a co-rotating twin screw extruder.

The matrix reinforcement was achieved by in-situ formation of silicate nanoparticles via exfoliation combined with simultaneous in-situ encapsulation of the nanosilicates in a thin shell of iPP-g-MA. The resulting aniosotropic core/shell type nanoparticles containing stacks of organohectorite layer as a core and iPP-g-MA as a shell, take a role as very effective new classes of nucleating agents for sPP crystallization, which can be a tailor made material for compatibility requirements of other semicrystalline polymers.

Garcia-Lopez, *et al.* [105] prepared the PPCNs with two different coupling agents, diethyl maleate grafted PP (PP-g-DEM) and commercial maleic anhydride grafted PP (PP-g-MA). Two different clays, a commercial montmorrilonite and sodium bentonite modified with octadecylammonium ions have also been used. This study reveals that mechanical properties of PPCN containing commercial clay are much better than those of PPCNs bearing bentonite. Similarly, the PP-g-MA coupling agent makes PPCNs better performance than PP-g-DEM. This is a consequence of the comparatively low degree of compatibilization between the polymer matrix and layered silicates. The results indicated that clay dispersion and interfacial adhesion were greatly affected by the type of matrix modification.

A novel approach for synthesizing PP-clay nanocomposites by melt intercalation was reported by Y Tang, *et al.* [106]. The authors used only pristine clay and PP instead of organically modified silicate or maleic anhydride modified PP or oligomer PP. This method utilizes a cationic surfactant, such as long chain alkyl ammonium salt. The resulting nanocomposites show the exfoliated structures and improved thermal stability comparing with that of pure PP.

In another approach, Kim, *et al.* [107] proposed a new method of nanocomposite formation by applying electric field on the melts of PP/clay hybrid systems without using any additives such as maleic anhydride or hydroxyl functionalized PP. In this method, electric field was applied on PP/clay melts between the parallel plates of a rheometer. The electrically activated PPCN show enhanced rheological properties with the disappearance of the characteristic peak in X-RD pattern. Electric field was found to facilitate the destruction of the layer-stacking and the separation of the silicate layers with the polymer chains penetrating into the gallery. But, the precise mechanism has not been clarified yet. The process can be applied to conventional extrusion processes with small additional cost and to other polymer/clay nanocomposite systems.

Kato, *et al.* [108] suggested a new method of producing pp/clay nanocomposites without the pre-treatment of clay with an organo cation. This method owed much attention on the nature of the clay which was exfoliated in water. In order to achieve the exfoliation of the clay mineral, water vapour at controlled pressure was injected into a complex of melted polypropylene and clay in a twin screw extruder.

Kim, *et al.* [109] studies the role of an antioxidant in the formation of PPCN. They prepared nanocomposites via simple melt mixing of three components, that is, PP, layered silicates modified with octadecylamine, and antioxidant. PPCN prepared without antioxidant showed immiscible state because PP and clay were quite incompatible each other. On the other hand, PPCN with antioxidant showed the intercalated states with the increased basal spacing of layered silicate. This indicates the role of antioxidant in enhancing the compatibility between PP and organically modified silicates. In rheological behavior, the nanocomposite with antioxidant showed higher storage modulus and complex viscosity than the unfilled PP and they exhibited a monotonous increase in tensile modulus and dynamic storage modulus. These enhanced mechanical properties demonstrate the reinforcing effect of nano-scale dispersed layered silicates in PP matrix.

As a part of these studies on polymer nanocomposites, Lee, *et al.* [110–111] reported the formation of PPCN by *in-situ* grafting-intercalating in melt. The organo-clay was first modified with maleic anhydride in melt with small quantity of a co-swelling agent and an initiator. It was then blended with PP in melt to obtain PP/clay grafting-intercalating composites (GIC). Due to the synergistic effects of graft-intercalation, the clay layers were dispersed well on a nano-scale in the PP matrix. The introduction of clay into the PP matrix improved the thermal stability remarkably and increased the storage modulus (stiffness) of the PP considerably. The narrow space surrounded by the dispersed clay layers and the interaction between the clay layers and macromolecules restricted the motion and enhanced mechanical properties.

More recently, Kotek, *et al.* [112] have prepared the PPCN by melt mixing. In order to increase polypropylene polarity, Cl and $SOCl_2$ groups were introduced by reaction with sulfuryl chloride under UV irradiation. Chlorosulfonated polypropylene was subsequently melt-compounded with organophilized montmorillonite clay to produce a master batch. The master batch was then blended with commercial *isotactic* polypropylene. An organophilized silicate and chlorosulfonated polypropylenes with different degrees of functionalization were examined. The presence of such a chlorosulfonated polypropylene in polypropylene/organoclay nanocomposites strongly promotes the clay dispersion in the matrix. The presence of dispersed clay improves the stiffness and tensile strength of the nanocomposites, but markedly reduces their strain at break.

In order to achieve homogeneous dispersion of clay layers without the use of compatibalizers, Shao, *et al.* [113] introduced new equipment called solid state shear compounding (S^3C) pan mill for the preparation of PPCN. Exfoliation of the silicate layers as well as formation of nanocomposites could be realized as a result of the weak interlayer structure of silicate and the fairly strong shear forces offered by pan-milling. Consequently, the characteristic peak disappeared in the X-RD pattern of the composite. The thickness of the milled silicates was decreased to the nanoscale level, and the exfoliated silicate layers were well dispersed in the PP matrix as confirmed by SEM and TEM. The PPCNs prepared by S^3C have higher melting temperature, thermal degradation temperature, and heat distortion temperature comparing to that of pure PP and the addition of silicates accelerates the crystallization considerably.

3. Characterization Methods

3.1. *X-Ray diffraction (X-RD) analysis*

X-RD is widely employed for the characterization of polymer-layered nanocomposites. The diffraction patterns produced by different types of hybrids are illustrated in Figure 8. The influence of intercalation of the

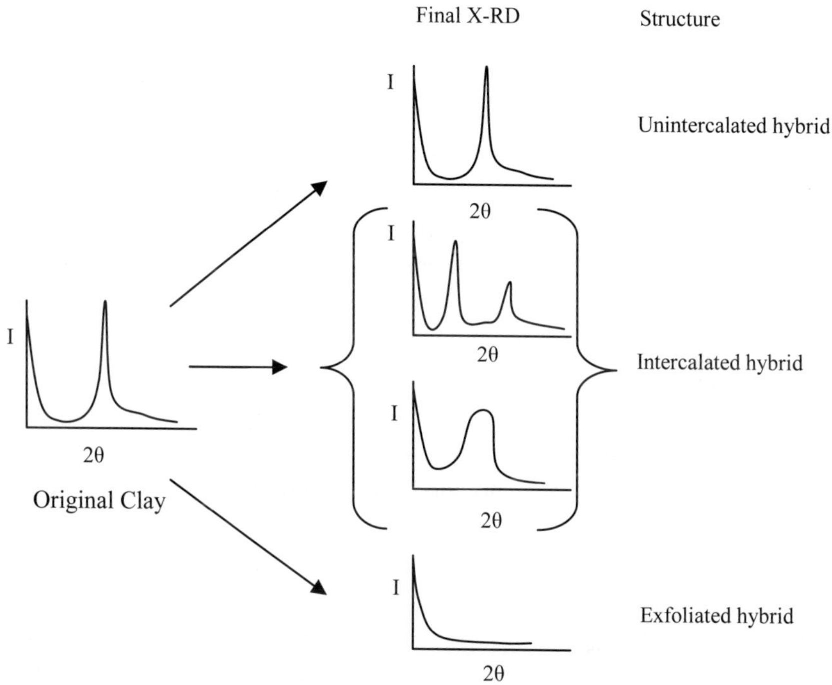

Fig. 8. The schematic explaining the expected X-ray diffraction patterns for various types of hybrid structures.

polymer between the silicate layers is indicated by changes in the intensity and angle of the peak of the basal reflections. Increased order is shown by a decrease in band amplitude. This provides an index of the degree of co-planarity of the silicate layers in an intercalated hybrid.

By monitoring the position of the peak angle and the intensity of the basal reflections from the distributed silicate layers, the structures of nanocomposites (either intercalated or exfoliated) may be identified. For example, in an exfoliated nanocomposite, the extensive layer separation associated with the delamination of the original silicate layers in the polymer matrix results in the eventual disappearance of any peaks in x-ray diffraction patterns. On the other hand, for intercalated nanocomposites, the finite layer expansion associated with the polymer intercalation results in the appearance of a new basal reflection corresponding to the larger basal space. X-RD offers a convenient method to determine the interlayer spacing of the silicate layers in the original layered silicates and in the intercalated nanocomposites (within 1~4 nm).

X-RD data was used to follow changes in the PP/clay nanocomposite at various nanofiller loading levels during the in-situ polymerization with metallocene/clay catalysts [80]. In Figure 9, X-RD result shows that the organoclay precursor has an increased d_{001}- spacing compared with the starting clay. After the polymerization, the X-RD pattern of the organoclay precursor has completely vanished, and a new diffraction peak appears at 5.5° (16Å) along with several weak diffraction peaks from the α-phase of crystalline PP. The intensity of the diffraction peak around 16 Å decreases with the polymerization time and eventually disappears as the polymerization progresses, which indicate the loss of long range order in the organo clay along the c-axis and/or the exfoliation of the stacked organo clay structure.

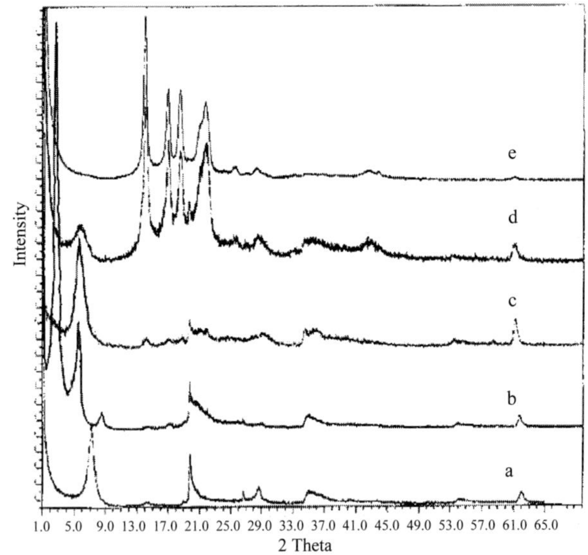

Fig. 9. X-ray diffraction patterns of (a) original clay, (b) organoclay, (c) organoclay with catalysts, (d) clay-PP composites after 2 min of introduction of propylene monomer and (e) clay propylene nanocomposites with about 5 wt% of filler loading [80]. Reproduced from reference [80] by T. Sun and J.M. Garces, *Advanced Materials*, 14, 128 (2002).

X-RD's potential, however, does not extend to the provision of formation concerning the spatial distribution of the silicate in the polymer matrix, nor the shape of the hybrid since all its data are "averaged" over the whole of the sample. Furthermore, some layered silicates do not display well defined basal reflections and it is very difficult to determine the intensity pattern and the shape of the relative peaks. Therefore, conclusions regarding the formation mechanisms and nanostructure of hybrids solely based on X-RD data cannot be other than approximations.

3.2. *Transmission electron microscopy (TEM)*

Transmission electron microscopy allows a qualitative understanding of the internal structure, spatial distribution of the various phases, and defect structure of nanocomposites through direct visualization. However, since a TEM image is obtained from the very small spot in the cross section, a special care must be exercised to ensure that the image can represent general circumstance. Analytical methods to probe the morphology of polypropylene/clay nanocomposites were discussed by Morgan and Gilman [114]. They recommended TEM as the best means to qualitatively assess the degree of dispersion of clay in polymer matrices and to observe the nano-structure of the silicates. Their TEM images demonstrated the structures of layered silicates which will be one of or combination of unintercalated, intercalated, and exfoliated silicates forms. They recommended X-RD to measure d-spacing in ordered-unintercalated stacks and ordered-intercalated tactoids combined with TEM, since X-RD was insufficient for determining disordered and exfoliated clays because these exhibited no reflection peak [115–117].

X-RD patterns and corresponding TEM images of PPCN are shown in Figure 10 and Figure 11 respectively [110]. Figure 10(a) shows the peak near 2.6°, which indicates that the layered silicates are intercalated and flocculated. The same results are observed in the TEM image in Figure 11(a). Many of the silicate layers dispersed with some partially ordered and intercalated stacks. X-RD pattern and TEM image of PPCN prepared by *in-situ* grafting-intercalating in melt are shown in

Figures 10(b) and 11(b). No peaks are observed in Figure 10(b) and it is evidenced by TEM image in Figure 11(b). This proves that both tools are helpful to observe the structures of the nanocomposites of layered silicates.

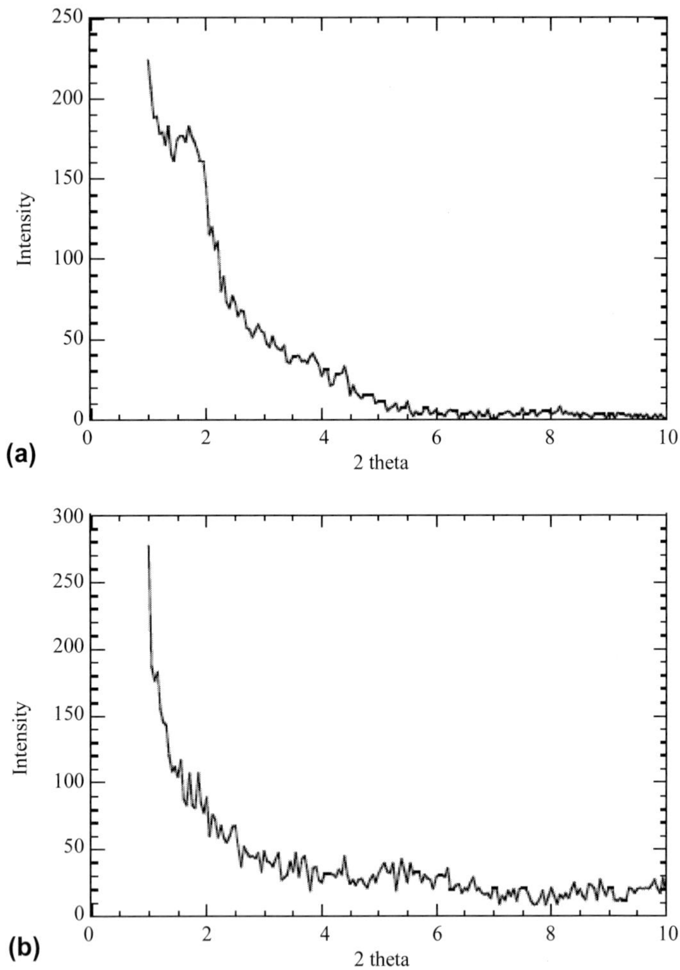

Fig. 10. X-RD Patterns of (a) GIC and (b) PPCN2 [110]. Reproduced from reference [110] by Yu-Qing Zhang, Joong-Hee Lee, John M. Rhee and Kyung Y. Rhee, *Composites Science and Technology*, 64(9), 1383 (2004).

(a)

(b)

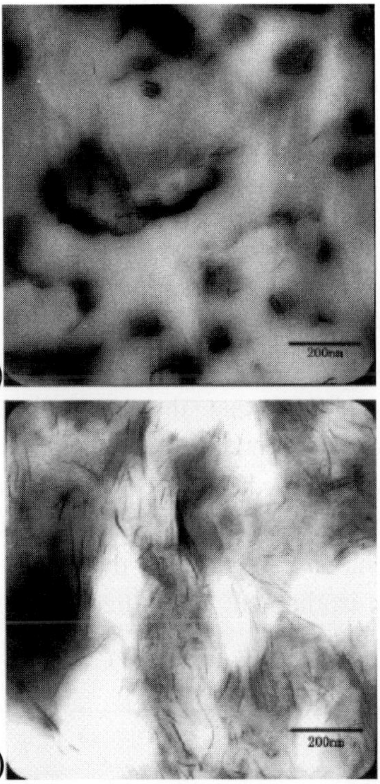

Fig. 11. TEM images of (a) PP/layered silicates and (b) PP-clay nanocomposites prepared by in-situ grafting-intercalating in melt [110]. Reproduced from reference [110] by Yu-Qing Zhang, Joong-Hee Lee, John M. Rhee and Kyung Y. Rhee, *Composites Science and Technology*, 64(9), 1383 (2004).

3.3. *Differential scanning calorimetry (DSC)*

Differential scanning calorimetry (DSC) provides further information concerning intercalation, reaction heat, crystallization and the processing window of the material. The many interactions in the intercalated chains of the polymer formed with the host species greatly reduce its rotational and translational mobility. This situation is similar to that in a segmented polymer, where restrictions on its mobility increase its glass transition temperature (Tg). A similar increase occurs in a nanocomposite due to the elevation of the energy threshold needed for the transition. This effect is readily detected by DSC [58]. DSC thermograms of PPCN with various clay contents are shown in Figure 12. With increasing clay content, the melting point of the composite decreases slightly, comparing to that of pure PP. The melting temperature of PPCN is thought to decrease due to the effects of the compatabilizer on the crystal integrality of PP [111].

In addition, PP is a semi-crystalline polymer. The final properties of composites based on PP in an engineering application are critically dependent on the extent of crystallinity and the nature of crystalline morphology of PP. Therefore, it is necessary to understand the nature, the degree of crystallinity, and the crystallization behavior of PPCNs. It has been recently found that the crystallization behavior and the crystalline morphology of nanocomposites are strongly affected by the presence of the layered silicates. Heterogeneous nucleation was observed in PPCN by several researchers [117,118]. In most cases, the crystallization behavior of these materials was investigated by performing a non-isothermal crystallization test and in a restricted composition range [119–120].

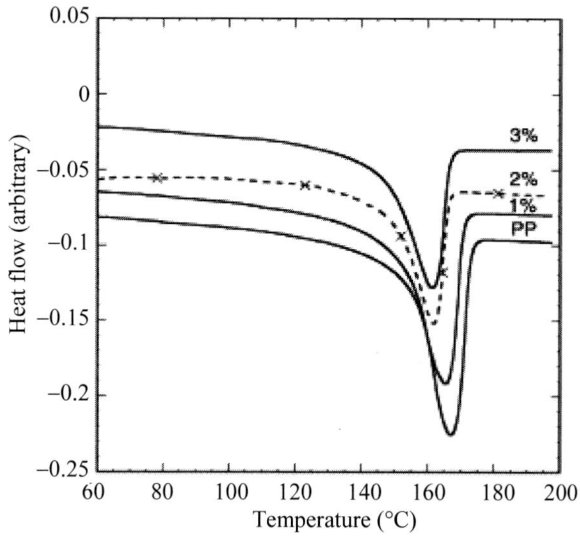

Fig. 12. DSC thermograms of PP and the composites for various clay contents [111]. Reproduced from reference [111] by Yu-Qing Zhang, Joong-Hee Lee, Han-Jong Jang and Chang-Woon Nah, *Composites Part B: Eng*, 35(2), 133 (2004).

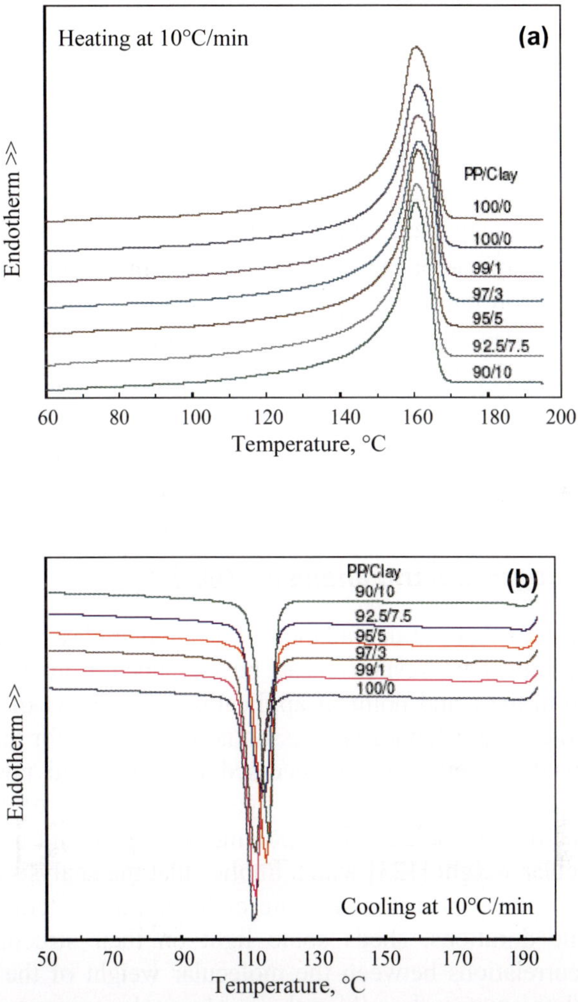

Fig. 13. DSC curves for the (a) heating and (b) cooling of PP/clay nanocomposites [121]. Reproduced from reference [121] by S. L. Liu, X. H. Lu, F. Y. Liew, S. H. Lim and M. S.Yong, On Line Report.

DSC analysis was adopted to evaluate the effects of clay addition on the melting and crystallization behavior of PP [121]. Figure 13 shows the heating and cooling curves of the PPCN. It also shows that melting temperature is not influenced with the presence of organoclay.

However, the crystallization temperature is shifted to higher temperature with clay content. The crystallization temperature of neat PP is about 114.5°C, while the crystallization temperature of 10 wt% clay filled nanocomposite is about 117.8°C. The increase in crystallization temperature in PPCN is due to the presence of clay, which promotes the crystallization of PP matrix. Similar to the melting behavior of PPCN, the addition of clay has little influence on the melting processes of PP/PP-g-MA/clay nanocomposites as shown in Figure 14. However, the crystallization temperature is further increased with the presence of PP-g-MA and clay. The crystallization temperature of 1 wt% PP-g-MA filled PPCN is 119.4°C; while 10 wt% PP-g-MA filled PPCN is about 121.5°C. The higher crystallization temperatures in PP/PP-g-MA/clay nanocomposites over PP/clay nanocomposites could be caused by the high crystallization temperature of PP-g-MA, which was about 121.6°C.

3.4. *Thermogravimetric analysis (TGA)*

The influence of thermal degradation of a polymer/clay nanocomposites on its time-dependent response can be explained from two points of view; (i) from the stand point of applications, the effect of degradation on the visco-elastic behavior is of essential importance for the prediction of mechanical properties of reprocessed industrial and post-consumer plastic wastes [122] and (ii) from the point of view of fundamental research, thermal degradation of a polymer is tantamount to a reduction in its molecular weight [123] which implies that the analysis of the time-dependent response of nanocomposite melts at various temperatures for various time durations, sheds some light on their structure–property relations (correlations between the molecular weight of the matrix and the material parameters describing the mechanical response).

The kinetics of thermal and thermo-oxidative degradation of polypropylene has attracted substantial attention in the past decade [124–127]. Thermal degradation and stability of nanocomposites with a polypropylene matrix and MMT filler are recently reported [128–133]. Some relations between the viscoelastic response of *isotactic* PP–MMT nanocomposites and their structure have been established by many researchers [134–141]. The reported literature indicates that the

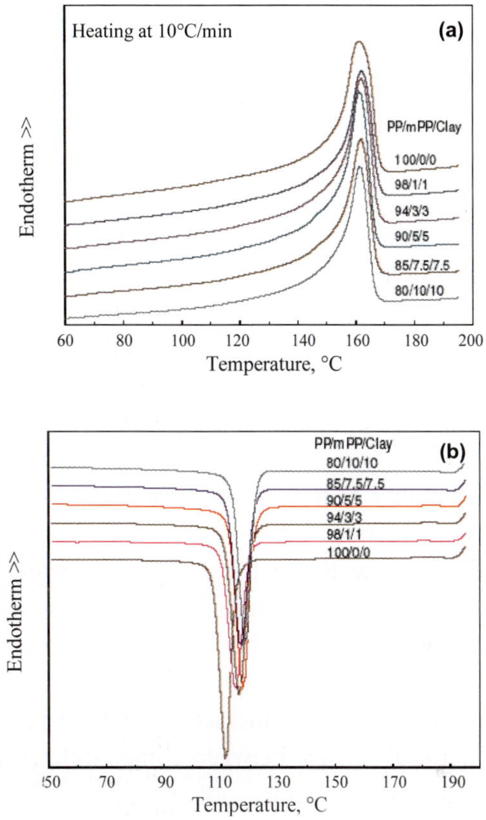

Fig. 14. DSC curves for the (a) heating and (b) cooling of PP/MA-g-PP/clay nanocomposites [121]. Reproduced from reference [121] by S. L. Liu, X. H. Lu, F. Y. Liew, S. H. Lim and M. S. Yong, On Line Report http://www.simtech.astar.edu.sg/Research/Technical Reports/TR04FT04.pdf

incorporation of clay into the polypropylene matrix results in a significant improvement in the thermal stability of the composites. The enhancement in the thermal stability of the polymer/clay composites has been attributed to the decreased permeability of oxygen caused by the exfoliated clay in PPCN.

The thermal degradation behaviour of PPCN was studied by Zanetti, *et al.* [141] using an isothermal and dynamic thermogravimetry. They suggested that the oxygen charring action and scavenging effect in the nanocomposite increase as the volatilization proceeds. In the nanocomposite, a catalytic role is played by the intimate polymer–silicate contact, which may further favor the oxidative dehydrogenation–crosslinking–charring process, and this would act as an efficient barrier to oxygen diffusion towards the bulk of the polymer. Surface polymer molecules trapped within the silicate are thus brought into close contact with oxygen to produce the thermally and oxidatively stable charred material providing a new char-layered silicate nanocomposite acting as an effective surface shield [141].

The thermal decomposition behavior of virgin PP and PPCN is represented in Figure 15. From the reported results [142], it is clear that the thermal stability of the base polymer is increased from the initial decomposition temperature of 225°C to 268°C because of nano-clay loading, with final weight loss of 95% and 90%, respectively. The increased thermal stability of nanocomposites might be attributed to the organic-inorganic interaction between the polymer and nano-clay. Comparative studies on the thermal stability and weight loss revealed that the weight loss of the uncompatibilized nanocomposite at final decomposition temperature was higher than that of the compatibilized nanocomposite. The enhanced thermal stability of the compatibilized hybrid is attributed to the strong interaction of base polymer and clay surface through chemical linkage between compatibilizer and nano-clay, which in turns mediates the surface polarity of the clay and the host polymer at the clay-polymer interface.

Thermogravimetric analysis of neat PP and PP/MA-g-PP/clay nanocomposite shows that PP volatilizes completely at 400°C through a radical chain process propagated by carbon-centered radicals originated by carbon–carbon bond scission as shown in Figure 16. But, PP/MA-g-PP/clay nanocomposites are more stable than pure PP by 50°C [143]. This effect is due to the barrier effect of the silicate nanolayers which operate in the nanocomposite level against oxygen diffusion, shielding the polymer from its action. The silicate nanostructure is an efficient barrier to oxygen diffusion towards the native polymer. The surface polymer molecules trapped within the silicate are thus brought into close contact with oxygen and the catalytic silicate layers to produce thermally and oxidative stable carbonized structures.

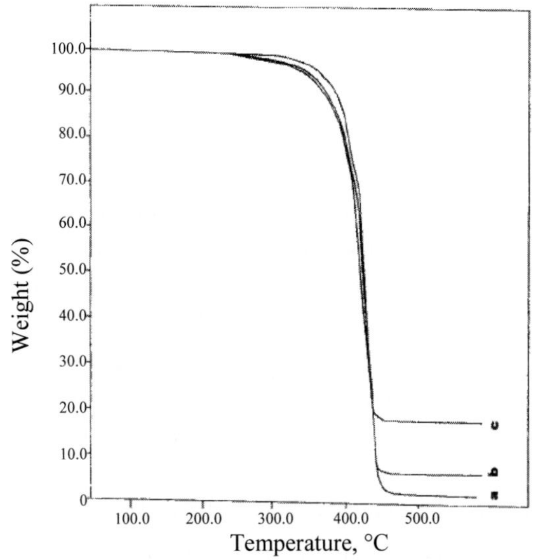

Fig. 15. TGA thermogram of (a) pure polypropylene, (b) polypropylene/clay, and (c) polypropylene/PP-g-MA/ clay [142]. Reproduced from reference [142] by S. Parija, S. K. Nayak, S. K. Verma and S. S. Thripaty, *Polymer Composites*, 25 (6), 646 (2004).

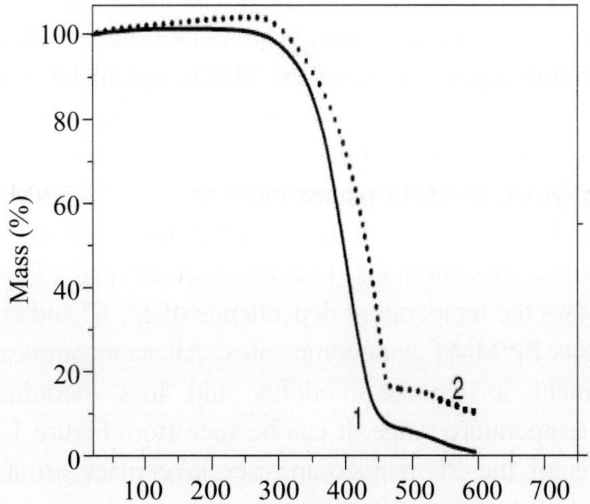

Fig. 16. TGA curves of (1) PP and (2) PP/MA-g-PP/Clay nanocomposite in air at a heating rate of 10 K min^{-1} [143]. Reproduced from reference [143] by S. M. Lomakin, I. L. Dubnikova, S. M. Berezina and G. E.Zailkov, *Polymer International*, 54(7), 999 (2005).

4. Physical and Mechanical Properties

4.1. *Dynamic mechanical analysis (DMA)*

DMA results are expressed by three main parameters; (1) the storage modulus (E') corresponding to the elastic response to the deformation, (2) the loss modulus (E''), corresponding to the plastic response to the deformation, and (3) tan δ (i.e., the (E'/E'') ratio), which is useful for determining the occurrence of molecular mobility transitions such as the glass transition temperature (T_g).

Early studies by Toyota workers, led by Hasegawa, focused on the increase of the dynamic storage modulus by introducing layered silicates to PP (95). PP/PP-g-MA/C18-MMT nanocomposites were formed, e.g. by blending 5 wt% MMT and 22 wt% PP-g-MA with isotactic PP. It was found that only large concentrations of PP-g-MA afforded reasonable clay dispersion, however, even in these type blends partial exfoliation and partial intercalation (as intercalated tactoids of MMT) were observed in TEM measurements. In the nanocomposite blend cited above the storage modulus was 1.8 times higher than that of the corresponding PP at 80°C. Hasegawa and coworkers prepared PP-g-MA (0.2 wt% MA)/C18-mmt (5.3 wt% MMT) nanocomposites, which exhibited apparently complete exfoliation of the MMT, and had 1.9 times higher tensile modulus than the PP-g-MA matrix polymer [96]. However, a simple PP/C18-MMT composite had a slightly higher tensile modulus than the PP-g-MA/C18-MMT nanocomposite.

DMA has been used to study the temperature dependence of the storage modulus, loss modulus, and tan δ of PP and PP/MMT [144]. Figure 17 shows the temperature dependence of E', E'' and tan δ for pure PP and various PP/MMT nanocomposites. All nanocomposites showed an enhancement in storage modulus and loss modulus over the investigated temperature range. It can be seen from Figure 17(c) that the T_g relaxations of the PP nanocomposites take place around 13–14°C compared with that of pure PP (20°C). In addition, the magnitudes of tan δ of the PP nanocomposites are decreased substantially compared with that of pure PP. The above observations may also be correlated with the intercalated nanostructures formed in the composites. The intercalated

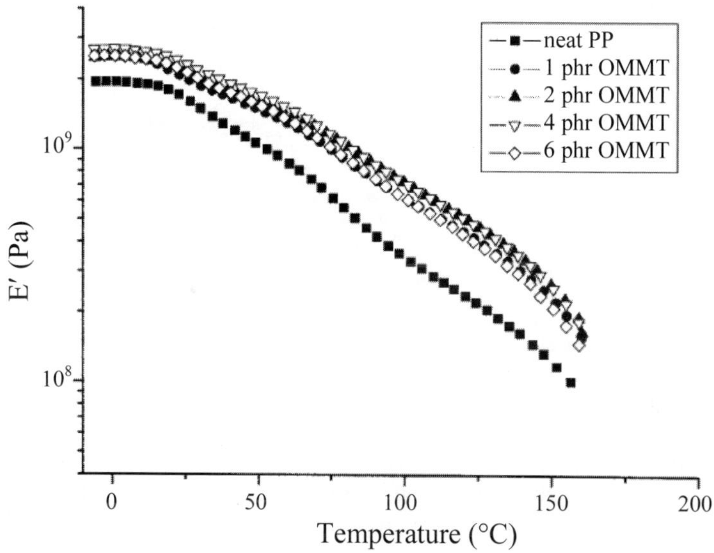
E′ (Pa)
10^9
10^8
0
50
100
150
200
Temperature (°C)
neat PP
1 phr OMMT
2 phr OMMT
4 phr OMMT
6 phr OMMT

(a)

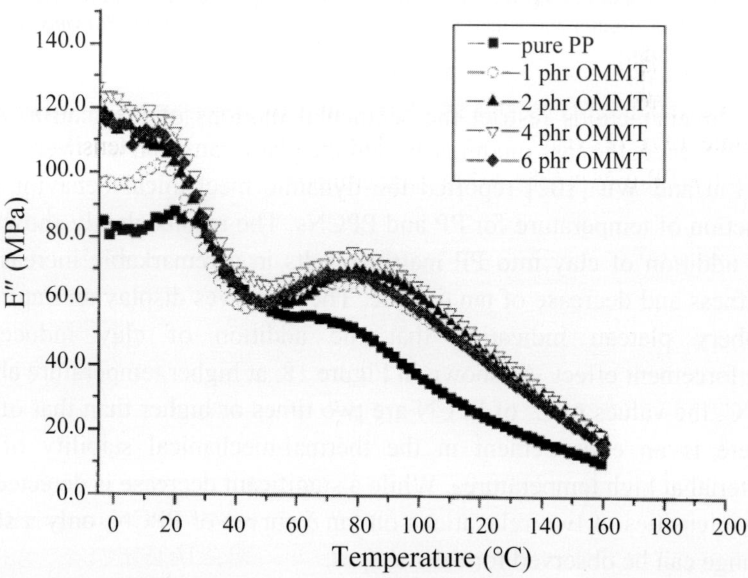
E″ (MPa)
140.0
120.0
100.0
80.0
60.0
40.0
20.0
0.0
0
20
40
60
80
100
120
140
160
180
200
Temperature (°C)
pure PP
1 phr OMMT
2 phr OMMT
4 phr OMMT
6 phr OMMT

(b)

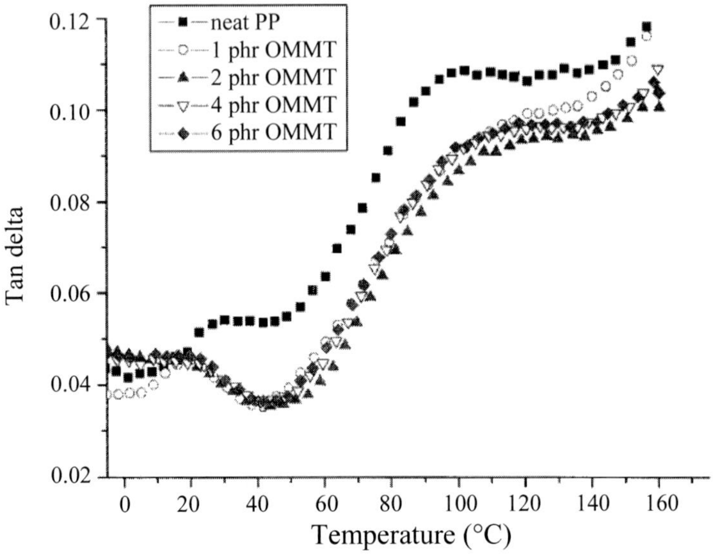

(c)

Fig. 17. Temperature dependence of (a) storage modulus E', (b) loss modulus E'' and (c) tan δ for PP and PP/OMMT nanocomposites [144]. Reproduced from reference [144] by Chao Ding, Demin Jia, Hui He, Baochun Guo and Haoqun Hong, *Polymer Testing*, 24(1), 94 (2005).

clay layers/tactoids restrict the segmental motions of the matrix. As a consequence, the loss modulus around the glass transition is lowered.

Liu and Wu [102] reported the dynamic mechanical behavior as a function of temperature for PP and PPCNs. The results clearly show that the addition of clay into PP matrix results in a remarkable increase of stiffness and decrease of tan δ value. The E' curves display an improved rubbery plateau indicating that the addition of clay induces a reinforcement effect. As shown in Figure 18, at higher temperature above 80°C, the values of E' of PPCN are two times or higher than that of PP. There is an enhancement in the thermal-mechanical stability of the material at high temperatures. While a significant decrease is detected for the intensities of two relaxations on tan δ curves of PPCN, only a slight change can be observed for their shapes.

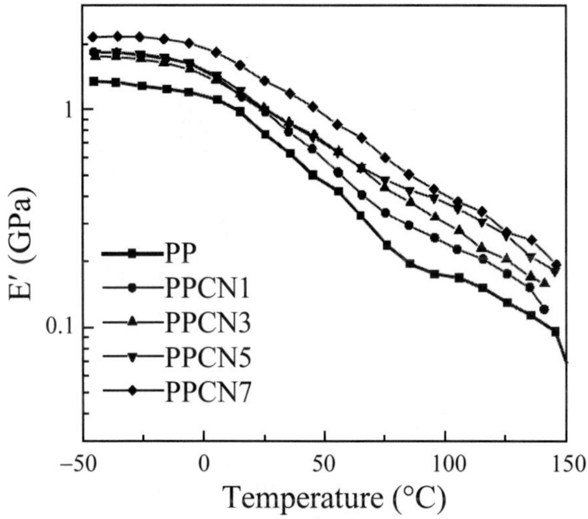
E′ (GPa)
1
0.1
PP
PPCN1
PPCN3
PPCN5
PPCN7
–50
0
50
100
150
Temperature (°C)

(a)

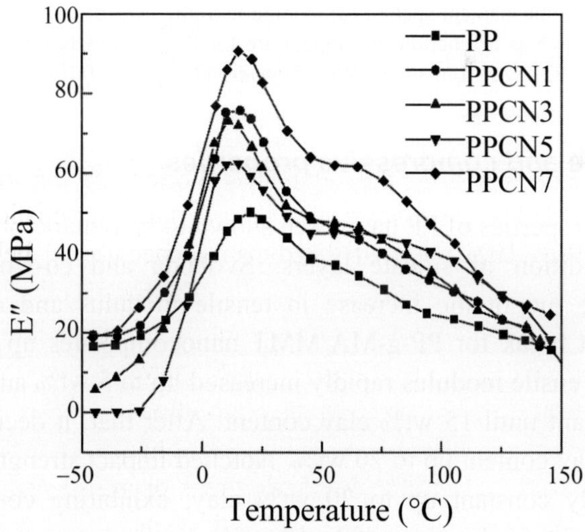
E″ (MPa)
100
80
60
40
20
0
PP
PPCN1
PPCN3
PPCN5
PPCN7
–50
0
50
100
150
Temperature (°C)

(b)

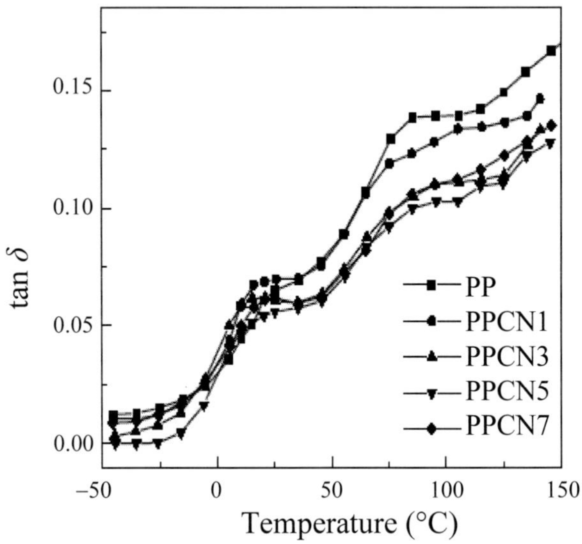

(c)

Fig. 18. Dynamic mechanical spectra (a) storage modulus E', (b) loss modulus E'', and (c) loss factor tan δ as a function of temperature for PP and PPCN [102]. Reproduced from reference [102] by X. Liu and Q. Wu, *Polymer*, 42, 10013 (2001).

4.2. *Tensile and compressive properties*

The tensile properties of PP have been shown to be remarkably improved with the addition of silicate layers. Svoboda and co-workers have observed the monotonic increase in tensile modulus and decrease in elongation at break for PP-g-MA/MMT nanocomposites up to 20 wt% clay [137]. Tensile modulus rapidly increased up to 5 wt% and remained almost constant until 15 wt% clay content. After that, it decreased with increasing clay content up to 20 wt%. Notched impact strength remained approximately constant up to 20 wt% clay, exhibiting comparatively modest increases in the range of 3 ~10 wt% clay.

Similar results were also reported by Liu and Wu [102]. The tensile strength of PPCN rapidly increases with increasing the clay content from 0 to 5 wt% as shown in Figure 19(a). The tensile strength of the nanocomposite containing 5 wt% clay (PPCN5) is 32.7 MPa. But the increasing trend is less when the clay content increases beyond 5 wt%. The tensile strength of the nancomposite for 7 wt% clay loading (PPCN7) is 33.1 MPa. Thus, 27% improvement in tensile strength is obtained in PPCN7 compared with pure PP. A similar trend is observed in tensile modulus as shown in Figure 19(b). Where 42% increase in tensile modulus is obtained with 7 wt% clay loading.

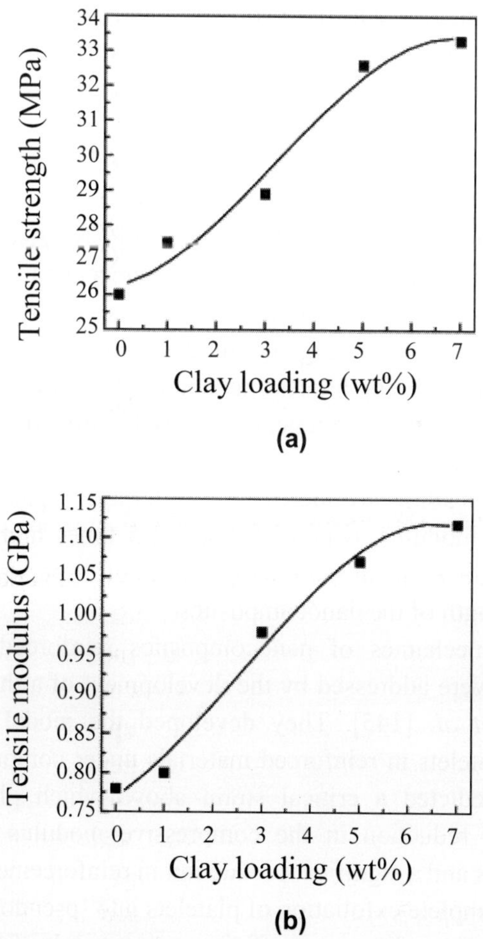

Fig. 19. Tensile properties of PPCNs as a function of clay content; (a) tensile strength and (b) tensile modulus of PPCN [102]. Reproduced with permission from reference [102] by X. Liu and Q. Wu., *Polymer*, 42, 10013 (2001).

Zhou, *et al.* [120] studied the uniaxial tensile behavior for the polypropylene and clay filled polypropylene nanocomposites at different strain rates, and the results are shown in Figure 20. It can be observed in these figures that the stress–strain relationships of polypropylene and its composites are non-linear even at lower strains than the yield strain. As shown in the figure, the effects of the strain rate in both materials on the tensile modulus and yield strength are similar, but the nanocomposite has little more strain rate sensitivity on the yield strength. Generally, the elastic modulus and yield strength increase with increasing strain rate. However, the effect of strain rate on the yield strain is relatively small for the range of strain rates tested.

Most recently, the extensive work by Kotek, *et al.* systematically defines the dependencies on compatibalizer functionality and organic modification on the tensile properties of PPCN. According to their studies, a considerable tensile property enhancement can be achieved only when appropriate compatibilizers are used to retreat the clay in conjugation with specific organic modification for the clay. Similar materials under different processing conditions showed much smaller improvements in the practical material properties [112]. The nanocomposites prepared using higher degrees of functionalization in compatibilizer leading to increase in interlayer spacing exhibit the highest tensile modulus, which is almost 1.5 times higher than that of pure polypropylene. In addition, the presence of dispersed clay improves the tensile strength of the nanocomposites.

The micromechanics of nanocomposites reinforced with platelet-shaped fillers were addressed by the development of a theoretical model by Douglas, *et al.* [145]. They developed the model to predict the buckling of platelets in reinforced materials under compressive loading. The model predicted a critical strain above which platelet buckling would lead to reduction in the compressive modulus relative to the tensile modulus and a significant reduction in reinforcement efficiency as a result of incomplete exfoliation of platelets into 'pseudoparticle' stacks. The reduction of reinforcement efficiency as a result of the deviation of the platelet from perfect biaxial in-plane orientation was also predicted by the model.

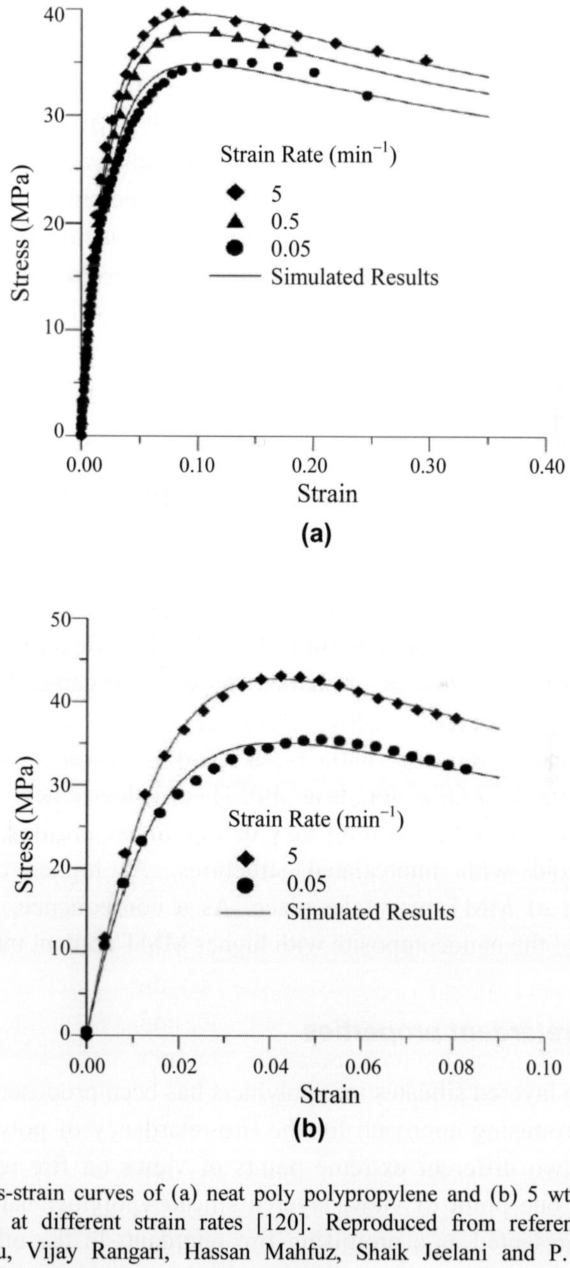

Fig. 20. Stress-strain curves of (a) neat poly polypropylene and (b) 5 wt% clay filled polypropylene at different strain rates [120]. Reproduced from reference [120] by Yuanxin Zhou, Vijay Rangari, Hassan Mahfuz, Shaik Jeelani and P. K. Mallick, *Materials Science & Engineering A*, 402, 109 (2005).

The common finding from many researches on the mechanical properties of polypropylene nanocomposites is that clay loading into PP makes somewhat improved properties, but does not improve them to the expected level. A large concentration of PP-g-MA is necessary to produce intercalated and exfoliated compounds with polypropylene. Otherwise, nanocomposites exhibit imperfect dispersion of the clay with morphologies containing partial intercalated tactoids and partial exfoliated clay platelets. The improvements of tensile properties of the nanocomposites are typically not sufficient to justify the additional costs to fabricate the composites until now.

4.3. *Flexural properties*

Reports on the flexural properties of PP/layered silicate nanocomposites are very limited. Many researchers found that flexural properties increased with increase in clay content in PP matrix [146–148].

Ding, *et al.* [146] reported the detailed measurement of flexural properties of neat PP and various PPCNs. The flexural modulus of nanocomposites increases remarkably with organically modified montmorillonite content as shown in Figure 21. The flexural modulus of PP/MMT nanocomposite with 6 phr MMT increases to 2.41 GPa compared to 1.27 GPa for neat PP. The enhancement in flexural properties at low MMT loading may be due to the uniformly dispersed MMT tactoid with intercalated structures. At higher content, the aggregation of MMT may take place. As a consequence, the flexural properties of the nanocomposite with higher MMT content may decrease.

4.4. *Fire retardant properties*

Introducing layered silicates into polymers has been proposed as a totally new and promising approach for the fire retardancy of polymers [149]. There are two different extreme points of views on fire retardancy of PPCNs. In one point of view, layered silicates/polymer nanocomposite has been presented as a promising fire retardant. In the other point of view, layered silicates nanocomposites should not be called fire retardant

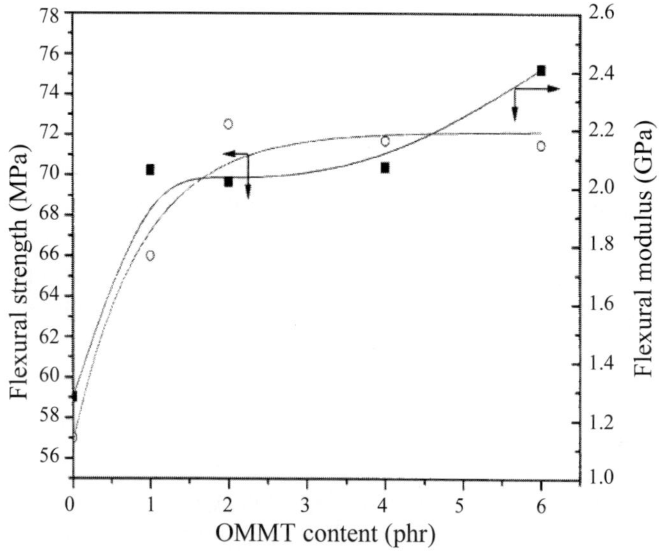

Fig 21. Effects of OMMT loading on flexural properties of PP/OMMT nanocomposites [146]. Reproduced from reference [146] by Chao Ding, Demin Jia, Hui He, Baochun Guo and Haoqun Hong, *Polymer Testing*, 24(1), 94 (2005).

at all. It should be noted that both views come from the similar studies and results, but interpreted with respect to different tasks and standards [150]. A fire retardant is anything that delays the start or decreases the propagation of a fire, whatever the size of the effect is. In other words, a fire retardant is a commercial product which improves a polymer material so that it can pass fire regulations and fire tests without too much influence on other properties.

The cone calorimeter is one of the most effective bench-scale methods for studying the flammability properties of materials. The cone calorimeter measures the fire-relevant properties such as heat release rate (HRR) and carbon monoxide yield. In heat release rate curve, in particular, the peak HRR has been found to be the most important parameter to evaluate fire safety [151].

According to Gilman and co-workers [152], the mechanism of flame-retardant behavior has been reported to be based on the formation of a carbonaceous-char layer on the outer surface of the composite. The surface char layer was found to have a high concentration of MMT layers and was claimed to become an excellent insulator and mass transport barrier, slowing oxygen diffusion into the surface and the escape of combustion products out of the surface. They reported that the heat release rate (HRR) was 75% lower for PP-g-MA/MMT (4 wt%) nanocomposite compared to the neat PP-g-MA matrix. Also, the mass loss rate (MLR) of the same PP-g-MA/MMT (4 wt%) nanocomposite was about one quarter of that of the PP-g-MA. They further found that the type of layered silicate, degree of dispersion and the processing conditions affected the magnitude of the flammability reduction in the nanocomposites.

Similar results were obtained and explained by a similar mechanism, based on the formation of a char layer by Wagenknecht, *et al.* [153]. A comparison was made between nanocomposites, based on a PP-g-MA matrix, containing 2.5 to 10 wt% clay, with talc composites having the same PP-g-MA matrix. It was found that the limiting oxygen index (LOI) was increased for the PP-g-MA/clay nanocomposites over the talc composites. Further, the time from ignition to the first occurrence of dripping was shown to increase with clay loading. Burning was reported to be much less intensive in the nanocomposites in comparison with pure polymer. Also, ash layers of ~1 mm in thickness were formed on the nanocomposites, which were correlated with the minimization of burning intensity and dripping.

In other studies, it was similarly reported that HRR and time-to-ignition were reduced in PP/clay nanocomposites [154,155]. The detailed thermal behavior and the formation of char layers on PP/PP-g-MA/clay nanocomposites were reported [156)]. The thermal oxidation process of the polymer was significantly slowed in the nanocomposites exhibiting high char yield. The inhibition of oxidation was explained on the basis of a physical barrier effect of the surface char layer, by ablative reassembling of the silicate and by a chemical catalytic action, due to the silicate and to the amine-modifier on the silicate. Fire retardant studies on polypropylene-layered silicate nanocomposites [157,158] have become even more attractive because of recent demonstrations of their flame retardant properties, namely a significant decrease in the peak heat release rate (peak HRR) and a decrease in the rate of mass loss rate (MLR) during combustion in a cone calorimeter.

Table 1. Cone calorimeter results for PP reactive blending (159).

Sample	t_{ign}(s)	PHRR(kW/m^2)	t_{PHRR} (s)	SEA(m^2/kg)	MLR(g/s m^2)
Brabender mixer					
PP	49	1642	103	290	22
PP-VB16-clay	27	1246(24)	102	283	22
PP-10A-clay	32	1084(34)	108	281	21
PP-MA	27	1091(34)	100	233	24
PS-Na-clay-MA-VB16 salt	30	1163(29)	102	271	22
PS-Na-clay-MA-10A salt	26	1136(31)	98	300	20
PP-Na-10A salt	23	1101(33)	103	303	22
Internal mixer					
PP-VB16-clay	48	1420(14)	108	386	22
PP-VB16-clay-MA	46	1456(11)	106	354	23
PP-Na-clay-MA-VB16-salt	46	1278(22)	114	380	20
Twin-screw extruder					
PP-VB16-clay	32	1344(18)	108	316	21
PP-VB16-clay-MA	34	1197(27)	100	327	20
PS-Na-clay-MA-VB16 salt	34	1351(18)	105	327	20

Reproduced from reference (159) by Dongyan Wang and Charles A. Wilkie., *Polymer Degradation and Stability*. 80(10), pp 171-182, (2003).

The cone calorimetric results for the polypropylene/layered silicate nanocomposites processed in two different techniques (i.e. internal mixer and the twin-screw extruder) are shown in Table 1 [159]. The peak HRR for this graft copolymer is routinely lower than that of pure PP. The value for pure PP is 1600 kW/m^2, while PP-g-MA gives 1046 kW/m^2, both at 35 kW/m^2 irradiance. Recently, same researchers report an authentic, mixed intercalated-exfoliated nanocomposite of polypropylene. The reduction in peak HRR for PP-g-MA is 54%, while the best value that has been obtained for PP itself is 20%. In this work, the typical values are in the range from 11% up to 34%. The mass loss rate does not change for all samples and the smoke is also relatively constant but there is a variation in time for ignition [160]. The greatest decrease in time for ignition occurs for systems which show the greatest reduction in peak HRR. This may be a criterion for nanocomposite formation, a significant reduction in time to ignite, and a significant reduction in peak HRR.

Wilkie, *et al.* [161] reported the HRR curves for the pure polymer PP and its nanocomposites are presented in Figure 22. The peak HRR of PPCN shows around 10% reduction compared with the pure PP, when there is less than 5% inorganic clay presence. When the loading of inorganic clay reaches 5%, PPCNs show about 40% reduction in peak HRR. These values obtained for the PPCNs are an indication of good dispersion, in agreement with the X-RD patterns and TEM images.

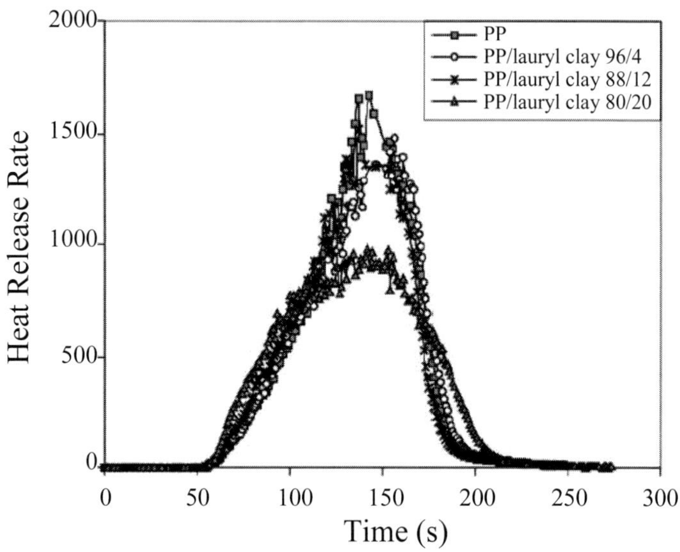

Fig. 22. Heat release rate curves for polypropylene and polypropylene/lauryl clay nanocomposites at 35 kW/m^2 heat flux [161]. Reproduced from reference [161] by Jinguo Zang, David D. Jiang and Charles A. Wilkie, *Thermochimica Acta*, 430, 107 (2005).

4.5. *Impact properties*

The available literature on impact strength of polypropylene/layered silicate nanocomposites indicates that the impact strength of nanocomposites is roughly doubled when compared with that of pure PP. This results mainly from the good dispersion of layered silicates in the composites and its ultra-fine morphology [162–164].

Mulhaupt and coworkers made an extensive study of PP/PP-g-MA/synthetic clay compounds prepared with fluoromica-modified synthetic clays, organically modified with alkyl chains ranging in length from C4 to C18 [64]. They found that the combination of 10 wt% of modified clay with C16 organic-modifier and 20 wt% of PP-g-MA with 4.2 wt% MA was sufficient to achieve an effective reinforcement. This

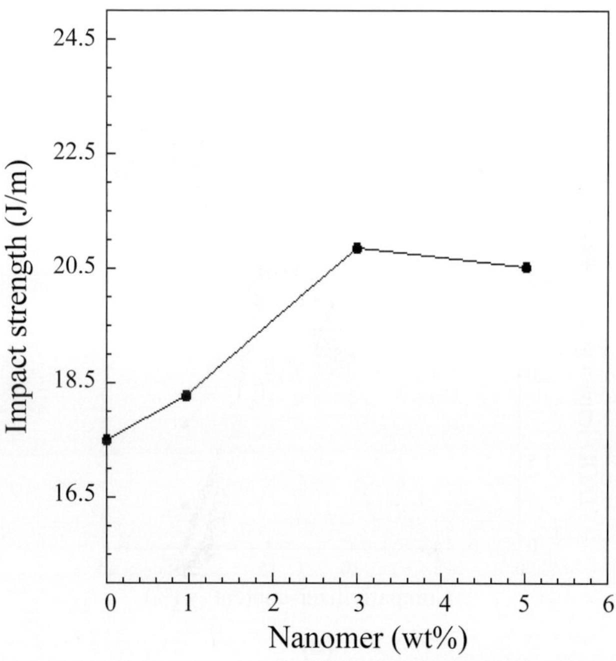

Fig. 23. Effect of nanomer loading on impact strength [142]. Reproduced from reference [142] by S. Parija, S. K. Nayak, S. K. Verma and S. S. Thripaty, *Polymer Composites*, 25(6), 646 (2004).

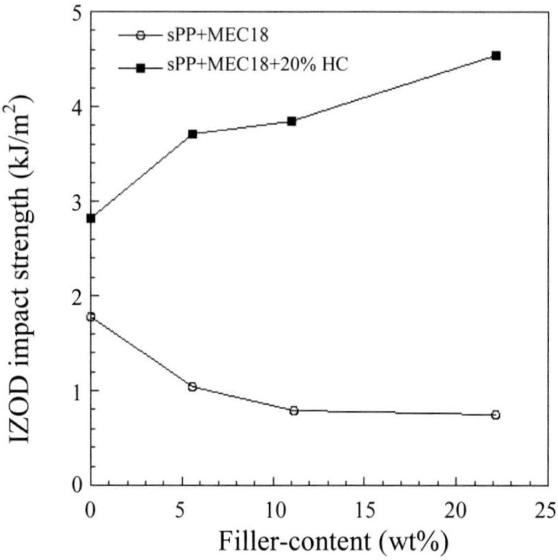

(a)

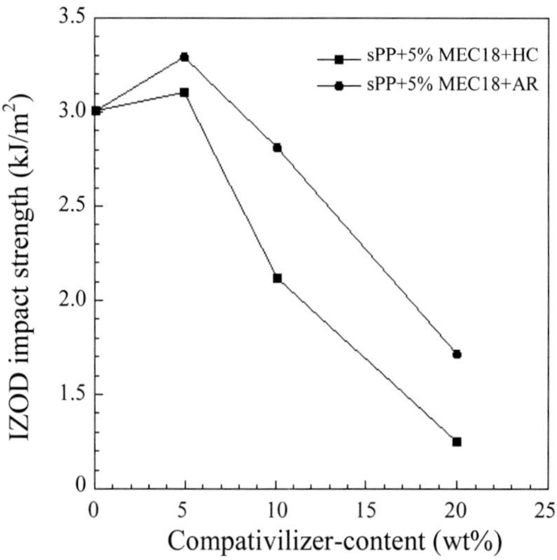

(b)

Fig. 24. Notched IZOD impact strength of sPP/organohectorite compounds with low (●) and high (■) molecular weight PP-*g*-MA compatibilizer [104]. Reproduced from reference [104] by D. Kaempfer, R. Thomann and R. Mulhaupt., *Polymer,* 43, 2909 (2002).

nanocomposite exhibited exfoliated and well-dispersed silicate layers in the TEM images and an increase in Young's modulus from 1,490 to 3,460 MPa and yield stress from 33 to 44 MPa with respect to bulk PP. However, notched Izod impact strength decreased from 1.7 to 1.4 KJ/m^2 with respect to bulk PP.

Figure 23 shows the effect of clay content on the impact strength of PPCNs. The impact strength of nanocomposites is increased by 4.81%, 19.51% and 17.48% for 1%, 3% and 5% clay loadings respectively in comparison to that of pure PP [142]. This increment in impact strength of nanocomposites may be due to the good dispersion of nano-scale fillers in the matrix network owing to inter diffusion of polymer matrix through the interlayer galleries of modified montmorillonite. The decreased impact strength at higher clay loading (5%) might be attributed to the detrimental effect of nanomer agglomerate due to the non-uniform dispersion of clay layers in the polymer matrix [165].

The influence of the molecular weight of the compatibilizer on the izod impact strength was studied by Kaempfer, *et al.* [104] with two different molecular weight PP-g-MA compatibilizers. As shown in Figure 24, better impact resistance was observed with lower molecular weight PP-g-MA compatibilizer (denoted as AR) than that of higher molecular weight (denoted as HC).

4.6. *Gas barrier properties.*

Giuliana, *et al.* [166] reported the transport properties of organic vapors in nanocomposites of organophilic layered silicates and *syndiotactic* polypropylene. The sorption of dichloromethane and *n*-pentane was reduced compared to the unfilled sPP, but there was no significant difference between the samples having different clay concentrations. The permeability for both vapors was reduced. In the case of dichloromethane, it showed a linear decrease until 10 wt% of the inorganic phase and a plateau from 10 to 20 wt% of organically modified layered silicates (OLS). The permeability for *n*-pentane showed a significant reduction up to 5% then a linear decrease at 5–20 wt% of OLS. In both cases, the improvement of barrier properties, in terms of

Table 2. Permeability (in Barrier) to CO_2, O_2 and N_2 gases of pure PP-g-MA and PP-g-MA-nanocomposite at different temperatures [167].

Temperature(K)	CO_2		O_2		N_2	
	PP-g-MA	PP-g-MA-nanocomposite	PP-g-MA	PP-g-MA-nanocomposite	PP-g-MA	PP-g-MA-nanocomposite
308	6.4	3.2	1.9	1	0.6	0.3
318	9.9	1.8	3.1	–	1.1	0.8
328	16.5	7.9	5.3	3.5	4.1	1.5
338	23.9	12.9	8.1	4.3	2.8	1.8

Reproduced from reference [167] by Adams Tidjani., *Polymer Degradation and Stability*., 87(1), pp 43-49, (2005).

permeability, was mainly caused by the reduced diffusion. Table 2 indicates the permeability of different gases (N_2, O_2 and CO_2) carried out on the investigated samples, where the permeability is given as barrier properties [167]. At the temperature of 308 K, a significant reduction (50%) of permeability of the nanocomposite compared to PP-*g*-MA was observed with acceptable reproducibility. Increasing the experimental temperature leads to the same observation. For each sample, an increase of the working temperature, which facilitates the molecular motion, increases the permeability. It is clear from the results obtained that the nanocomposites provide a good barrier to gases because of the presence of the clay layers. These silicate layers are responsible for delaying the molecular movement by making the diffusion path more tortuous.

5. Rheological Behaviour

The effects of clay and the orientation of the clay platelets on the rheology of polypropylene nanocomposites has been an area of particular interest, secondary to mechanical properties [130,131]. Lele and coworkers found that the zero-shear viscosity of the compatibilized nanocomposites (PP/PP-g-MA/O-MMT) was at least 3 orders of magnitude higher than that of PP and the uncompatibilized hybrids [130]. The strong dependence of melt viscosity on clay concentration was claimed to arise from frictional interactions between the silicate layers, but not from the immobilization of confined chains between silicate layers. At low shear rates, a 3-D percolated structure of dispersed clay tactoids having a high zero-shear viscosity was postulated. At higher shear rates, this network was speculated to break, imparting a solid-like behavior to the melt, which exhibited yielding and decreasing viscosity as shear rate increased. Rheo-XRD experiments gave direct evidence of flow-induced orientation of the clay tactoids. Yielding was linked to the cooperative orientation of the clay. The rheological data for the uncompatibilized hybrid exhibited a much lower zero-shear viscosity and much less pronounced yield behavior, and the corresponding rheo-XRD data showed a smaller orientation of the clay tactoids. This dramatic increase in low shear rate viscosity has been utilized to prepare polypropylene homopolymers and copolymers/clay compounds with improved melt strength in extrusion for the production of films, fibers and a variety of molded articles [65]. The rheology of such clay compounds was shown to be controllable by varying the composition ratio of PP-g-MA compatibilizer/clay.

Solomon and Almusallam [136] conducted the linear and non-linear rheological measurements on a nanocomposite of PP and organically modified montmorillonite with a maleic anhydride functionalized PP compatibilizer. They suggested a hybrid structure of weakly agglomerated network of clay multiplatelets. The reforming of the structure in quiescent conditions after rupture is due to attractive interparticle interactions and not due to Brownian motion. Galgali, *et al.* [130] exploited creep measurements on PP nanocomposites to confirm the origin of the solid-like behavior inferred from the dynamic oscillatory response. They have shown a dramatic decrease in the creep compliance for the nanocomposites prepared with compatibilized PP and 6 wt% montmorillonite. Furthermore, for nanocomposites of a compatibilized PP and 9 wt% montmorillonite, they have reported a dramatic three-order of magnitude drop in the zero shear viscosity beyond the apparent yield stress, suggesting that the solid-like character in the quiescent state is a result of the percolated structure of the layered-silicates.

Solomon, *et al.* reported the strain amplitude sensitivity of the visco-elastic response for PP/clay nanocomposites [136], they observed that for the case of layered silicate nanocomposites, shear thinning began to occur at significantly smaller strain amplitude than that in the unfilled polymer. Furthermore, the strain amplitude for the onset of shear thinning decreases with increasing silicate loading and has been attributed to the increased filler–filler interactions and the meta-stable nature of the silicate network structure. Okamoto, *et al.* [168] have reported that the perpendicular orientation of silicate layers is observed in PP based nanocomposites in the melt state subjected to extensional flow, where the extensional flow rate is the square of the converging flow rates in the orthogonal directions. The perpendicular orientation is observed by solid-state TEM measurements and is attributed for the observed unique strain-hardening and rheopexy in these hybrids.

Rheology curves for PP/clay nanocomposites are shown in Figure 25. The composites were heated to 200°C and then pressured at 20, 30, 40, 50, and 60 Kg/cm^2 [169]. The shear stress (τ_w) and apparent shear rate ($\dot{\gamma}_a$) curve is plotted and each group of the plotted curve satisfies the power law equation, i.e. $\tau_w = K\dot{\gamma}_a^n$ (where n is the power law index and K is the consistency index) in a certain range of shear rates. The values of slope determined from the linear regression of rheology curves are non-Newtonian exponent n. All data are listed in Table 3. All the values of n are less than 1, implying that PP and modified composites exhibit typical shear-thinning behavior. The shear rate $\dot{\gamma}$ can be calculated from the Equation (1)

$$\dot{\gamma} = \frac{(3n+1)\dot{\gamma}_a}{4n} \tag{1}$$

The apparent viscosity η of composites can be calculated by the following equation according to various n values:

$$\eta = \frac{\tau_w}{\gamma} = K \cdot \dot{\gamma}^{(n-1)} \tag{2}$$

The curves of log η verses log $\dot{\gamma}_a$ (shown in Figure 26) show that the higher shear rate carries less apparent viscosity for all of the composites, which indicates that molecular chains deviate from equilibrium

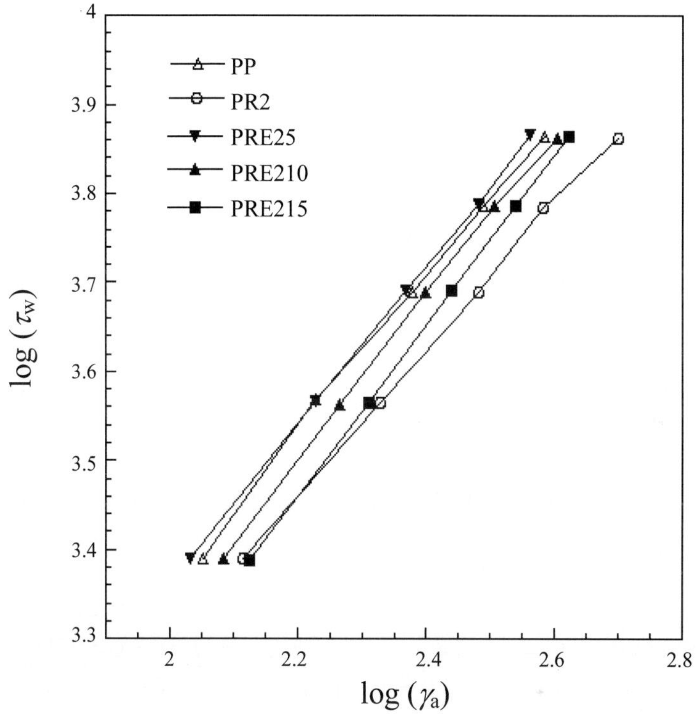

Fig. 25. Plots of log τ_w verses log $\dot{\gamma}_a$ for PP and the composites [169]. Reproduced from reference [169] by Xiao-Yan Ma, Guo-Zeng Liang, Hai-Lin Liu, Jing-Yin Fei and Yun Huang, *Journal of Appied Polymer Science*, 97(5), 1915 (2005)

Table 3. The values of non-Newton exponent *n* for PP and the composites [164].

Systems	Shear rate(s^{-1})	*n*
PP	113.2–386.0	0.89
PR2	130.9–504.2	0.83
PRE25	108.4–366.5	0.90
PRE210	121.2–404.4	0.92
PRE215	134.1–420.3	0.96

Reproduced from reference [164] by Xiao-Yan Ma, Guo-Zeng Liang, Hai-Lin Liu, Jing-Yin Fei and Yun Huang, *Journal of Applied Polymer Science*, 97(5), 1915 (2005).

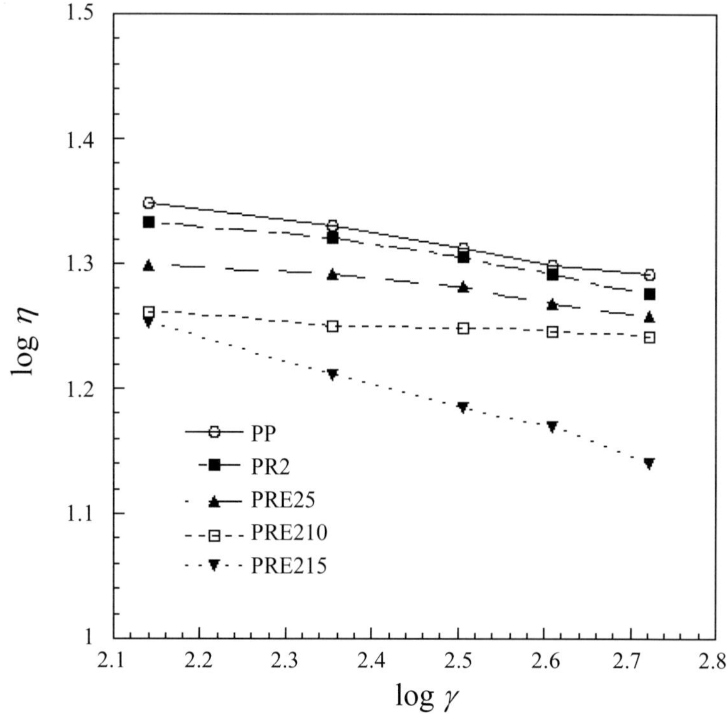

Fig. 26. Plots of log η verses log $\dot{\gamma}_a$ for PP and the composites [169]. Reproduced from reference [169] by Xiao-Yan Ma, Guo-Zeng Liang, Hai-Lin Liu, Jing-Yin Fei and Yun Huang, *Journal of Applied Polymer Science*, 97(5), 1915 (2005).

configuration to molecular orientation along the flow direction under high temperatures and pressures.

6. Challenges and Opportunities

Polypropylene is one of the most widely used thermoplastics, not only because of its balance of physical and mechanical properties, but also due to its environmental friendliness (e.g. recyclability) and low cost. Although the addition of mineral fillers to polypropylenes enhances the stiffness properties, the concomitant effect is typically a decrease in toughness properties, since the inorganic filler particles act as stress-concentrators. Therefore, the formation of propylene-based polymer/clay nanocomposites is an area of intensive investigation.

Efforts were made to improve both the dispersion of clay in PP matrix by using functional oligomers as compatibilizers and the mechanical properties of clay nanocomposites to be used for industrial applications. One of the attractive application fields of clay nanocomposites is automotive parts. The use of nanocomposites in automobile parts or systems is expected to reduce weight and to promote recycling. Applying these nanocomposites to structurally non critical parts such as front and rear fascia, cowl top cover, waist line molding, and valve cover could offer a 25 wt% weight saving on average over highly filled plastics and further savings could be expanded into structural components, such as interior parts and body panels. The enhanced physical properties that clay nanocomposites offer, such as stiffness, strength, and dimensional stability, are as important as the weight and energy savings. Therefore, the state of development of polypropylene/clay nanocomposites is currently in an early stage. From the technical stand point, the enhanced properties of nanocomposites with uniformly dispersed clays do not compensate the increased cost of production and this problem also remains a daunting challenge. Certainly, progressive researches on polypropylene/clay nanocomposites continue to spur an interest in commercial development and to make these materials feasible.

7. Conclusions

The enormous efforts to improve physical performance of PPCNs as well as easy production technique will make them very promising new class of materials. As we reviewed in previous sections, PPCNs exhibited somewhat reasonable increases in thermal stability and mechanical properties. Flammability of pure PP could be significantly reduced by introducing layered silicates into PP.

The tightly bound stacks of clay platelets in natural clay may be expanded to certain levels by simple dissolution in a polar solvent. However, the separation of clay platelets in a non-polar polymer is comparatively difficult. Therefore, organic modification of natural clay minerals is required to disperse the clay in PP. This approach involves cations exchange with alkyl ammonium chains. Moreover, many experimental studies and commercial developments suggested that using PP-g-MA or functionalized PP as a compatibilizer would help greatly to disperse the clays in PP.

Nevertheless, more research needs to be carried out in order to fully understand certain factors such as exfoliation verses intercalation driving forces yielding the different nanocomposites structures and structure/property relationship with respect to fire retardancy and physico-mechanical properties. From present point of view, a cost effective fabrication of PP/clay nanocomposite with high performance is viable with little more efforts in near future.

References

1. P. C. LeBron, Z. Wang and T. J. Pinnavaia, *Appl. Clay. Sci.*, 15, 11 (1999).
2. M. Alexandre and P. Dubois, *Mater. Sci. Eng.*, 12, 1 (2000).
3. R. Mulhaput, T. Engelhardt and N. Schall, *Kunststof*, 9(10), 178 (2001).
4. E. Manias, A. Touny, L. Wu, K. Strawhecker, B. Lu and T. C. Chung, *Chem. Mater.*, 13, 3516 (2001).
5. D. Schmidt, D. Shah and E. P. Giannelis, *Current Openion in Solid State and Mater. Sci.*, 6, 205 (2002).
6. A. Leuteritz, D. Pospiech, B. Kretzschmar, M. Willeke, D. Jehnichen, U. Jentzschrundke and A. Janke, *Adv. Eng. Mater.*, 5(9), 678 (2003).
7. A. Korakianiti, V. Papaefthimiou, T. Daflou, S. Kennou and V. G. Gregoriou, *Macromol. Symp.*, 205, 71 (2004).
8. Y. Kojima, A. Usuki, M. Kawasumi, Y. Fukushima, T. Karauchi and O. Kamigaito, *J. Poly. Sci. A: Poly. Chem.*, 31, 983 (1993).
9. E. P. Giannelis, *Adv. Mater.*, 8, 29 (1996).
10. E. P. Giannelis, R. Krishnamoorty and E. Manias, *Adv. Polym. Sci.*, 138, 107 (1999).
11. R. A. Vaia, G. Price, P. N. Ruth, H. T. Nguyen and J. Lichtenhan, *Appl. Clay. Sci.*, 15, 67 (1999).
12. M. Biswas and S. Sinha Roy, *Adv. Polym. Sci.*, 155, 167 (2001).
13. S. Parija, S. K. Nayak, S. K. Verma and S. S. Tripathy, *Polym. Compos.*, 25(6), 646 (2004).
14. J. K. Pandey, K. R. Reddy, A. Pratheep Kumar and R. P. Singh, *Polym. Deg. Stab.*, 88(2), 234 (2005).
15. F. G. Ramos Filho, J. A. Tomas, S. M. Marcelo, S. Rabello and S. M. L. Silva, *Polym. Degra. Stab.*, 89(3), 383 (2005).

16. W. H. Awad, J. W. Gilman, M. Nyden, R. H. Harris, Jr.Thomas, E. Sutto, J. Callahan, P. C. Trulove, H. C. DeLong and D. M. Fox, *Thermochim. Acta*, 409(1), 3 (2004).
17. T. D. Fornes, P. J. Yoon and D. R. Paul, *Polymer*, 44(24), 7545 (2003).
18. E. P. Giannelis, *Appl. Organomet. Chem.*, 12, 675 (1998).
19. Z. F. Wang, B. Wang, N. Qi, H. F. Zhang and L. Q. Zhang, *Polymer*, 46(3), 719 (2005).
20. R. Xu, E. Manias, A. J. Snyder and J. Runt, *Macromolecules*, 34, 337 (2001).
21. R. K. Bharadwaj, *Macromolecules*, 34, 1989 (2001).
22. P. B. Messersmith and E. P. Giannelis, *J. Polym. Sci. A: Polym. Chem.*, 33, 1047 (1995).
23. K. Yano, A. Usuki, A. Okada, T. Kurauchi and O. Kamigaito, *J. Polym. Sci. A: Polym. Chem.*, 31, 2493 (1993).
24. Y. Kojima, A. Usuki, M. Kawasumi, Y. Fukushima, A. Okada, T. Kurauchi and O. Kamigaito, *J. Mater. Res.*, 8, 1179 (1993).
25. S. Keszei, Sz. Matkó, Gy. Bertalan, P. Anna, Gy. Marosi and A. Tóth, *Eur. Polym. J.*, 41(4), 697 (2005).
26. C. H. Hong, Y. B. Lee, J. W. Bae, J. Young Jho, B. Uk Nam, G. J. Nam and K. J. Lee, *J. Appl. Polym. Sci.*, 97(6), 2375 (2005).
27. M. Zanetti, L. Costa, *Polymer*, 45(13), 4367 (2004).
28. A. B. Morgan, R. H. Harris Jr., T. Kashiwagi, L. J. Chyall and J. W. Gilman, *Fire Mater.*, 26(6), 247 (2002).
29. J. W. Gilman, C. L. Jackson, A. B. Morgan, R. Harris, Jr, E. Manias, E. P. Giannelis, M. Wuthenow, D. Hilton and S. H. Phillips, *Chem. Mater.,* 12, 1866 (2000).
30. S. Bourbigot, M. LeBras, F. Dabrowski, J. W. Gilman and T. Kashiwagi, *Fire Mater.*, 24, 201 (2000).
31. J. W. Gilman, *Appl. Clay. Sci.*, 15, 31 (1999).
32. J. W. Gilman, T. Kashiwagi and J. D. Lichtenhan, *SAMPE J.*, 33, 40 (1997).
33. M. Okamato, Rapra Review Report, Rapra Technology Ltd., London. 163 (2003).
34. G. M. Odegard, T. C. Clancy and T. S. Gates, *Polymer*, 46(2), 553 (2005).
35. J. Gou, Z. Liang, C. Zhang and B. Wang, *Compos. B: Eng. Sci.*, 36, 6–7, 524 (2005).
36. J. Gou, B. Minaie, B. Wang, Z. Liang and C. Zhang, *Compu. Mater. Sci.*, 31, 3–4, 225 (2004).
37. R. Toth, A. Coslanich, M. Ferrone, M. Fermeglia, S. Pricl, S. Miertus and E. Chiellini, *Polymer*, 45(23), 8075 (2004).
38. M. Pospisil, A. Kalendova, P. Capkova, J. Simoník and M. Valaskova, *J. Colloid and Inter. Sci.*, 277(1), 154 (2004).
39. N. Sheng, M. C. Boyce, D. M. Parks, G. C. Rutledge, J. I. Abes and R. E. Cohen, *Polymer*, 45(2), 487, (2004).
40. L. Zhu and K. A. Narh, *J. Polym. Sci., B: Polym. Phys.*, 42(12) 2391 (2004).
41. M. Fermeglia, M. Ferrone and S. Pricl, *Fluid Phase Equilibria*, 212(1–2), 315 (2003).
42. G. Tanaka and L. A. Goettler, *Polymer*, 43(2), 541 (2002).

43. B. Minisini and F. Tsobnang, *Comp. A: Appl. Sci. and Manufact.*, 36(4), 539 (2005).
44. K. Wang, S. Liang, R. Du, Q. Zhang and Q. Fu, *Polymer*, 45(23), 7953 (2004).
45. E. Manias and V. Kuppa, *Colloids & Surf., A* 187–188, 509 (2001).
46. V. Kuppa and E. Manias, *Chem. Mater.*, 14, 2171 (2002).
47. D. B. Zax, D. K. Yang, R. A. Santos, H. Hegmann, E. P. Giannelis and E. Manias, *J. Chem. Phys.*, 112, 2945 (2000).
48. S. H. Anastasiadis and K. Karatasos, *Phys. Rev. Lett.*, 84, 915 (2000).
49. V. V. Ginsburg and A. C. Balazs, *Adv. Mater.*, 12, 1805 (2000).
50. E. Manias, E. Chen, R. Krishnamoorti, J. Genzer, E. J. Kramer and E. P. Giannelis, *Macromolecules*, 33, 7955 (2000).
51. C. Singh and A. C. Balazs, *Polym. Int.*, 49, 469 (2000).
52. V. V. Ginzburg and A. C. Balazs, *Macromolecules*, 32, 5681 (1999).
53. A. Blumstein, *Bull. Chem. Soc.*, 889 (1961).
54. D. J. Greenland, *J. Colloid. Sci.*, 18, 647 (1963).
55. K. Tanihara and M. M. Nakagawa, *Nippin Kagaku Kaishi*, 5, 782 (1975).
56. B. K. G. Theng, *Elsevier*, Amsterdam 1979.
57. R. A. Vaia, H. Ishii and E. P.Giannelis, *Chem. Mater.*, 5, 1694 (1993).
58. A. Okada, Y. Fukoshima, S. Inagaki, A. Usuki, S. Sugiyama, T. Kurashi and O. Kamigaito, 1988, *U.S. Patent* 4,739,007.
59. Y. Fukoshima, A. Okada, M. Kawasumi, T. Kurauchi and O. Kamigaito. *Clay Miner.*, 23, 27 (1988).
60. A. Usuki, Y. Kojima, M. Kawasumi, A. Okada, Y. Fukishima, T. Karauchi and O. Kamigaito, *J. Mater. Res.*, 8, 1179 (1993).
61. J. W. Gilman, C. L. Jackson, A. B. Morgan, R. Harris Jr, E. Manias, E. P. Giannelis, M. Wuthenow, D. Hilton and S. H. Phillips, *Chem. Mater.*. 12, 1886 (2000).
62. M. Zanetti, G. Canimo, D. Canavese, A. B. Morgan, F. J. Lamelas and C. A. Wilkie, *Chem. Mater.*, 14, 180 (2002).
63. E. Manias, A. Touny, L. Wu, K. Strawhecker, B. Lu and T. C. Chung, *Chem. Mater.*, 13, 3516 (2001).
64. P. Reichert, H. Nitz, S. Klinke, R. Brandsch, R. Thomann and R. Mulhaupt, *Macromol. Mater. Eng.*, 274, 8 (2000).
65. M. J. Solomon, A. S. Almusallam, K. F. Seefeldt, A. Somwangathanaroj and P. Varadan, *Macromolecules*, 34, 1864 (2001).
66. P. H. Nam, P. Maiti, M. Okamato, T. Kotaka, N. Hasegawa and A. Usuki, *Polymer*, 42, 9633 (2001).
67. G. Galgali, C. Ramesh and A. Lele, *Macromolecules*, 34, 852 (2001).
68. J. Wu and M. Lerner, *Chem. Meter.*, 5, 835 (1993).
69. G. Lagaly, *Appl. Clay Sci.*, 15, 1 (1999).
70. A. Usuki, M. Kawasumi, Y. Kojima, A. Okada, T. Krauchi and O. Kamigaito, *J. Mater. Res.*, 8, 1174 (1993).
71. A. Akelih and A. Moet, *J. Mater. Sci.*, 31, 3589 (1996).
72. P. Reichert, J. Kressler, R. Tomann, R. Mulhaupt and G. Stoppelmann, *Acta Polym.*, 49, 116 (1998).

73. T. K. Chen, Y. I. Tien and K. H. Wei, *J. Polym. Sci.: Part A Polym. Chem.*, 37, 2225 (1999).
74. S. L. C. Hsu and K. C. Chang, *Polymer*, 43, 4097 (2002).
75. A. Pattanayak and S. C. Jana, *Polymer*, 46(10), 3394 (2005).
76. J. Heinemann, P. Reichert, R. Thomann and R. Mülhaupt, *Macromol. Rapid Commun.*, 20, 423 (1999).
77. J. Tudor, L. Willington, D. O'Hare and B. Royan, *Chem. Commun.*, 2031 (1996).
78. J. S. Bergman, G. W. Coates, H. Chen, E. P. Giannelis and M. G. Thomas, *Chem. Comm.*, 21, 2179 (1999).
79. J. Heinemann, P. Reichert, R. Thomann and R. Mülhaupt, *Macromol. Rapid. Commun.*, 20, 423 (1999).
80. T. Sun and J. M. Garces, *Adv. Mater.*, 14 128 (2002).
81. J. M. Hwu and G. J. Jiang, *J. Appl. Polym. Sci.*, 95. 1228 (2005).
82. A. Arada and E. Ruiz-Hitzky, *Adv. Mater.*, 2, 545 (1990).
83. J. Wu and M. Lerner, *Chem. Mater.*, 4, 1395 (1993).
84. J. J. Tunney, Y. C. Liu and K. H. Wei, *Chem. Mater.*, 8, 927 (1996).
85. H. R. Fischer, L. H. Gielgens and T. P. M. Koster, *Acta. Polym.*, 50, 122 (1999).
86. M. Rodlert, C. J. G. Plummer, L. G. Y. Leterrier, H. J. M. Grünbauer and J. A. E. Manson, *Polymer*, 45(3), 949 (2004).
87. B. Pourabas and V. Raeesi, *Polymer*, 46(15), 5533 (2005).
88. A. Oya, Y. Kurokawa and H. Yasuda, *J. Mater. Sci.*, 35, 1045 (2000).
89. R. A. Vaia, K. D. Jandt, E. J. Krammer and E. P. Giannelis, *Macromolecules*, 28, 8080 (1995).
90. R. A. Vaia, K. D. Jandt, E. J. Krammer and E. P. Giannelis, *Chem. Mater.*, 28, 2628 (1996).
91. R. A. Vaia and E. P. Giannelis, *Macromolecules*, 30, 7990 (1997).
92. R. A. Vaia and E. P. Giannelis, *Macromolecules*, 30, 8000 (1997).
93. A. C. Balazas, C. Singh and E. Zhulina, *Macromolecules*, 31, 8370 (1998).
94. A. Usuki, M. Kato, A. Okada and T. Kauruchi, *J. Appl. Polym. Sci.*, 63, 137 (1997).
95. N. Hasegawa, N. Kawasumi, M. Kato, A. Usuki and A. Okado, *J. Appl. Polym. Sci.*, 67, 87 (1998).
96. M. Kawasumi, N. Hasegawa, M. Kato, A. Usaki and A. Okado, *Macromolecules*, 30, 6333 (1997).
97. M. Kato, A. Usuki and A. Okado, *J. Appl. Polym. Sci.*, 66, 1781 (1997).
98. A. Oya, Y. Kurokawa and H. Yasuda, *J. Mater. Sci.*, 35, 1045 (2000).
99. P. Reichert, H. Nitz, S. Klinke, R. Brandsch, R. Thomann and R. Mulhapt, *Macrmol. Mater. Eng.*, 275, 8 (2000).
100. T. T. Chastek, E. L. Que, J. S. Shore, R.J. Lowy III, C. Macosko and A. Stein, *Polymer*, 46 (12), 1421 (2005).
101. P. H. Nam, P. Maiti, M. Okamato, T. Kotaka, T. Hasegawa and A. Usuki, *Polymer*, 42, 9633 (2001).
102. X. Liu and Q. Wu, *Polymer*, 42, 10013 (2001).
103. E. Manias, *Mater. Res. Soc. Bull.*, 26, 862 (2001).
104. D. Kaempfer, R. Thomann and R. Mulhaupt, *Polymer*, 43, 2909 (2002).

105. D. Garcia-Lopez, O. Picazo, J. C. Merino and J. M. Pastor, *Eur. Polym. J.*, 39, 945 (2003).
106. Y. Tang, Y. Hu, S. Wang, Z. Gui, Z. Chen and W. Fan, *J. Appl. Polym. Sci.*, 89, 2586 (2003).
107. D. H. Kim, J. U. Park, K. H. Ahn and S. J. Lee., *Macromol. Rapid. Commun.*, 24, 388 (2003).
108. M. Kato, M. Mitsushita, K. Fukumori, *Polym. Eng. Sci.*, 44, 1205 (2004).
109. J. H. Kim, C. M. Koo, Y. S. Choi, K. H. Wang and I. J. Chung, *Polymer*, 45, 7719 (2004).
110. Y. Zhang, J. H. Lee, J. M. Rhee and K. Y. Rhee, *Comp. Sci. Tech.*, 64(9), 1383 (2004).
111. Y. Zhang, J. H. Lee, H. J. Jang and C. W. Nah, *Comp B: Eng.*, 35(2), 133(2004).
112. J. Kotek, I. Kelnar, M. Studenovský and J. Baldrian, *Polymer*, 46, 4876 (2005).
113. W. Shao, Q. Wang and H. Ma, *Polym. Intl.*, 54, 336 (2005).
114. A. B. Morgan and J. W. Gilman, *J. Appl. Polym. Sci.*, 87, 1329 (2003).
115. K.-N. Kim, H. Kim and J.-W. Lee, *Polym. Eng. Sci.*, 41, 1963 (2001).
116. Q. Zhang, Y. Wang and Q. Fu, *J. Polym. Sci.: Part B: Polym. Phys.*, 2003, 41, 1.
117. E. Manias, A. Touny, L. Wu, B. Lu, K. Strawhecker, J. W. Gilman and T. C. Chung, *Polym. Mater. Sci. Eng.*, 82, 282 (2000).
118. Jian Li, Chixing Zhou and Wang Gang, *Polym. Test*, 22(2), 217 (2003).
119. V. P. Privalko, R. V. Dinzhos and E. G. Privalko, *Thermochim. Acta*, 432, 76 (2005).
120. Y. Zhou, V. Rangari, H. Mahfuz, S. Jeelani and P. K. Mallick, *Mater. Sci. & Eng. A* XXX, XXX(2005) (article in press and available online).
121. S. L. Liu, X. H. Lu, F. Y. Liew, S. H. Lim and M. S. Yong, *On Line Report* http://www.simtech.astar.edu.sg/Research/TechnicalReports/TR04FT04.pdf.
122. L. Incarnato, P. Scarfato, D. Acierno, M. R. Milana and R. Feliciani, *J. Appl. Polym. Sci.*, 89, 1768 (2003).
123. N. Grassie and G. Scott, *Polym. Degr. & Stab.*, Camb. Univ. press, Cambridge 1985.
124. J. H. Chan and S. T. Balke, *Polym. Degrad. Stab.*, 57, 113–125, 127–134, 135–149 (1997).
125. J. D. Peterson, S. Vyazovkin and C. A. Wight, *Macromol. Chem. Phys.* 202, 775 (2001).
126. B. Fayolle, L. Audouin and J. Verdu, *Polym. Degrad. Stab.*, 75, 123 (2002).
127. Z. Gao, T. Kaneko, I. Amasaki and M. Nakada, *Polym. Degrad. Stab.*, 80, 269 (2003).
128. P. Kodgire, R. Kalgaonkar, S. Hambir, N. Bulakh and J. P. Jog, *J. Appl. Polym. Sci.*, 81, 1786 (2001).
129. J. Ma, Z. Qi and Y. Hu, *J. Appl. Polym. Sci.*, 82, 3611 (2001).
130. G. Galgali, C. Ramesh and A. Lele, *Macromolecules*, 34, 852 (2001).
131. C. M. Koo, M. J. Kim, M. H. Choi, S. O. Kim and I. J. Chung, *J. Appl. Polym. Sci.*, 88, 1526 (2003).
132. J. Li, C. Zhou, G. Wang and D. Zhao, *J. Appl. Polym. Sci.*, 89, 318 (2003).

133. J. D. He, M. K. Cheung, M. S. Yang and Z. Qi, *J. Appl. Polym. Sci.*, 89, 3404 (2003).
134. G. Florêncio, R. Filho, T. Jeferson, A. Mélo, M. S. Rabello, M. L. Suédina and M. L. Silva, *Polym. Degrad. Stab.*, 89(3), 383 (2005).
135. M. Diagne, M. Guèye, L. Vidal and A. Tidjani, *Polym. Degrad.Stab.*, 89(3), 418 (2005).
136. M. J. Solomon, A. S. Almussallam, K. F. Seefeldt, A. Somwangthanaroj and P. Varadam, Macromolecules, 34, 1864 (2001).
137. P. Svoboda, C. Zeng, H. Wang, L. J. Lee and D. L. Tomasko, *J. Appl. Polym. Sci.*, 85, 1562 (2002).
138. T. S. Ellis and J. S. D'Angelo, *J. Appl. Polym. Sci.*, 90, 1639 (2003).
139. F. C. Chiu, S. M. Lai, J. W. Chen and P. H. Chu, *J. Polym. Sci. B: Polym. Phys.*, 4(22), 4139 (2004).
140. J. Zhang, D. Jiang, A.Charles and A. Wilkie, *Thermochim. Acta*, 430, 107 (2005).
141. M. Zanetti, G. Camino, P. Reichert and R. Mülhaupt, *Macromol. Rapid. Commun.*, 22, 176 (2001).
142. S. Parija, S. K. Nayak, S. K. Verma and S. S. Thripaty, *Polym. Comp.*, 25(6), 646 (2004)
143. S. M. Lomakin, I. L. Dubnikova, S. M. Berezina and G. E. Zaikov, *Polym. Intl.*, 54(7), 999 (2005).
144. C. Ding, D. Jia, H. He, B. Guo and H. Hong, *Polym. Test*, 24(1), 94 (2005).
145. D. A. Brune and J. Biccerano, *Polymer*, 43, 369 (2002).
146. Chao Ding, Demin Jia, Hui He, Baochun Guo and Haoqun Hong, *Polym. Test*, 24(1), 94 (2005).
147. E. M. Benetti, V. Causin, C. Marega, A. Marigo, G. Ferrara, A. Ferraro, M. Consalvi and F. Fantinel, *Polymer*, 46(19), 8275 (2005).
148. M. N. Bureau, F. Perrin-Sarazin, M.-T. Ton-That, *Polym. Eng. & Sci.*, 44(6), 1142 (2004).
149. J. W. Gilman, T. Kashiwagi and J. D. Lichtenhan, *SAMPE. J.*, 33, 40 (1997).
150. M. Bartholmai and B. Schartel, *Polym. Adv. Technol.*, 15, 355 (2004).
151. V. Babrauskas and R. D. Peacock, *Fire Safety J.*, 18, 255 (1992).
152. J. W. Gilman, C. L. Jackson, A. B. Morgan, R. Harris, Jr., E. Manias, E. P. Giannelis, M. Wuthenow, D. Hilton and S. H. phillips, *Chem. Mater.*, 2000, 12, 1866.
153. U. Wagenknecht, B. Kretzschmar and G. Reinhardt, *Macromol. Symp.*, 2003, 194, 207.
154. D. Wang and C. A. Wilkie, *Polym. Degrad. Stabil.*, 80, 171 (2003).
155. A. B. Morgan and J. D. Harris, *Polymer*, 44, 2313 (2003).
156. M. Zanetti, G. Camino, P. Reichert and R. Mulhaupt, *Macromol. Rapid Commun.*, 22, 176 (2001).
157. Gy. Marosi, P. Anna, A. Márton, Gy. Bertalan, A. Bóta, A. Tóth, M. Mohai and I. Rácz, *Polym. Adv. Technol.*, 13. 1103 (2002).
158. Jeffrey W. Gilman., *Appl. Clay Sci.*, 15, 31 (1999).
159. D. Wang and C. A. Wilkie, *Polym. Deg. Stab.*, 80(10), 171(2003).

160. J. W. Gilman, T. Kashiwagi, M. Nyden, J. E. T. Brown, C. L. Jackson and S. Lomakin. In: Al-Maliaka S., Golovoy A. and Wilkie C.A., editors. *Chemistry and Technology of Polymer Additives. London*: Blackwell Scientific; 249 (1998).
161. J. Zhang, D. David Jiang and C. A. Wilkie, *Thermochim. Acta*, 430, 107 (2005).
161. Y. Tang, Y. Hu, B. Li, L. Liu, Z. Wang, Z. Chen and W. Fan, *J. Polym. Sci. A: Polym. Chem.* 42 (23), 6163 (2004).
162. Q. Zhang, Q. Fu, L. Jiang and Y. Lei, *Polym. Intl.*, 49(12), 1561 (2000).
163. M. T. Ton-That, F. Perrin-Sarazin, K. C. Cole, M. N. Bureau and J. Denault, *Polym. Eng. Sci.*, 44(7), 1212 (2004).
164. M. Xiaoyan, L. Guozheng, L. Haijun, L. Hailin and H. Yun, *J. Appl. Polym. Sci.* 97(5), 1907 (2005).
165. Y. Kurokawa, H. Yusuda and A. Oya, *J. Mater. Sci. Lett.*, 15, 1481 (1996).
166. G. Giuliana, T. Mariarosaria, V. Vittoria, K. Dirk and R. Mülhaupt, *Polymer*, 44(13), 3679 (2003).
167. A. Tidjani, *Polym. Degrad. & Stab.*, 87(1), 43, 2005.
168. M. Okamoto, P. H. Nam, P. Maiti, T. Kotaka, N. Hasegawa and A. Usuki, *Nano. Lett.*, 1, 295 (2001).
169. X. Y. Ma, G. Liang, H. Liu, J. Fei and Y. Huang, *J. Appl. Polym. Sci.*, 97(5), 1915 (2005).

CHAPTER 9

Clay Nanocomposites of Polyurethanes and Epoxies: Preparation Methods and Properties

Sadhan C. Jana

Department of Polymer Engineering
University of Akron
Akron, OH 44325-0301, USA
janas@uakron.edu

1. Introduction

This chapter presents a detailed treatment on nanocomposite products of thermoplastic polyurethanes and thermosetting epoxies in view of widespread applications in industry. The general understanding of nanoparticle dispersion in thermoplastic polyurethanes and epoxies can be easily extended to other thermosetting and thermoplastic polymers and elastomers.

A recent market survey predicted 13.3% annual growth of thermosetting polymer industry in the US alone due to increased usage in semi-conductors and other polymer-intensive technologies [1]. Industrial trade magazines also portray positive outlook for thermoset materials in conventionally thermoplastic industries [2]. Most thermosetting resins, such as epoxies, polyimides, unsaturated polyesters, and cyanate esters are used in making polymeric composite articles. Thus, new nanocomposite products of thermosetting resins have an immediate market utility in aerospace, automotive, and coatings industry. This has triggered strong research activities on epoxy-clay nanocomposites [3–28] starting with the pioneering work of Pinnavaia *et al.* in the early 1990s [3–5,7–9] and recently on clay composites of unsaturated polyesters [29–31] and PMR-type polyimide nanocomposites [32–40].

Epoxy nanocomposites have been developed in association with nanoclay [3–28], carbon nanofibers [41–43] and nanotubes [44–49], and nanoscale silica and fumed silica particles [50–54]. Some investigators also studied composites of small quantities of nanoparticulate fillers and continuous glass and carbon fibers [55–57]. The presence of nanofillers offers increased stiffness and toughness, a reduction in water absorption and permeability to water vapor, and an increase in electrical conductivity and coefficient of thermal expansion.

Polyurethanes have niche market in shoe and other industries, such as coatings, adhesives, foams, rubbers, and as thermoplastic elastomers [58–60], sports track, and biomedical. These applications can immensely benefit from large improvements in properties and performance brought about by nanotechnology research. The advent of nanotechnology adds versatility in the design of such products. An array of products with superior properties can be made using nanoscopic filler particles, such as layered silicates, carbon nanotubes and nanofibers, often with the possibility of reactions and energetic interactions between polymer chains and nanoparticles [61–100]. In addition, the weight of finished articles can be greatly reduced as superior properties are obtained with small quantities of nanofillers, in the range of 3–5% by weight.

Nanocomposites of polyurethanes can be prepared by mixing nanofillers with pre-made polyurethanes in conventional mixing devices, such as twin- and single-screw extruders (Figure 1a) or by incorporating nanofillers during condensation polymerization from isocyantates, long chain diols, and chain extenders, as in Figure 1(b, c). Dispersion of nanoparticles in epoxy is much more challenging, although an array of techniques are used in an effort to produce good dispersion, e.g., by sonication [101], high intensity mixing and highly specific chemical treatment [25] before curing of epoxy. As epoxy curing is carried out under quiescent conditions, nanoparticles must be dispersed well in epoxy before curing. Shear flows and sonication applied to achieve good dispersion of clay agglomerates in low viscosity monomeric liquid epoxy do not lead to exfoliation of clay particles owing to low levels of shear stress. This leaves only one possibility — exfoliation of clay particles must be achieved during curing.

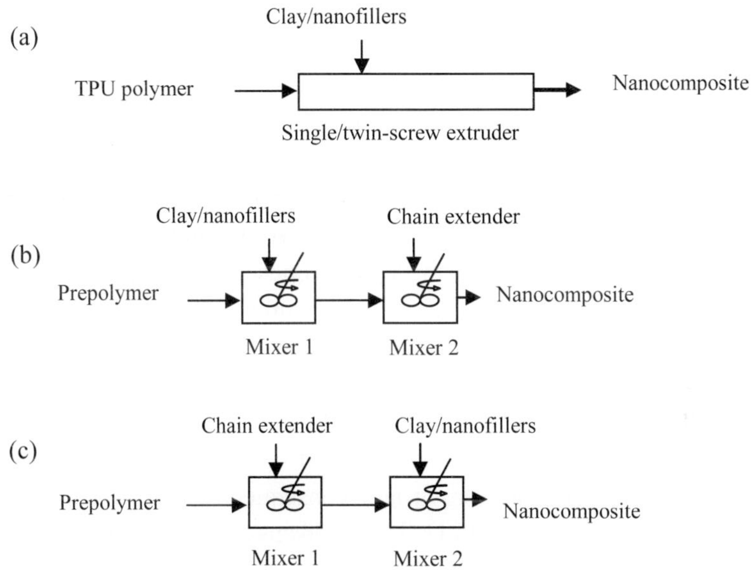

Fig. 1. Schemes for TPU-nanocomposite preparation. (a) Simple mixing scheme. Simultaneous polymerization and filler-polymer mixing, (b) Method I, (c) Method II.

2. Scopes for Dispersion of Nanofillers

The magnitude of electrostatic attractive forces holding together individual nanoscopic filler particles is much higher due to small gap between the particles — of the order of a few nanometers [102–105] — and is often greater than the shear forces produced in typical mixing equipment. Shear forces are limited by low viscosity monomers in cases where filler particles are pre-dispersed before polymerization. Consequently, nanoscale dispersion of filler particles cannot be easily produced in such nanocomposites. Nevertheless, other forms of dispersion forces are routinely used. For example, filler-monomer mixtures are sonicated for long periods of time in order to break particle agglomerates. Although somewhat successful in achieving dispersion, these methods often cause breakage of high aspect ratio filler particles and sometimes trigger premature polymerization of monomers. In addition, nanoparticles may undergo settling from well-dispersed monomer-nanoparticle mixtures due to higher particle density.

In many systems, polymerization rate in nano-confined space between filler particles is significantly higher than those in the bulk. This has been used successfully in polyamides to produce nanocomposites of organically treated nanoclay by in situ polymerization methods [106–110]. For example, in polyamide 6-nanoclay system, intra-gallery caprolactam polymerization occurs at much faster rate due to catalytic effects of alkylammonium ions present inside organically treated clay galleries. A similar mechanism was proposed by Pinnavaia and coworkers [3–5,7–9] for epoxy-nanoclay mixtures. In this case, the intra-gallery reactions between epoxide and amine groups are catalyzed by alkylammonium ions. As a result, epoxy polarity is reduced inside the clay galleries, which induces transport of additional epoxy molecules into clay galleries (Figure 2). Although intuitive, this mechanism cannot explain exfoliation in many cases and consequently new mechanisms were proposed in literature [20,21]. This is discussed in Section 4.1.

The *in situ* polymerization scheme — successful in explaining exfoliation of clay galleries in polyamide and epoxy systems — cannot be readily extended to thermoplastic polyurethanes, although polyurethane chains can be synthesized in the presence of nanoparticles [61–64,68,92]. In this case, linear chains of polyurethane polymer do not exert enough entropic force on clay layers to produce exfoliation. Section 3.1 discusses relative merits of several methods for exfoliation of clay particles in TPU-clay systems.

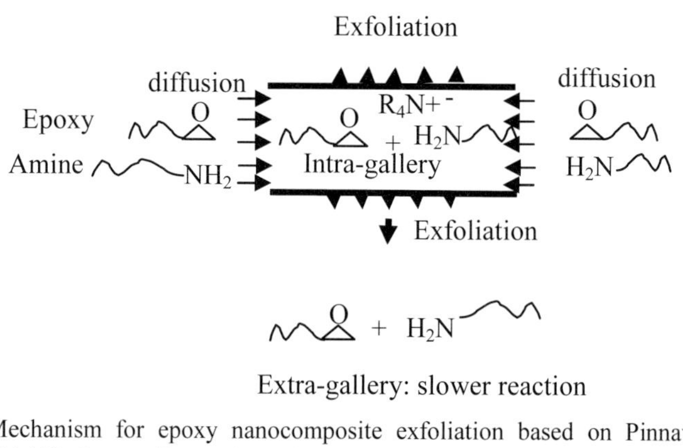

Fig. 2. Mechanism for epoxy nanocomposite exfoliation based on Pinnavaia *et al.* [3–5,7–9,11,12]. Intra-gallery polymerization is catalyzed by alkylammonium ion, R_4N^+.

3. Clay Nanocomposites of Polyurethanes

3.1. *Methods of preparation*

A general preparation method for polyurethane nanocomposites involves the following steps. First, nanoparticles are pre-dispersed in polyols. Second, the polyol-nanoparticle suspension is reacted with diisocyanates to produce polyurethane nanocomposites. This methodology was followed by several investigators [61–64]. Wang and Pinnavaia [61] were first to prepare polyurethane-clay nanocomposites by intercalating clay galleries with glycerol propoxylate polyol and reacting the clay-polyol mixture with methylene diphenyl diisocyanate. These authors treated clay particles with alkylammonium ions containing 12- and 18-carbon atoms in the alkyl chains. Zilg *et al.* [62] prepared polyurethane nanocomposites of organophilic mica, trihydroxy terminated oligo(propyleneoxide) polyol, diphenyl methane diisocyante in the same manner. Petrovic *et al.* [65] prepared nanocomposites of nano-silica by reacting a dispersion of polyol and nanosilica with diisocyanate Hu *et al.* [70] treated montmorillonite clay with trimethyl hexadecyl ammonium ion, prepared a dispersion of treated clay in poly(propylene oxide) glycol, and reacted clay-polyol mixture with toluene diisocyanate to obtain polyurethane nanocomposites. Chen *et al.* [63], on the other hand, modified montmorillonite clay with protonated 12-aminolauric acid and polymerized ε-caprolactone in the presence of treated clay. The resultant material was then mixed with a solution of separately prepared polyurethane to yield the nanocomposite product.

A second approach for obtaining polyurethane nanocomposites was developed for reactive nanoclay [68,92–95]. Nanoclay particles, e.g., montmorillonite and bentonite can be organically treated with alkyl ammonium ions containing hydroxylated alkyl groups, which participate in reactions with polymer chains carrying –NCO groups or diisocyanates. Montmorrilonite clay has been treated with alkyl ammonium ions carrying 1-OH, 2-OH, and 3-OH groups [68]. In these cases, the alcoholic –OH groups present in alkylammonium ions account for a part of –OH groups in the overall stoichiometric ratio of isocyanate and hydroxyl groups. Tien and Wei [68] prepared polyurethane nanocomposites by allowing clay-polymer tethering reactions in solution under isothermal, well-stirred conditions. In this case, $-CH_2OH$ groups of alkylammonium ions of organic treatment of clay were assumed to react with –NCO groups on prepolymer chains. As an evidence of clay-polymer reactivity, these authors presented urethane linkage formation in the reactions between diphenyl methane diisocyanate and organically treated clay. Osman *et al.* [80] prepared polyurethane nanocomposite adhesives in solution in the same manner as Tien and Wei [82] but did not find evidences of clay-prepolymer reactions.

A third approach was developed by Pattanayak and Jana [92–95], whereby organically treated clay and polymer were mixed in bulk as depicted in Figure 1(b–c) and clay-polymer reactions took place during mixing. Pattanayak and Jana [92–95] used commercially available clay, Cloisite 30B (Southern Clay Products) as reactive clay and exploited the $-CH_2CH_2OH$ groups in the structure of the quaternary ammonium ion, $N^+(CH_2CH_2OH)_2(CH_3)T$, where T represents an alkyl group with approximately 65% $C_{18}H_{37}$, 30% $C_{16}H_{33}$, 5% $C_{14}H_{29}$ [111] to promote clay-polymer reactions. In Method I (Figure 1b), the reactive clay particles, containing hydroxy-alkyl groups in the alkylammonium ions, were allowed to react with isocyanate groups in pre-polymer chains during clay-prepolymer mixing. In Method II (Figure 1b), the extended chain polymers with residual –NCO groups were allowed to react with clay during clay-polymer mixing. The bulk polymerization methods, as depicted in Figure 1, avoid the use of solvents. Some processing related advantages of bulk polymerization methods are as follows. (1) Larger shear forces during mixing can aid clay dispersion. (2) Products can be easily finished into useful articles by injection molding, film casting, or other techniques. (3) Conventional mixers, such as single- and twin-screw extruders can be used to produce such nanocomposites. Yet another method, Method III (Figure 3), can be developed in which all ingredients, including isocyanate, polyol, chain extender, and nanoclay are mixed and allowed to react. Note that these methods do not require prior intercalation of clay particles by polyols or polymer chains. Nevertheless, bulk polymerization methods have some limitations. For example, diffusional constraints hinder the rates of polymerization as well as clay-polymer tethering reactions. In addition, reaction conditions

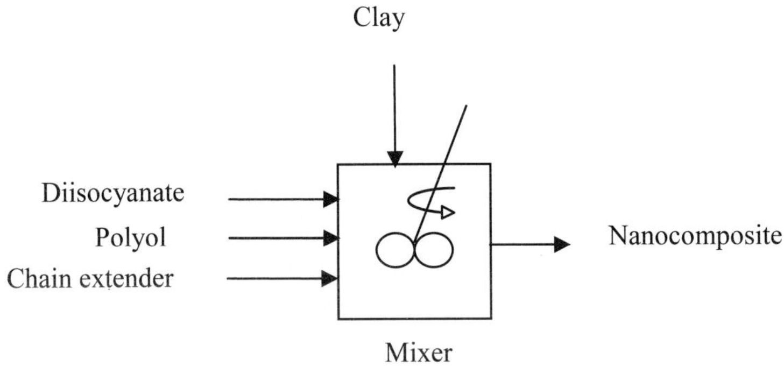

Fig. 3. TPU nanocomposite preparation by simultaneous mixing of ingredients with clay (Method III).

are seldom isothermal. A concomitant rise in temperature may trigger many side reactions, e.g., formation of biurets and allophanates [112,113].

Other avenues of clay-polymer tethering via bulk polymerization have been attempted. Cao *et al.* [96] treated nanoclay with hydrochloride salt of organotin catalyst carrying $-NH_2$ endgroups. In this manner, the chain extension reaction inside clay galleries occurred at equivalent rates as in outside the galleries.

3.2. *Clay-polymer reactions*

Clay-polymer reactions can be studied using infrared spectroscopy. The intensities of stretching of –NCO groups at 2270 cm^{-1} and of carbonyl groups at 1705–1715 cm^{-1} in Fourier transform infra-red (FT-IR) spectrum are monitored. Tien and Wei [68] inferred clay-polymer reactions by observing formation of hydrogen bonded urethane carbonyl and hydrogen-bonded urethane –NH groups in the reactions between reactive clay and diphenyl methane diisocyanate. Pattanayak and Jana [92] studied urethane linkage formation between $-CH_2CH_2OH$ groups on alkylammonium ions of Cloisite 30B clay and –NCO end groups of polymer chains. These authors also considered reactivity of isocyanates with structural –OH groups, e.g., of Al-OH bonds and hydrogen-bonded water [92,114]. A gradual reduction of –NCO peaks with time can be attributed to (a) reactions of –NCO groups with structural –OH, water molecules in clay and on KBr discs used in FTIR spectroscopy, (b) $-CH_2CH_2OH$ groups in clay, and (c) dimerization and trimerization reactions [115]. The conversion (α) of –NCO groups can be defined as $\alpha \equiv \frac{A_{NCO,0} - A_{NCO}}{A_{NCO,0}}$, where A_{NCO} is the area under the peak at 2270 cm^{-1} due to –NCO stretching at any time t and $A_{NCO,0}$ is the initial area of –NCO peak. A constant area under the peaks between 2860 and 2940 cm^{-1} due to –CH stretching is used as internal standard. In the work of Pattanayak and Jana [92], a much higher value of $\alpha \sim 0.38$ for Cloisite 30B clay than for prepolymer ~0.08, untreated clay ~0.12, and clay treated with alkyl ammonium ions not carrying $-CH_2CH_2OH$ groups, ~0.09 after 60 minutes of reaction indicated substantial urethane linkage formation during Cloisite 30B-polymer reactions. A non-zero value of α for the pre-polymer indicated reactivity with trace amounts of moisture on KBr discs, which is known to yield urea [112,113,116,117]. A corresponding increase in the area under the carbonyl peak (A_{CO}) at 1733 cm^{-1}, provides support that –NCO groups are consumed in urethane formation from the reactions between –NCO and $-CH_2CH_2OH$ groups (Figure 4). In Figure 4(b), $\beta \equiv \frac{A_{CO} - A_{CO,0}}{A_{CO,0}}$ with $A_{CO,0}$ as the initial area under the CO peak. Note that clay-polymer reactions must be carried out in dry conditions as one molecule of H_2O present in clay galleries can effectively react with two –NCO groups and form one urea carbonyl group. Pattanayak and Jana [92] showed that the rates of reactions between structural –OH groups in clay galleries and isocyanate groups of polymer chains are negligibly small.

3.3. *State of clay particle dispersion*

Most studies to date reported intercalated and partially exfoliated states of dispersion of clay particles in polyurethane nanocomposites. These states were characterized by wide angle X-ray diffraction (WAXD) patterns, where the absence of clay peaks or a reduction of scattering

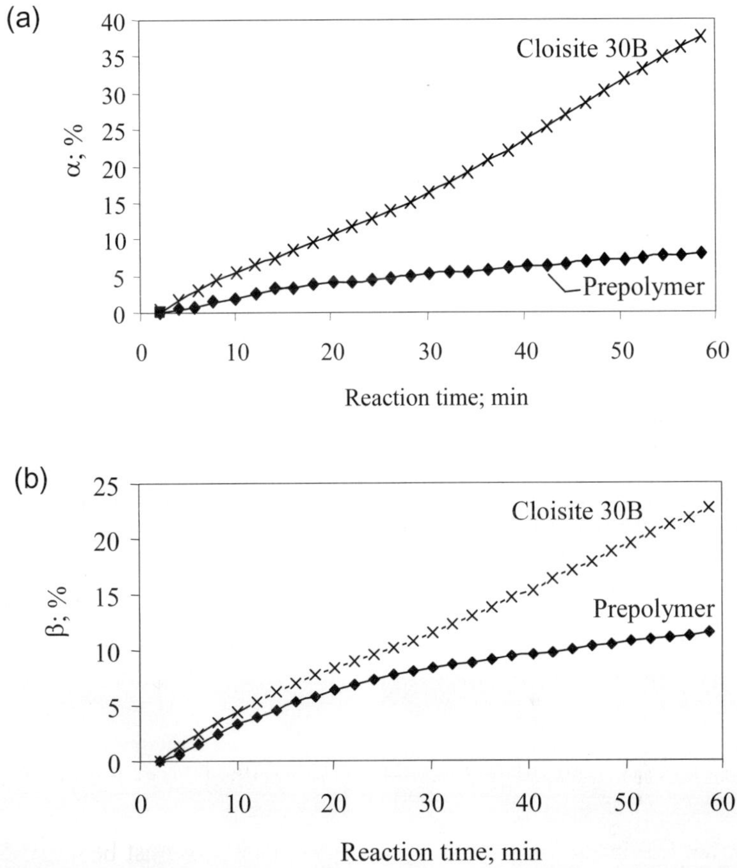

Fig. 4. Urethane formation from reactions between $-CH_2CH_2OH$ groups in Cloisite 30B clay and –NCO groups in pre-polymer chains at 80°C [92]. (a) Conversion of –NCO groups and (b) formation of urethane linkage.

intensity are taken as signs of better dispersion of clay particles. A representative set of WAXD patterns are presented in Figure 5. The patterns presented in Figure 5 show peaks due to residual organically treated clay, e.g., at $2\theta = 5.2°$ and due to polymer intercalated clay, e.g., at $2\theta = 3–3.75°$. These patterns are typical for polyurethane-clay systems where clay particles are mixed with pre-made polyurethanes or when clay particles do not have favorable interactions with polyurethane chains. Only in a handful of cases, WAXD patterns do not show any

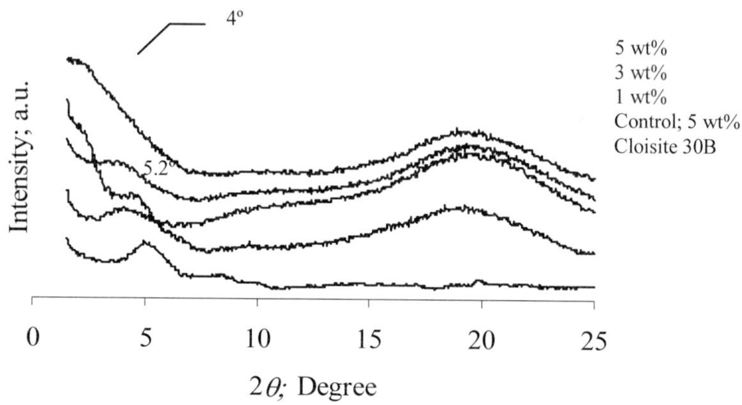

Fig. 5. WAXD patterns of nanocomposites of TPU and Cloisite 30B prepared by Method II (Figure 1c). TPU was synthesized from polyesterpolyol, methylenediisocyanate, and 1,4 butanediol, with –NCO:OH ratio maintained at 1:1 after taking into account –OH groups in clay. Chain-extended polymer was mixed with clay at 130°C for 6 minutes [92]. Control material was obtained by mixing pre-made TPU for 6 minutes at 130°C.

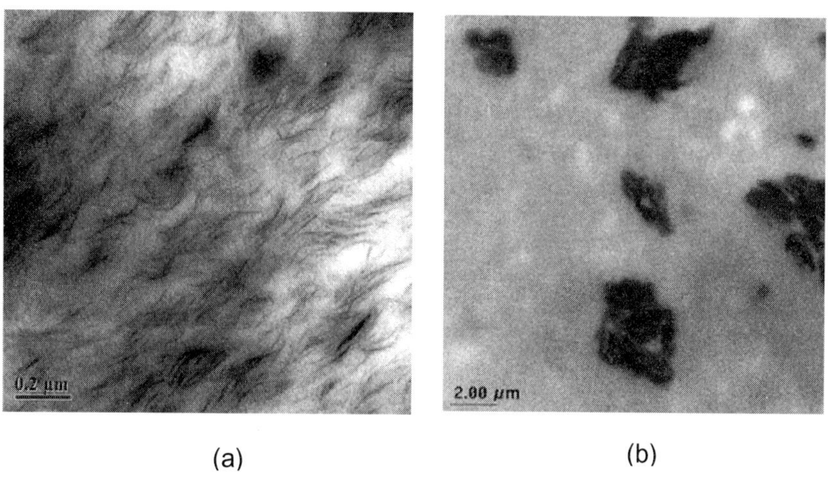

Fig. 6. TEM images of nanocomposites of 5 wt% Cloisite 30B produced by (a) Method II and (b) Method I. The image in Figure 6(a) corresponds to WAXD patterns presented in Figures 5 [92]. Methods I and II are described in Figure 1.

major peak and indicate clay exfoliation. One such case is seen in Figure 5 for 5 wt% reactive clay, which has no distinguishable peak for $2\theta > 1.5^{\circ}$, indicating that the separation between two adjacent clay layers is at least 6 nm. Many authors present WAXD patterns and companion transmission electron microscope images to corroborate good separation of individual clay layers in the polymer. A representative TEM image showing exfoliated state of clay particles is shown in Figure 6(a). Figure 6(b) shows that methods of preparation significantly influence the state of clay particle dispersion.

3.4. *Effect of viscosity on clay particle dispersion*

It was shown by some authors that both clay-polymer reactions and shear stress generated during mixing are needed for good dispersion of clay particles [92]. The level of shear stress depends upon viscosity of the polymer and the shear rate during mixing. The peak shear rate in most batch mixers and some extrusion equipment is below 100 s^{-1}, while the average shear rate falls in the range of 10–50 s^{-1}. Consequently, very little shear stress is generated during mixing of low viscosity monomers, such as polyol and prepolymer. However, a recent study [87] reported that clay particles can be fully exfoliated by dispersing in polypropylene glycol, although no direct evidence of exfoliation was presented. Figure 6(b) presents TEM image of clay particles in polyurethanes showing aggregates of a few micrometer in size [92,95]. Clay particles in this case were dispersed in prepolymer by stirring in a batch mixer before chain extension reactions were carried out with butanediol. In this case, the zero-shear viscosity of the prepolymer was low, ~500 Pa-s at 80°C. On the other hand, chain-extended polymer offered much higher zero-shear viscosity ~4500 Pa.s at 130°C and helped produce clay particle exfoliation [92].

The polymer viscosity can be changed by using polyol molecular weight and polyol type. Pattanayak and Jana [92] found that the viscosity of polyurethanes prepared from 2 mole MDI, 1 mole polyol, and 1 mole 1,4 butanediol changed from ~4500 Pa.s for polyether polyol of molecular weight 1000 to ~2500 Pa.s for molecular weight 2000 at 130°C. On the other hand, polyurethane of polyester polyol of molecular

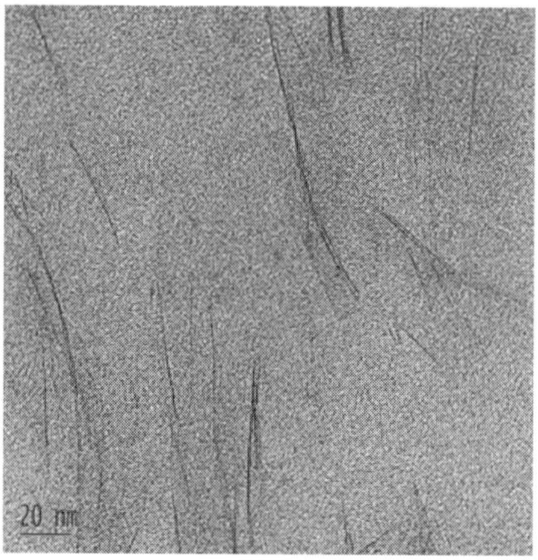

Fig. 7. TEM image of nanocomposites of 5 wt% Cloisite 30B produced by Method II. Polyesterpolyol of molecular weight 2000 was used.

weight 2000 provided a viscosity of 3800 Pa.s at 130°C. In the latter case, the clay particles were found to be very well exfoliated as revealed by TEM image in Figure 7 [92–95]. It was established by Pattanayak and Jana [92] that clay-polymer reactions must occur during composite preparation for efficient exfoliation to take place. On the other hand, shear mixing of pre-made polyurethanes with clay particles leads to only intercalated composites [66,69,73,83,88,89,97]

3.5. *Hydrogen bonding on clay dispersion*

Hydrogen bond formation between organically treated clay and polymer chains have been found useful in producing efficient clay particle dispersion in many polymer nanocomposite systems. Han *et al.* [118–119] as well as Paul and coworkers [120] established that $-CH_2CH_2OH$ groups on organically treated clay particles promotes exfoliation of clay particles during shear mixing with polar polymers due to hydrogen bonding with polar groups of polymer chains. However, Jana and Pattanayak [92–95] observed the following effects of clay particles on hydrogen bonding in polyurethane-clay systems. (1) Hydrogen bonding among hard segments is impeded when natural montmorillonite clay or clay particles treated with alkyl ammonium ions of non-polar alkyl groups are used. (2) Additional hydrogen bonds form between hydroxy-alkyl groups of alkyl ammonium ions on clay particles and –CO and –NH groups of urethane linkages on polymer chains. This is schematically presented in Figure 8. Some polymer chain ends with –NCO groups can diffuse to the vicinity of the clay galleries during nanocomposites preparation and react with $–CH_2CH_2OH$ group of the quaternary ammonium ions to produce urethane linkage, –CO–NH–. The urethane linkages, in turn, can form hydrogen bonds with the second $–CH_2CH_2OH$ group residing on the same quaternary ammonium ion. The extent of additional hydrogen bonding is faciliated by exfoliated clay particles, as many unreacted $–CH_2CH_2OH$ groups are exposed. It was seen in literature that materials prepared by mixing Cloisite 30B clay and pre-made polyurethanes exhibits the same ratio of A_{HCO}/A_{FCO} (Table 1), as materials prepared by Method II as described in Figure 1(c) [92]. WAXD patterns and TEM images present evidences that clay particles remained in intercalated state in the former and in exfoliated state in the latter. In view of this, one can infer that clay-polymer hydrogen bonding does not have much influence on clay particle dispersion in polyurethane systems.

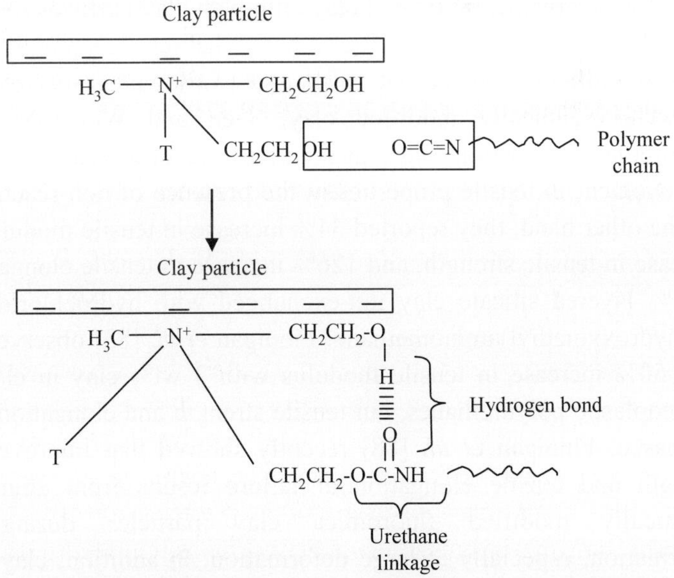

Fig. 8. Hydrogen bonding by clay-tethered polyurethane chain [92].

Table 1. Ratio of the area under the peak of hydrogen-bonded –NH (A_{NH}), total C=O (A_{CO}), and –CH stretching (A_{CH}) of FTIR spectra and the ratio of area under the peak of hydrogen bonded C=O (A_{HCO}) and free C=O (A_{FCO}) groups (92).

Material		A_{HCO}/A_{FCO}	A_{CO}/A_{CH}	A_{NH}/A_{CH}
PU		1.0	0.54	0.41
Composite of Cloisite 30B	1 wt%	0.8	0.63	0.26
	3 wt%	0.8	0.63	0.25
	5 wt%	0.8	0.63	0.25
Control material with 5 wt% Cloisite 30B		0.9	0.54	0.39

3.6. Properties

3.6.1. *Tensile properties*

An increase of tensile modulus with the addition of nanoclay has been a norm in all studies reported to date. Many studies also reported an increase of tensile strength. The extent of increase, however, depends on the state of exfoliation of clay particles. Wang and Pinnavaia [61] observed respectively 28% and 100% increase in tensile modulus with 5 and 10 wt% organically treated clay, although clay particles remained in intercalated state. These authors also noted that tensile strength increased linearly with clay content, e.g., 67% and 133% with respectively 5 and 10 wt% organically modified clay. Tien and Wei [68] prepared polyurethane nanocomposites in solution and observed only marginal improvement in tensile properties in the presence of non-reactive clays. On the other hand, they reported 34% increase in tensile modulus, 165% increase in tensile strength, and 126% increase in tensile elongation with 1 wt% layered silicate clay ion-exchanged with hydrochloride salt of tris(hydroxymethyl)aminomethane. Finnigan *et al.* [83] observed greater than 50% increase in tensile modulus with 3 wt% clay in elastomeric thermoplastic polyurethanes, but tensile strength and elongation at break decreased. Finnigan *et al.* [88] recently showed that improved tensile strength and tensile elongation at failure results from alignment of organically modified fluoromica clay particles during tensile deformation, especially at large deformation. In addition, clay-polymer interactions via secondary bonding also contribute to such improved properties. It was shown that properties improvement is more dramatic in the presence of smaller size clay particles, as they tend to orient more than larger size tactoids. Xia *et al.* [87] showed that tensile strength increased, while tensile elongation decreased linearly with the degree of clay particle exfoliation.

Figure 9 presents typical stress-strain diagrams of polyurethane and its composites with reactive clay particles. Pattanayak and Jana [93] showed that the modulus and tensile strength increased by 110% and 170% respectively over pristine PU and the strain at break decreased slightly, irrespective of whether reactive clay or non-reactive clay particles were used.

It was reported in some studies [92–95] that hydrogen bonding does not have much impact on tensile strength in polyurethane-clay composites, especially if hard segment content is kept low, e.g., less than 36 wt%. Pattanayak and Jana [92,93] compared the extent of soft- and hard-segment hydrogen bond in materials prepared by *in-situ* bulk polymerization with those obtained by mixing of pre-made polyurethanes and reactive layered silicate clay. It was found that the extent of hydrogen bonding by urethane –NH and –CO groups were similar in both materials, although the tensile strength of former materials was much larger, e.g., ~12.8 MPa compared to ~3.4 MPa in the presence of 5 wt% clay. However, this observation has not been confirmed for higher hard segment content, although it was generally observed that clay particles reduced the extent of hard segment hydrogen bonding [85,92].

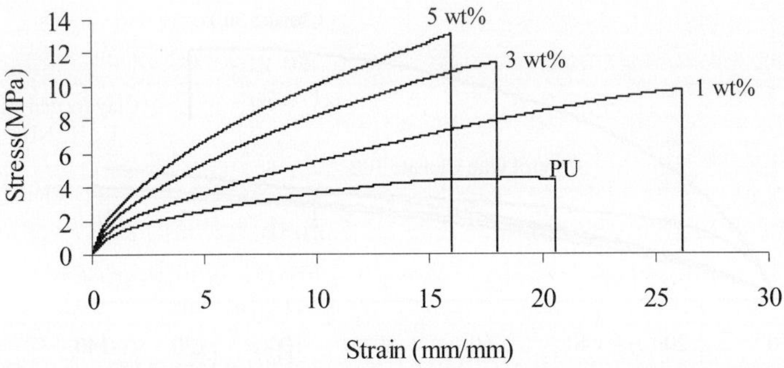

Fig. 9. Stress-strain diagram of TPU nanocomposites of Cloisite 30B clay prepared by Method II. TPU nanocomposite was synthesized from polyetherpolyo of molecular weight 1000 [93].

3.6.2. *Tear and fracture properties*

Pattanayak and Jana [93–95] showed that the values of tear stress, strain, modulus, and energy for break all increased over pristine PU in the presence of reactive clay. A typical stress-strain diagram in tear tests is presented in Figure 10. In all cases, tear modulus increased with clay content, although larger increase is observed in the case of reactive clay. The stress at break, a measure of resistance to growth of fracture, increases with reactive clay content indicating that polymer-tethered clay particles delayed crack propagation. On the other hand, microdispersed clay particles in composite materials of non-reactive clay or in materials obtained by mixing reactive clay with pre-made polyurethanes made marginal contributions to tear properties. The energy required to break, which is measured from the area under the curve of tear stress vs. tear strain curves increased with the increase of clay loading (Table 2).

The presence clay particles also improves fracture toughness. The values of plane-stress fracture toughness (K_c) of the nanocomposites increases with clay content, although larger improvement is obtained when clay particles are well-dispersed and possibly with clay-polymer tethering. For example, approximately 120% increase over TPU is observed with 5 wt% reactive clay, while the control material and the composites of 5 wt% unmodified montmorillonite clay showed either lower values or almost no improvement [93].

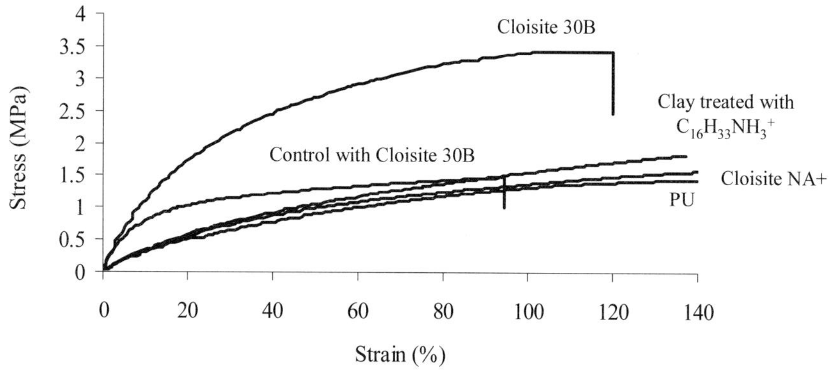

Fig. 10. Tear stress-strain diagram of composites with 5 wt% clay. TPU nanocomposite was synthesized from polyetherpolyo of molecular weight 1000 [93].

Table 2. Fracture toughness of polyurethane nanocomposites with Cloisite 30B [93].

Material		K_c MPa $(m)^{1/2}$
PU		0.032 ± 0.002
With Cloisite 30B	1 wt%	0.045 ± 0.002
	3 wt%	0.066 ± 0.007
	5 wt%	0.071 ± 0.008

3.6.3. *Abrasion resistance*

Clay particles can potentially offer resistance to abrasion, although the extent of improvement in abrasion resistance greatly depends on the type of nanoclay, its modification, and the extent of dispersion. Pattanayak and Jana [93] found that loss of materials during abrasion test is greatly reduced if clay particles are finely dispersed and if clay-polymer tethering is allowed. They found 40% increase in abrasion resistance with 5 wt% Cloisite 30B clay. Almost no improvement in abrasion resistance is observed when clay particles are micro-dispersed [93].

3.6.4. *Thermal properties*

Thermal properties of interests in polyurethanes are glass transition temperature of soft segments, melting transition of hard segments, dynamic mechanical properties, and thermal decomposition behavior of polymer chains. One study reported a small decrease [63], while another reported a small increase [74] of soft-segment glass transition temperature, although a majority revealed that glass transition temperature of soft-segment phase does not change much in the presence clay particles [64,68,72,81,83,89,92–95]. The melting transitions corresponding to hard-segment phases are usually relatively small, especially in systems with lower hard segment content and are not affected by the clay particles.

Thermal degradation of polyurethanes occurs in two steps as revealed from tests on thermal stability of polyurethanes by thermo-gravimetric analyzer. Figure 11 presents typical differential thermo-gravimetric analysis data. The first degradation temperatures (T_1), i.e. the temperatures at 5% weight loss of composites, is mostly dominated by

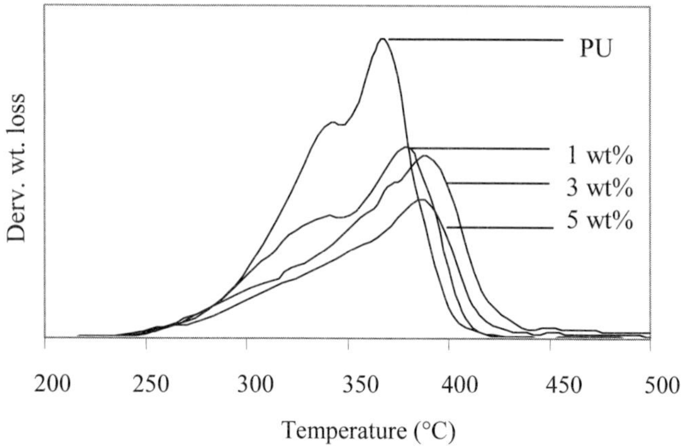

Fig. 11. Differential thermo-gravimetric analysis curves of PU and its nanocomposites with Cloisite 30B clay. TPU nanocomposite was synthesized from polyetherpolyo of molecular weight 1000 [93].

the degradation of organic modifier present in the clay and falls in the same neighborhood of pristine PU [93]. The derivative of the weight loss curves gives distinct degradation temperature peak (T_2) (Figure 11), where the presence of clay particles increases the values of T_2 by about 20°C over pristine PU. This result indicates that thermal degradation behavior is not influenced by clay-tethered polymer chains.

3.6.5. *Rheological properties*

The linear viscoelastic properties of pristine PU and its composites with nanoclay were studied by Pattanayak and Jana [93] and Plummer *et al.* [85]. These studies showed that the storage modulus increased monotonously and the viscosity decreased due to break-up of isotropic network structure with the increase of frequency. The values of storage modulus (G′) are much higher in systems with exfoliated clay particles and with clay particles tethered to polymer chains, e.g, in the case of Cloisite 30B in Figure 12. A micro-scale dispersion of clay particles in the case of untreated clay (Cloisite NA+) resulted in lower value of G′. A plateau-like behavior in G′ vs. ω curves at low values of ω in the case of reactive clay is similar to other clay-polymer nanocomposites [121,122] and is a result of clay-polymer tethering, uniform nanoscale dispersion, and much larger surface area of clay particles exposed to polymer chains. The dependence of G′ and G″ on ω in the terminal zone holds the following relationships: $G' \sim \omega^{a}$; $G'' \sim \omega^{b}$; $G' \sim (G'')^{\alpha}$ It was shown by Pattanayak and Jana [93,94] that values of *a* (~1.9) and *b* (~1.0) for pristine polyurethane are close to the ideal homopolymer values of respectively 2.0 and 1.0, which reduce significantly in the presence of clay. Their study also showed that tethered clay particles significantly retarded the motion of polymer chains as reflected from much smaller values of *a* (~0.24) and *b* (~0.57) in the case of Cloisite 30B.

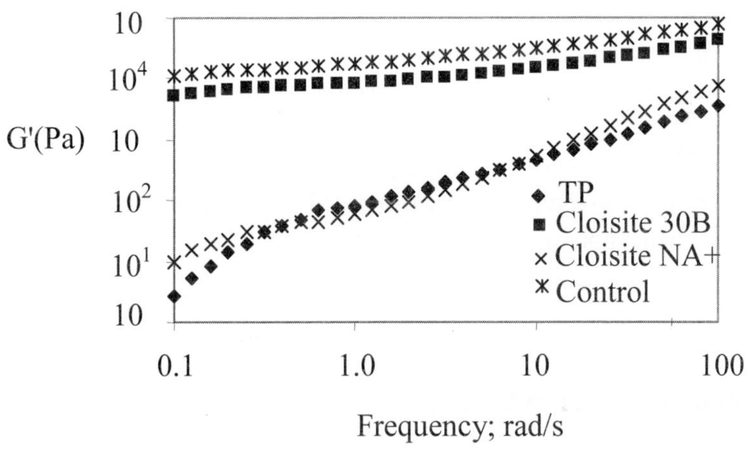

Fig. 12. Storage modulus of TPU and its nanocomposites with 5 wt% clay. TPU nanocomposite was synthesized from polyetherpolyo of molecular weight 1000 [93].

3.6.6. *Barrier properties*

The incorporation of layered silicate clay reduced diffusion rate of small molecules, such as water and oxygen, through polyurethane films. Osman *et al.* [82] found exponential decay of permeability of oxygen with layered silicate content in the polymer up to a limit of 3% by volume, beyond which the permeability values did not change, apparently due to reduced particle-particle distance. Xu *et al.* [123] found a five-fold reduction of water permeability with the introduction of 6%

by volume organically treated layered silicate particles in biomedical polyurethane nanocomposites. In the works of both Osman *et al.* [82] and Xu *et al.* [123], the clay particles remained mostly in intercalated states, as revealed by peaks in X-ray diffraction diagrams. In another study, Suter and coworkers [80] observed reduction of permeability in polyurethane-clay adhesive nanocomposites by approximately 40% and 60% respectively for oxygen and water vapor with 5% by volume clay particles. The presence of 5wt% organically clay particles also reduced water-induced swelling of polyurethanes by half [124] and caused a reduction of permeability of dichloromethane vapor [74].

4. Epoxy-clay Nanocomposites

Epoxy-clay systems have been widely studied in the past decade and several review articles have been written on these research activities (28,125). Therefore, in this section we present two issues not covered in review articles and other research publications. First, the mechanisms of exfoliation of clay particles are discussed. Second, the usefulness of epoxy as dispersion aid of layered silicate particles in thermoplastic polymers is discussed. Both issues are important in the design of composite materials.

4.1. *Mechanism of exfoliation*

Pinnavaia [3–5,7–9] and later Berglund and coworkers [11–12] proposed that in clay-epoxy system monomeric epoxy gradually diffuses into clay galleries as a consequence of imbalance in epoxy curing rate between inside and outside of clay galleries. Consequently, individual clay layers are pushed out of clay tactoids, leading to exfoliation. It was also noted that epoxy cures at a faster rate inside clay galleries due to catalytic activity of the alkyl ammonium ions [5,7,8]. The catalytic activity in turn is a strong function of acidity of the alkyl ammonium ion, and decreases in the following order, primary > secondary > tertiary. Later, several authors used time-resolved small and wide angle X-ray diffraction experiments and showed that d-spacing of clay structures changes in steps, as in Figure 13(a) [15,18], rather than continuously as proposed by

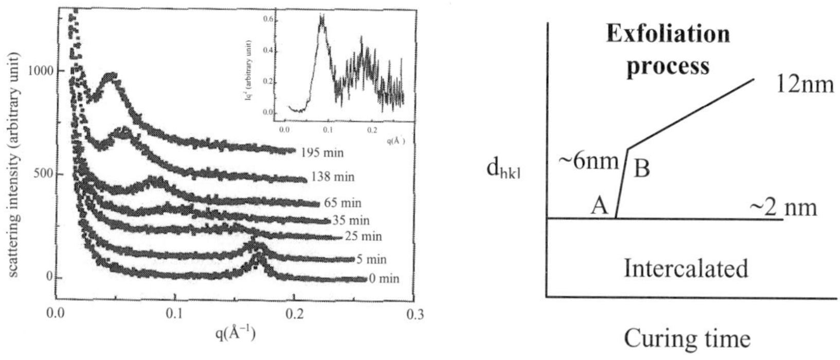

(a) (b)

Fig. 13. (a) Time-resolved small angle synchroton X-ray scattering patterns in curing of bisphenol A epoxy and montmorillonite clay mixture at 130°C. The inset shows the data after 75 minutes. [15]. (b) Schematic showing clay layer exfoliation with curing time as basal spacing of clay (d_{hkl}) expands from ~2 nm to more than 12 nm.

Pinnavaia *et al.* [3–5,7–9] and Berglund *et al.* [11,12]. The trend in Figure 13(b) indicates that the transition from intercalated to exfoliated state may take place over a very short period of time.

Jana and Park [20,21] hypothesized that elastic forces developed inside clay galleries are responsible for exfoliation of clay structures. In this case, cross-linking epoxy molecules build network structures and store energy to recoil with growing molecular weight. This energy, measured in terms of storage modulus of cross-linking epoxy molecules, can push clay layers apart against the attractive forces holding clay particles together. The attractive forces originate due to electrostatic attraction between the alkyl ammonium ions and the negatively charged clay particles and the van der Waals forces between the organic fragments of the alkyl ammonium ions. An additional force due to viscosity of extra-gallery epoxy also acts against clay layer separation. The entropic elastic forces must outweigh the attractive forces and viscous forces, which happens beyond a critical conversion. Consequently, adjacent clay layers move away from each other. This scenario is presented in Figure 14. Experimental data on epoxy-clay systems supports such scenario, as reflected in Figure 15.

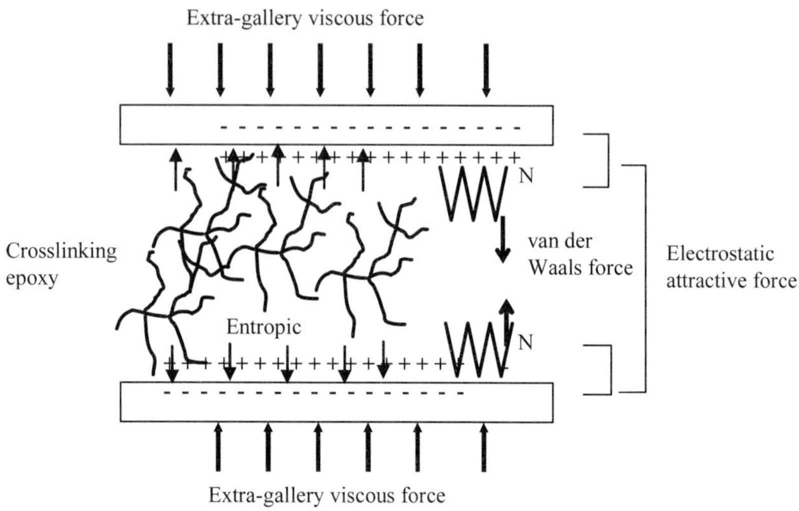

Fig. 14. Force balance on clay particles forming the clay gallery [20].

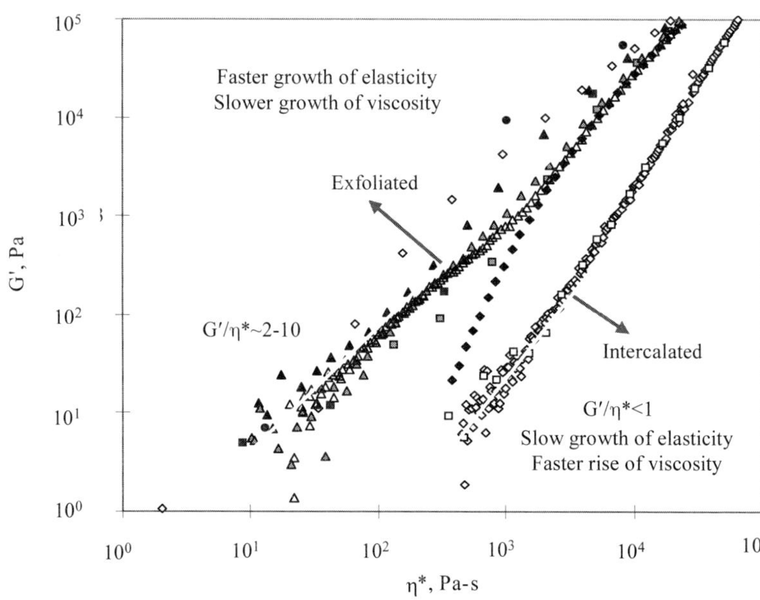

Fig. 15. Trend showing relationship between viscosity and storage modulus in intercalated and exfoliated epoxy clay systems [20].

The clay layer separation from tactoids starts with the outermost layers as the inner layers have a higher ionic bonding energy than the surface layers [126], as shown schematically in Figure 16. The removal of clay layers from the tactoids continue until the epoxy networks turn into gel, at which point the viscosity rises sharply. In this case, clay exfoliation stops as the extra-gallery viscous force overcomes the entropic force. Thus gel point indicates an upper bound of time available for complete exfoliation of all tactoids.

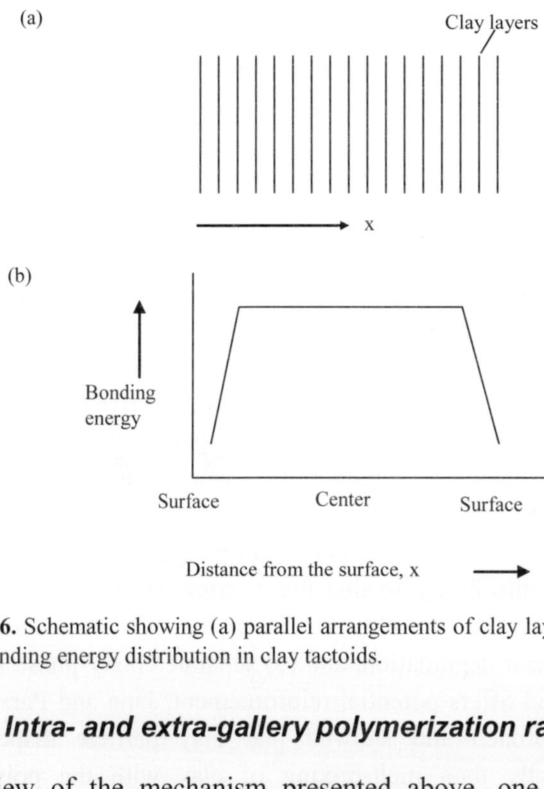

Fig. 16. Schematic showing (a) parallel arrangements of clay layers in clay tactoids and (b) bonding energy distribution in clay tactoids.

4.2. *Intra- and extra-gallery polymerization rates*

In view of the mechanism presented above, one might argue if an imbalance of intra- and extra-gallery polymerization is at all needed. A faster intra-gallery polymerization rate originating from the acceleration of epoxy curing reactions by the organic modifiers [5] can also account for faster growth of intra-gallery storage modulus. Despite such differences in the rate of polymerization between intra- and extra-gallery, exfoliation was observed to occur only in some selected cases. It was established that exfoliation of clay layers is possible even if the intra- and extra-gallery polymerization rates are matched, provided the storage modulus of intra-gallery polymer is much higher than the viscosity of the extra-gallery polymer [20,21]. Therefore, a disparity between the rates of intra- and extra-gallery polymerization is not a necessary condition for achieving fully exfoliated clay structures.

4.3. *Epoxy as dispersion aid of layered silicate clay*

Epoxy has been used in the past [127–132] as processing aid of intractable polymers, whereby, the temperature of processing of polymers such as polyphenyleneether can be reduced by almost 100°C with the addition of small quantities of epoxies. Short and continuous fiber-reinforced composite materials of thermoplastic-thermosetting hybrid matrix polymers [133–135] have been produced using this approach. In addition, low molecular weight epoxies facilitated dispersion of nanoscale fumed silica particles in high temperature thermoplastic polymers, such as polyethersulphone (PES) [136].

The same approach can be extended to achieve effective dispersion of layered silicate clay particles in thermoplastic polymers, which show miscibility with low molecular weight epoxies. Such an approach has the following wholesome attributes: (1) Epoxy molecules, intercalated in clay galleries, produce exfoliation of clay. (2) Epoxy reduces viscosity of the polymer due to miscibility so that the mixture can be processed at much lower temperature and clay particles can be homogeneously dispersed without shear degradation, and (3) separate epoxy-phase forms after crosslinking and offers potential reinforcement. Jana and Park [22] elaborated this approach and showed that clay particle dispersion improved significantly than melt-mixing of clay with the polymer. Figure 17 presents representative TEM images showing improved clay dispersion. A curious observation was made by these authors that the shape of clay particle agglomerates, whether fully exfoliated or intercalated, dictated the shape of a majority of crosslinked epoxy domains especially for low epoxy content, such as in Figure 17(b). The

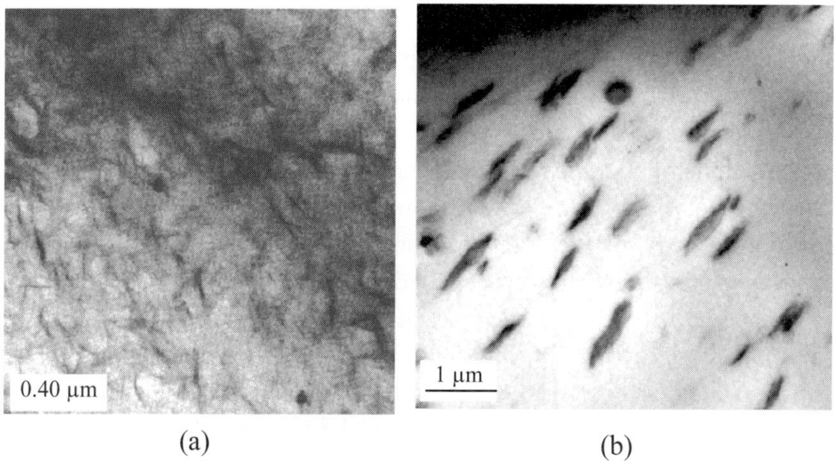

Fig. 17. (a) TEM image showing state of clay particle dispersion in PMMA. (b) TEM image of cured epoxy-PMMA-clay composite. The composition of PMMA, epoxy and clay in parts by weight were 90:10:1 [22].

shape of phase separated epoxy domains tend to become more spherical with the increase of epoxy content and contain much smaller fraction of clay particles.

5. Summary and Future Outlook

This chapter presented a summary of literature on synthesis methods and properties of polyurethane-clay and epoxy-clay nanocomposites. It is seen that the mechanisms of clay particle exfoliation in these two systems are different. While much larger values of storage modulus than viscosity produces exfoliation in clay-epoxy system, best case scenario in the case of polyurethanes is obtained when clay particles are tethered to polymer chains and shear viscosity during clay-polymer mixing is high. Mixing of pre-made polyurethanes with clay does not produce clay exfoliation even though shear viscosity is maintained high.

The existing literature on polymer nanocomposites and the results discussed in this Chapter thus far have taught us that several chemical and engineering issues must be considered while designing synthesis and processing routes for new polymer nanocomposite products. In addition, new research focus must be geared towards synthesis of multifunctional nanofillers to offer conductivity, optical activity, and shape memory actions on top of the usual mechanical reinforcement. In case of polyurethanes, multi-functionality can be achieved by using combinations of layered silicate clay and carbon nanofibers. In this case, layered silicate clay can cause a reduction of water uptake, increase of abrasion resistance, tear toughness, and tensile properties, while carbon nanofibers can augment electrical conductivity.

In some other cases, a whole set of desired mechanical properties can be obtained by using 'building block' concepts. For example, polyurethane-clay composites can be developed to offer substantial enhancements in tensile, compressive, and tear strengths without much increase of tensile modulus. These composites can be used as seat cushions in automobiles, as sport tracks, or as shoe soles. An avenue to achieve this is to design a composite with shorter polymer chains tethered to clay particles. The shorter chains will reduce the modulus, while the clay particles tethered to them will increase the modulus. Additionally, tethered clay particles will act as crosslink sites and contribute to tensile strength. These issues have been addressed in a recent study where excess isocyanates were used in conjunction with Method I (Figure 1a) to produce materials which provided substantial increase in tensile strength with small increase of tensile modulus [78]. Several building block scenarios are presented in Figure 18. A typical polymer chain with clay particles tethered to it consists of soft segments, hard segments, and tethered clay particles. The chain length between successive clay particles depends on the rates of reaction of –NCO groups with chain extender and with reactive clay particles. Thus the relative rates of chain extension reactions and reactions between clay and –NCO groups of polymer chains can be used as control to arrive at various chain configurations and hence properties. If chain extension reactions are faster than clay-polymer reactions, scenario presented in Figure 18(b) is realized.

Epoxy-clay composites also offer various possibilities. For example, mixtures of partially cured epoxy (below gel point) and nanoclay can be subjected to high shear in internal mixers in an effort to achieve good dispersion of clay particles. In this case, much higher values of shear

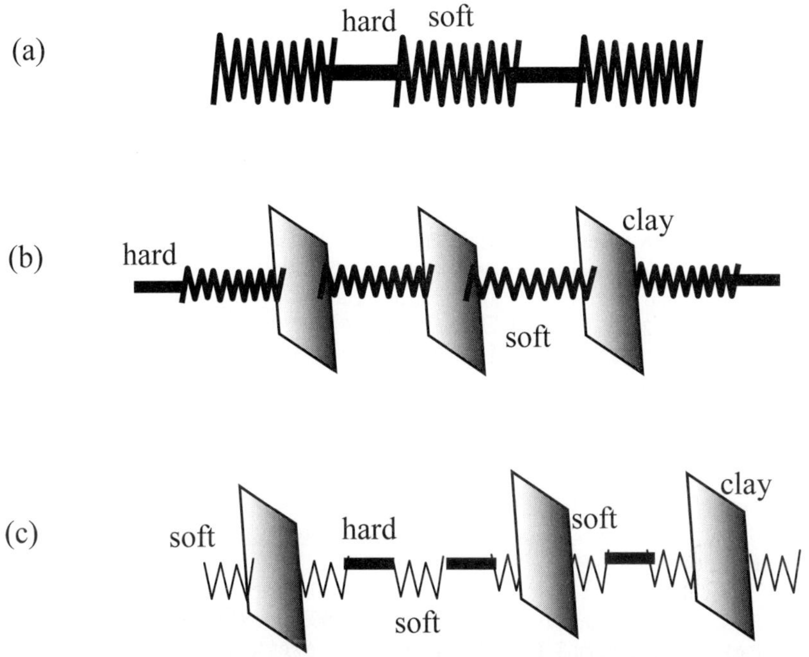

Fig. 18. Building blocks of polyurethanes and its nanocomposites. (a) Segmented polyurethanes with hard and soft segments. (b) Clay tethered chains with small hard segment content. (c) Clay tethered chains with large hard segment content. Distance between two clay particles is much greater than in (b). Cases (b) and (c) are obtained respectively for faster and slower clay-polymer reactions.

stress are obtained compared to low molecular weight epoxy, which aid clay dispersion. In addition, settling of clay particles is hindered due to much higher viscosity. Well-dispersed clay-epoxy mixtures can be processed by regular means, such as by resin transfer molding, wet lay up, or filament winding and cured.

References

1. www.freedoniagroup.com/Electronic-Polymers.html
2. Lewis, R. and Dorado, P. R. New thermoset can replace engineering thermoplastics. Plastics Engineering (Brookfield, CT, United States), 2001, 57, 58.
3. Lan, T. and Pinnavaia, T. J., *Chem. Mater.*, 1994, 6, 2216.

4. Lan, T., Kaviratna, P. D. and Pinnavaia, T. J., *Chem. Mater.*, 1994, 6, 573.
5. Wang, M. S. and Pinnavaia, T. J., *Chem. Mater.*, 1994, 6, 468.
6. Messersmith, P. and Giannelis, E. P., *Chem. Mater.*, 1994, 6, 1719.
7. Lan, T., Kaviratna, P. D. and Pinnavaia, T. J., *Chem. Mater.*, 1995, 7, 2144.
8. Lan, T., Kaviratna, P. D. and Pinnavaia, T. J., *J. Phys. Chem. Solids*, 1996, 57, 1005.
9. Shi, H., Lan, T. and Pinnavaia, T. J., *Chem. Mater.*, 1996, 8, 1584.
10. Brown, J. M., Curliss, D. and Vaia, R. A., *Chem. Mater.*, 2000, 12, 3376.
11. Kornmann, X., Lindberg, H. and Berglund, L. A., *Polymer*, 2001, 42, 1303.
12. Kornmann, X., Lindberg, H. and Berglund, L. A., *Polymer*, 2001, 42, 4493.
13. Chen, C. and Curliss, D., *SAMPE Journal,* 2001, 37, 11.
14. Zerda, A.S. and Lesser, A.J., *J. Polym. Sci., Part B: Polym. Phys.*, 2001, 39, 1137.
15. Chin, I. J., T.-A., T., Kim, H. C., Russell, T. P. and Wang, J., *Polymer*, 2001, 41, 591.
16. Becker, O., Varley, R. and Simon, G., *Polymer*, 2002, 43, 4365.
17. Ratna, D., Becker, O., Krishnamurthy, R., Simon, G. and Varley, R.J., *Polymer*, 2003, 44, 7449.
18. Kong, D. and Park, C. E., *Chem. Mater.*, 2003, 15, 419.
19. Becker, O., Cheng, Y-B., Varley, R. J. and Simon, G., *Macromolecules*, 2003, 36, 1616.
20. Park, J. H. and Jana, S. C., *Macromolecules*, 2003, 36, 2758.
21. Park, J. H. and Jana, S. C., *Macromolecules*, 2003, 36, 8391.
22. Park, J. H. and Jana, S. C., *Polymer*, 2004, 45, 7673–7679.
23. Frohlich, J., Thomman, R., Gryshchuk, Karger-Kocsis, J. and Mulhaupt, R., *J. Appl. Polym. Sci.*, 2004, 92, 3088.
24. Ma, J., Yu, Z.-Z., Zhang, O.-X., Xie, X.-L., Mai, Y.-W. and Luck, I., 2004, *Chem. Mater.*, 16, 757.
25. Wang, K., Wang, L., Wu, J., Chen, L. and He, C., *Langmuir*, 2005, 21, 3613.
26. Miyagawa, H., Foo, K. H., Daniel, I. M. and Drzal, L. T., *J. Appl. Polym. Sci.*, 2005, 96, 281.
27. Miyagawa, H., Misra, M., Drzal, L. T. and Mohanty, A. K., *Polym. Eng. Sci.*, 2005, 45, 487.
28. Becker, O. and Simon, G., *Adv. Polym. Sci.*, 2005, 179, 29.
29. Karger-Kocsis, J., Gryshchuk, O., Frohlich, J. and Mulhaupt, R., *Comp.Sci. Tech.*, 2003, 63, 2045.
30. Bharadwaj, R. K., Mehrabi, A. R., Hamilton, C., Trujillo, C., Murga, M., Fan, R., Chavira, A. and Thompson, A. K., *Polymer*, 2003, 43, 3699.
31. Lee, J. and Giannelis, E. P. *Polym. Pre., Amer. Chem. Soc.*, 1997, 38, 688; Kornmann, X., Berglund, L. A., Sterte, J. and Giannelis, E. P., *Polym. Eng. Sci.*, 1998, 38, 1351; Xu, L. and Lee, L. J., *Polym. Eng. Sci.*, 2005, 45, 496.
32. Agag, T., Koga, T. and Takeichi, T., *Polymer* v 42 n 8 2001 p. 3399–3408.
33. Chang, J.-H., Park, K. M. and Cho, D., *Polym. Eng. Sci.*, 2001, 41, 2001 p. 1514.
34. Chang, Jin-Hae and Park, Kwang Min, *Polym. Eng. Sci.*, 2001, 41 2226.
35. Delozier, D. M., Orwoll, R. A., Cahoon, J. F., Ladislaw, J. S., Smith Jr., J. G. and Connell, J. W., *Polymer*, 2003, 44, 2231.

36. Leu, C.-M., Reddy, G. M., Wei, K.-H. and Shu, C.-F., *Chem. Mater.*, 2003, 15, 2261.
37. Delozier, D. M., Orwoll, R. A., Cahoon, J. F., Ladislaw, J. S., Smith Jr., J. G. and Connell, J. W., *High Perf. Polym.* 2003, 15, 329.
38. Park, C., Smith, J. G., Connell, J. W., Lowther, S. E., Working, D. C. and Siochi, E. J., *Polymer*, 2005, 46, 9694.
39. Campbell, S. and Scheiman, D., *High Perf. Polym.* 2002, 14, 17.
40. Abdalla, M. O., Dean, D. and Campbell, S., *Mater. Res. Soc. Symp. Proc.* 2002, 726, 179.
41. Choi, Y.-K., Sugimoto, K., Song, S.-M., Gotoh, Y., Ohkoshi, Y. and Endo, M., *Carbon*, 2005, 43, 2199.
42. Prasse, T., Cavaille, J.-Y. and Bauhofer, W., *Comp. Sci. Tech.*, 2003, 63, 1835.
43. Zhong, W.-H., Li, J., Xu, L. R., Michel, J. A., Sullivan, L. M. and Lukehart, C. M., *J. Nanosci. Nanotech.*, 2004, 4, 794.
44. llaoui, A., Bai, S., Cheng, H. M. and Bai, J. B., *Comp. Sci. Tech.*, 2002, 62, 1993.
45. Allaoui, A., Bai, J. B. and Rieux, N., *Polym. Polym. Comp.*, 2003, 11, 171.
46. Sandler, J. K. W., Kirk, J. E., Kinloch, I. A., Shaffer, M. S. P. and Windle, A. H., *Polymer*, 2003, 44, 5893.
47. Gojny, F. H. and Schulte, K., *Comp. Sci. Tech.*, 2004, 64, 2303.
48. Kim, Y. J., Shin, T. S., Choi, H. D., Kwon, J. H., Chung, Y.-C. and Yoon, H. G., *Carbon*, 2005, 43, 23.
49. Zilli, D., Chiliotte, C., Escobar, M. M., Bekeris, V., Rubiolo, G. R., Cukierman, A. L. and Goyanes, S. *Polymer*, 2005, 46, 6090.
50. Ragosta, G., Abbate, M., Musto, P., Scarinzi, G. and Mascia, L. *Polymer*, 2005, 46, 10506.
51. Bandyopadhyay, A., De Sarkar, M. and Bhowmick, A. K., *J. Polym. Sci., Part B: Polym. Phys.*, 2005, 43, 2399.
52 Kang, S., Hong, S. I., Choe, C. R., Park, M., Rim, S. and Kim, J., *Polymer*, 2000, 42, 879.
53. Roscher, C., *Eur. Coat. J.* 2003, 138, 140.
54. Fujiwara, M., Kojima, K., Tanaka, Y. and Nomura, R., *J. Mater. Chem.*, 2004, 14, 1195.
55. Aktas, L., Hamidi, Y. K. and Altan, M. C., *Plastics, Rubber and Composites*, 2004, 33, 267.
56. Sandler, J. K. W., Pegel, S., Cadek, M., Gojny, F., van Es, M., Lohmar, J., Blau, W. J., Schulte, K., Windle, A. H. and Shaffer, M. S. P., *Polymer*, 2004, 45, 2001.
57. Timmerman, J. F., Hayes, B. S. and Seferis, J. C., *Comp. Sci. Tech.*, 2002, 62, 1249.
58. Woods, G., *The ICI Polyurethanes Book*, John Wiley & Sons: New York, 1990.
59. Lamba, N. M. K., Woodhouse, K. A. and Cooper, S. L., *Polyurethanes in Biomedical Applications*, CRC Press: Boca Raton, 1998.
60. Fabris, H. J., Advances in Urethane Science and Technology, Technomic: New York, 1976.
61. Wang, Z. and Pinnavaia, T. J., *Chem. Mater.*, 1998, 10, 3769–3771.
62. Zilg, C., Thomann, R., Muelhaupt, R. and Finter, J., *Adv. Mater.*, 1999, 11, 49–52.

63. Chen, T. K., Tien, Y. I. and Wei, K. H., *J. Polym. Sci. Part A: Polym. Chem.*, 1999, 37, 2225.
64. Chen, T. K., Tien, Y. I. and Wei, K. H., *Polymer*, 2000, *41*, 1345–1353.
65. Petrovic, Z.S., Javni, I., Waddon, A., Banhegyi, G., *J. Appl. Polym. Sci.*, 2000, *76*, 133–151.
66. Xu, R., Manias, E., Snyder, A. J. and Runt, J., *Macromolecules*, 2001, *34*, 337–339.
67. Ma, J., Zhang, S. and Qi, Z., *J. Appl. Polym. Sci.*, 2001, *82*, 1444–1448.
68. Tien, Y. I. and Wei, K. H., *Macromolecules*, 2001, 34, 9045–9052.
69. Tien, Y. I. and Wei, K. H., *Polymer*, 2001, *42*, 3213–3221.
70. Hu, Y., Song. L., Xu, J., Yang, L., Chen, Z. and Fan, W., *Colloid Polym. Sci.*, 2001, 279, 819–822.
71. Yao, K. J., Song, M., Hourston, D. J. and Luo, D. Z., *Polymer*, 2002, 43, 1017–1020.
72. Tien, Y. I. and Wei, K. H., *J. Appl. Polym. Sci.*, 2002, 86, 1741–1748.
73. Chang, J. H. and An, Y. U., *J. Polym. Sci.: Part B: Polym. Phys.*, 2002, 40, 670–677.
74. Tortora, M., Gorrasi, G., Vittoria, V., Galli, G., Ritrovati, S. and Chiellini, E., *Polymer*, 2002, 43, 6147–6157.
75. Zhang, X., Xu, R., Wu, Z. And Zhou, C., *Polym. Int.*, 2003, 52, 790–794.
76. Song, M., Hourston, D. J., Yao, K. J., Tay, J. K. H. and Ansarifar, M. A., *J. Appl. Polym. Sci.*, 2003, 90, 3239–3243.
77. Mishra, J. K., Kim, I. and Ha, C. S., *Macromol. Rapid Commun.*, 2003, 24, 671–675.
78. Chen, X., Wu, L., Zhou, S. and You, B., *Polym. Int.*, ***2003***, 52, 993–998.
79. Rhoney, I., Brown, S., Hudson, N. E. and Pethrik, R. A., *J. Appl. Polym. Sci.*, 2003, 91, 1335–1343.
80. Osman, M. A., Mittal, V., Morbidelli, M. and Suter, U. W., *Macromolecules*, 2003, 36, 9851–9858.
81. Dai, X., Xu, J., Guo, X., Lu, Y., Shen, D. and Zhao, N., *Macromolecules*, 2004, 37, 5615.
82. Osman, M. A., Mittal, V. and Lusti, H. R., *Macromol. Rapid Commun.*, 2004, 25, 1145.
83. Finnigan, B., Martin, D., Halley, P., Truss, R. and Campbell, K., *Polymer*, 2004, 45, 2249.
84. Rhoney, I., Brown, S., Hudson, N. E. and Pethrick, R. A., *J. Appl. Polym. Sci.*, 2004, 91, 1335
85. Plummer, C. J. G., Rodlert, M., Bucaille, J.-L., Grunbauer, H. J. M. and Manson, J.-A. E., *Polymer*, 2005, 46, 6543.
86. Gorrasi, G., Tortora, M. and Vittoria, V., *J. Polym. Sci. Part B: Polym. Phys.*, 2005, 43, 2454.
87. Xia, H., Shaw, S. J. and Song, M., *Polym. Int.*, 2005, 54, 1392.
88. Finnigan, B., Jack, K., Campbell, K., Halley, P., Truss, R., Casey, P., Cookson, D., King, S. and Martin, D., *Macromolecules*, 2005, 38, 7386.
89. Finnigan, B., Martin, D., Halley, P., Truss, R. and Campbell, K., *J. Appl. Polym. Sci.*, 2005, 97, 300.

90. Ma, C.-C. M., Huang, Y.-L., Kuan, H.-C. and Chiu, Y.-S., *J. Polym. Sci. Part B: Polym. Phys.*, 2005, 43, 345.
91. Ma, C.-C. M., Su, H.-Y., Kuan, H.-C., Kuan, C.-F. and Chiang, C.-L., *J. Polym. Sci. Part B: Polym. Phys.*, 2005, 43, 1076.
92. Pattanayak, A. and Jana, S. C., *Polymer*, 2005, 46, 3275.
93. Pattanayak, A. and Jana, S. C., *Polymer*, 2005, 46, 3394.
94. Pattanayak, A. and Jana, S. C., *Polymer*, 2005, 46, 5183.
95. Pattanayak, A. and Jana, S. C., *Polym. Eng. Sci.*, 2005, 45, 1532.
96. Cao, X., Lee, L. J., Widya, T. and Macosko, C., *Polymer*, 2005, 46, 775.
97. Ma, X., Lu, H., Lian, G., Zhao, J. and Lu, T., *J. Appl. Polym. Sci.*, 2005, 96. 1165.
98. Kuan, H.-C., Ma, C.-C. M., Chuang, W.-P. and Su, H.-Y., *J. Polym. Sci. Part B: Polym. Phys.*, 2005, 43, 1.
99. Kwon, J.-Y. and Kim, H.-D., *J. Appl. Polym. Sci.*, 2005, 96, 595.
100. Koerner, H., Liu, W., Alexander, M., Mirau, P., Dowty, H. and Vaia, R., *Polymer*, 2005, 46, 4405.
101. Lam, C.-K., Lau, K.-T., Cheung, H. and Ling, H., *Mater. Lett.*, 2005, 59, 1369.
102. Heinz, H., Vaia, R. and Farmer, B. L., Abstract PMSE-532, 230th ACS National Meeting, Aug. 28–Sept.1, 2005, Washington, DC.
103. Yoshioka, H., *Physica B*, 2000, 284–288, 1756.
104. Kim, B., Park, H. and Sigmund, W. M., *Langmuir*, 2003, 19, 2525.
105. Pailett, M., Poncharal, P. and Zahab, A., *Phys. Rev. Lett.*, 2005, 94, 186801.
106. Fukushima, Y., Okada, A., Kawasumi, M., Kurauchi, T. and Kamigaito, O., *Clay Miner.*, 1988, 23, 27.
107. Usuki, A., Kojima, Y., Kawasumi, M., Okada, A., Fukushima, Y., Kurauchi, T. and Kamigaito, O., *J. Mater. Res.*, 1993, 8, 1179.
108. Kojima, Y., Usuki, A., Kawasumi, M, Okada, A., Kurauchi, T. and Kamigaito, O. J., *Polym. Sci., Part A: Polym. Chem.*, 1993, 31, 983.
109. Yano, K., Usuki, A., Okada, A. Kurauchi, T. and Kamigaito, O., *J. Polym. Sci., Part A: Polym. Chem.*, 1993, 31, 2493.
110. Okada, A. and Usuki, A., *Mater. Sci. Eng.: C*, 1995, C3, 109.
111. www.nanoclay.com/data/30B.htm
112. Lyman, D. J., *Rev. Macromol. Chem.*, 1, 191–237 (1966).
113. Heintz, A. M., Duffy, D. J., Hsu, S. L., Suen, W., Chu, W. and Paul, C. W., *Macromolecules*, 2003, 36, 2695–2704.
114. Madejova, J., *Vib. Spectrosc.*, 2003, 31, 1.
115. Lin, I. S., Kresta, J. E. and Frisch, K. C., In Reaction injection molding and fast polymerization reactions, Polymer Science and Technology: Volume 18, 147–163, Plenum Press: New York, 1982.
116. Heiss, H. L., Combs, F. P., Gemeinhardt, P. G., Saunders, J. H. and Hardy, E. E., *J. Ind. Eng. Chem.*, 1959, 51, 929–934.
117. Lu, Q. W., Hoye, T. R. and Macosko, C. W., *J. Polym. Sci., Part A. Polym. Chem.*, 2002, *40*, 2310–2328.
118. Zhou, M. and Han, C. D., *Macromolecules*, 2005, 38, 9602.
119. Zha, W., Choi, S., Lee, K. M. and Han, C. D., *Macromolecules*, 2005, 38, 8418.

120. Fornes, T. D., Yoon, P. J., Hunter, D. L., Keskkula, H. and Paul, D. R., *Polymer*, 2002, 43, 5915.
121. Krishnamoorti, R. and Giannelis, E.P., *Macromolecules*, 1997, 30, 4097.
122. Ayyer, R. K. and Leonov, A. I., *Rheol. Acta*, 2004, 43, 283.
123. Xu, R., Manias, E., Snyder, A. J. and Runt, J., *J. Biomed. Mater. Res. Part A*, 2003, 64A, 114.
124. Kim, B. K., Seo, J. W. and Jeong, H. M., *Eur. Polym. J.*, 2003, 39, 85.
125. D'Souza, N. A., *Encycl. Nanosci. Nanotech.*, 2004, 3, 253–265.
126. Van vlack, L. H. In *Physical Ceramics for Engineers*, Addison-Wesley, New York, 1964.
127. Meijer, H. E. H., Venderbosch, R. W., Goossens, J. G. P. and Lemstra, P. J., *High Perf. Poly.* 1996; 8(1): 133–167.
128. Venderbosch, R. W., Meijer, H. E. H. and Lemstra, P. J., *Polymer*, 1994; 35(20): 4339–4357.
129. Venderbosch, R. W., Meijer, H. E. H. and Lemstra, P. J., *Polymer*, 1995; 36(6): 1167–1178.
130. Venderbosch, R. W., Meijer, H. E. H. and Lemstra, P. J., *Polymer*, 1995; 36(15): 2903–2913.
131. Ishi, Y. and Ryan, A. J., *Macromolecules*, 2000; 33(1): 158–166.
132. Ishi, Y. and Ryan, A. J., *Macromolecules*, 2000; 33(1): 167–176.
133. Saalbrink, A., Lorteije, A. And Peijs, T., *Composites Part A* 1998; 29A(9–10): 1243–1250.
134. Jana, S. C. and Prieto, A., *J. Appl. Poly. Sci.*, 2002, 86(9): 2159–2167.
135. Jana, S. C. and Prieto, A., *J. Appl. Poly. Sci.*, 2002, 86(9): 2168–2173.
136. Jana, S. C. and Jain, S. *Polymer*, 2001; 42(16): 6897–6905.